DESIGNING SOLDIER SYSTEMS

Human Factors in Defence

Series Editors:
Dr Don Harris, Managing Director of HFI Solutions Ltd, UK
Professor Neville Stanton, Chair in Human Factors at the
University of Southampton, UK
Dr Eduardo Salas, University of Central Florida, USA

Human factors is key to enabling today's armed forces to implement their vision to "produce battle-winning people and equipment that are fit for the challenge of today, ready for the tasks of tomorrow and capable of building for the future" (source: UK MoD). Modern armed forces fulfil a wider variety of roles thanever before. In addition to defending sovereign territory and prosecuting armed conflicts, military personnel are engaged in homeland defence and in undertaking peacekeeping operations and delivering humanitarian aid right across the world.

This requires top class personnel, trained to the highest standards in the use offirst class equipment. The military has long recognised that good human factors is essential if these aims are to be achieved.

The defence sector is far and away the largest employer of human factors personnel across the globe and is the largest funder of basic and applied research. Much of this research is applicable to a wide audience, not just the military; this series aims to give readers access to some of this high quality work.

Ashgate's *Human Factors in Defence* series comprises of specially commissioned books from internationally recognised experts in the field. They provide in-depth, authoritative accounts of key human factors issues being addressed by the defence industry across the world.

Designing Soldier Systems
Current Issues in Human Factors

EDITED BY

PAMELA SAVAGE-KNEPSHIELD

U.S. Army Research Laboratory – Human Research and Engineering Directorate, USA

JOHN MARTIN

U.S. Army Research Laboratory – Human Research and Engineering Directorate, USA

JOHN LOCKETT III

U.S. Army Research Laboratory – Human Research and Engineering Directorate, USA

and

LAUREL ALLENDER

U.S. Army Research Laboratory – Human Research and Engineering Directorate, USA

CRC Press
Taylor & Francis Group
Boca Raton London New York

CRC Press is an imprint of the
Taylor & Francis Group, an **informa** business

CRC Press
Taylor & Francis Group
6000 Broken Sound Parkway NW, Suite 300
Boca Raton, FL 33487-2742

First issued in paperback 2017

© 2012 by Pamela Savage-Knepshield, John Martin, John Lockett III, Laurel Allender and the contributors
CRC Press is an imprint of Taylor & Francis Group, an Informa business

No claim to original U.S. Government works

Version Date: 20160226

ISBN 13: 978-1-138-07697-6 (pbk)
ISBN 13: 978-1-4094-0777-5 (hbk)

Visit the Taylor & Francis Web site at
http://www.taylorandfrancis.com

and the CRC Press Web site at
http://www.crcpress.com

Contents

List of Figures

List of Tables

This book is dedicated to the men and women who have unselfishly and steadfastly served in the US Armed Forces to uphold the principles that our nation holds so dear and to all those that have made the ultimate sacrifice; especially remembered are Army Guardsmen SPC Bryan Rearick and SSG Jorge Oliveira.

About the Editors

Laurel Allender, Ph.D., was appointed as the Director of the Army Research Laboratory (ARL) Human Research and Engineering Directorate (HRED) at Aberdeen Proving Ground (APG), MD, in January 2011 to direct basic and applied research and advanced development in three core competency areas: human performance, simulation and training technologies, and human systems integration (HSI). Dr. Allender began her career in 1984 at the Army Research Institute for the Behavioral and Social Sciences at Fort Bliss, TX, where her research led to the development of an automated assessment capability for Patriot training systems. Dr. Allender received her Ph.D. in Applied Experimental Cognitive Psychology from Rice University.

John Lockett III is Chief of the Soldier Performance Division at ARL HRED. Prior to that, he led the MANPRINT Methods and Analysis Branch and the Integration Methods Branch. He received a Masters in Industrial and Systems Engineering from Virginia Tech and a B.S. in Engineering Psychology from Tufts University. He has over 25 years' research and development experience in Human Factors and has concentrated on application of workload analysis and human figure modeling technologies to MANPRINT, the US Army's HSI program.

John H. Martin is an operations research analyst working in the area of MANPRINT and human factors engineering. He is currently an Army Materiel Command (AMC) Field Assistance in Science Technology (FAST) Science and Technology Advisor at the Joint Readiness Training Center in Grafenwoehr, Germany. He has graduate degrees from Columbia University in Mathematics and Education. He has also done graduate work in Human Factors at Virginia Tech and the University of Michigan and is a member of the Army Acquisition Corps.

Pamela A. Savage-Knepshield, Ph.D., is a research psychologist and Chief of the Human Factors Integration Division at ARL HRED. She earned a doctorate in Psychology from Rutgers University and holds one patent assigned to AT&T Corp. In 2008, she received the Department of the Army's MANPRINT Practitioner of the Year award. In 2010, she returned from serving a tour in Iraq as an AMC FAST Science Advisor and is a recipient of the Superior Civilian Award and the Global War on Terrorism Civilian Service Medal.

List of Contributors

Petra E. Alfred is a research psychologist with ARL HRED's Army Medical Department Field Element at Fort Sam Houston, TX. Ms. Alfred received an M.A. in Industrial/Organizational Psychology from the George Mason University and an M.S. in HSI from the Naval Postgraduate School. Ms. Alfred's work focuses on applied human factors and macroergonomic research aimed to reduce academic attrition in military training programs.

Bruce E. Amrein attended the Baltimore Polytechnic Institute and graduated from Loyola University Maryland with a B.S. in Engineering/Physics. He also holds an MBA and Masters in Engineering Science. After serving as a commissioned officer in the US Army's Signal Corps, he has been with ARL since its inception in 1992. He currently serves as the Chief of ARL HRED's Perceptual Sciences Branch. He holds nine patents in technology areas ranging from the control of military vehicle drive trains to medical devices.

Michael J. Barnes is a research psychologist with ARL HRED's Fort Huachuca Field Element. Currently, he manages the human robot interaction (HRI) component of the Safe Operations for Unmanned Systems for Reconnaissance in Complex Urban Environments Army Technology Objective. In addition to HRI, his research interests include investigations of risk visualization, intelligence processes, and unmanned aerial vehicles crew systems. He was educated at the University of Arizona (B.A.) and the New Mexico State University (M.A.).

Raymond M. Bateman, Ph.D., is an operations research analyst at ARL HRED's Field Element at the US Army Medical Department in Fort Sam Houston, TX. He graduated with a doctorate from the Colorado School of Mines while assigned to the US Army Special Forces where he served in positions from Detachment Commander to Deputy Commander of a Special Forces Group conducting foreign internal defense missions in Central and South America. While assigned to ARL he has served two tours in Iraq as a Senior Science Advisor.

Andrew Bodenhamer is a human factors engineer at ARL HRED's Field Element at the US Army Maneuver Support Center in Fort Leonard Wood, MO. He holds degrees in Mechanical Engineering from the University of Missouri – Rolla and Industrial and Operations Engineering from the University of Michigan. He is the co-recipient of the 2004 HFES Practitioners Award from the Industrial Ergonomics group for his work on ergonomic analysis of fixed bridging.

Jorge Capo, Ph.D., was born and raised on the island of Puerto Rico. He obtained his bachelor's degree in biology and the M.D. from the University of Puerto Rico. His research experiences include prototypical psychotropic drug development and brain–computer interaction. Prior to returning to Puerto Rico to complete his residency, he completed a post-doctoral position with ARL HRED investigating brain–computer interaction technologies.

Jessie Y.C. Chen, Ph.D., is a research psychologist with ARL HRED's Field Element in Orlando, FL. She received her doctorate in Applied Experimental and Human Factors Psychology from the University of Central Florida. Prior to joining ARL, she was a post-doctoral fellow with the US Army Research Institute for Behavioral and Social Sciences. Dr. Chen is the recipient of the 2008 Army-wide Modeling and Simulation Award for the category of Individual Experimentation. Her research interests include human-robot interaction, human-vehicle interaction, supervisory control, and individual differences.

Keryl A. Cosenzo, Ph.D., is a research psychologist for ARL HRED. She received a doctorate in the Psychological Sciences (Biopsychology) from Virginia Tech. Her research focuses on the development of automation strategies to enable soldiers to perform optimally in the complex HRI environment, and human-in-the-loop simulation to evaluate and optimize HSI.

Bradley Davis is an industrial/human factors engineer at ARL HRED's Aviation and Missile Command (AMCOM) Field Element at Redstone Arsenal, AL. He received a B.S. in Mechanical Engineering and an M.S. in Engineering Management from the University of Missouri-Rolla. Among Mr. Davis's research interests are augmented reality, biomechanics, human performance, mental workload, situation awareness, and usability engineering.

Thomas W. Davis, Ph.D., is Chief of ARL HRED's Weapons Branch headquartered at the AMCOM Field Element. His responsibilities include MANPRINT and human factors engineering for missile, aviation, aviation support equipment, unmanned aerial vehicles, and unmanned ground vehicles. Dr. Davis was selected in 2006 as Department of the Army, MANPRINT Practitioner of the Year.

Linda R. Elliott, Ph.D., is a research psychologist at ARL HRED's Field Element at the US Army Infantry Center in Fort Benning, GA. She graduated with a degree in Organization Behavior and Industrial Psychology from Michigan State University while assigned to US Air Force Headquarters Staff, Air Force Research Laboratory, at Brooks Air Force Base, TX. She is a member of the NATO working group on tactile displays and the International Standards Committee to develop standards for tactile and haptic displays.

Lamar Garrett is Chief of ARL HRED's Edgewood Chemical Biological Field Element located in APG-Edgewood, MD. His responsibilities include MANPRINT and human factors engineering for chem-bio defense, biometric, integrated base defense, and persistent surveillance systems. He holds a B.S. in Engineering Management from Park University and an M.S. in System Engineering from Johns Hopkins University. Mr. Garrett was selected in 2010 as Department of the Army, MANPRINT Practitioner of the Year.

Stephen Gordon, Ph.D., is an electrical engineer who holds B.S. degrees in Mechanical and Electrical Engineering from Tennessee Technological University, and an M.S. and a Ph.D. in Electrical Engineering from Vanderbilt University. Since joining the DCS Corporation, he has applied his experience to the support of electrophysiological data acquisition and analysis and is currently working on methods for assessing structure–function couplings in the brain based on EEG, as well as methods for deriving and classifying operator state using both eye-tracking and EEG measures.

Ellen C. Haas, Ph.D., received her B.A. in Psychology from the University of Colorado at Boulder, her M.S. in Industrial Psychology from California State University at Long Beach, and her Ph.D. in Industrial and Systems Engineering from Virginia Polytechnic Institute and State University in Blacksburg, Virginia. She is a research engineer at ARL HRED and heads the ARL Multimodal Controls and Displays Laboratory. She has written and published extensively about auditory and tactile warnings and displays, multimodal displays and speech recognition controls.

Jeffrey T. Hansberger, Ph.D. is a research psychologist for ARL HRED at the AMCOM Field Element. He received his M.S. and Ph.D. in Human Factors and Applied Cognition from George Mason University. He supports and conducts experimentation with DARPA and numerous university and industry research partners in the areas of command and control, distributed cognition, strategy development, and adaptation within military teams and environments.

John K. Hawley, Ph.D., has worked as an applied psychologist for more than 25 years in a variety of government and private-sector organizations. He recently served as project leader on an Army effort to examine human performance contributors to fratricide and recommend potential solutions, and is now working with the air defense community to implement and evaluate selected recommendations. These recommendations involve changes to HSI practices, test and evaluation methods, personnel assignment practices, and operator and crew training.

Charles L. Hernandez is a human factors specialist at ARL HRED's Fort Sill Field Element, and a retired Lieutenant Colonel in the Field Artillery with 23 years of active duty service. He provides subject matter expertise in fire support tactics, techniques and command and control procedures for research programs that are of interest to the Army and the Fires Center of Excellence. Mr. Hernandez was deployed during Operation Iraqi Freedom as a senior scientist for an AMC FAST Science and Technology Advisory Team.

Michael Sage Jessee is an engineering psychologist at ARL HRED's AMCOM Field Element, specializing in eye-tracking as an assessment tool for Army aviation platforms. His analysis techniques for characterizing visual and mental workload have contributed to product improvements for cockpit enhancement. In 2009, he received his M.A. in Experimental Psychology from the University of Alabama in Huntsville and continues to maintain industry and academic relationships by serving on the board of the Tennessee Valley Chapter of the Human Factors and Ergonomics Society (HFES).

Kathy L. Kehring is an electronics engineer at ARL HRED. She served as an AMC FAST Science Advisor to the V Corps in Heidelberg Germany and is a recipient of the Superior Civilian Service Award. She holds an M.S. in Electrical Engineering from The Johns Hopkins University, and a B.S. in Electrical Engineering from The Pennsylvania State University. Currently she heads HRED's Tactical Environment Simulation Facility where she leads the development of an immersive simulator that incorporates a state-of-the-art mobility interface.

Richard W. Kozycki is a graduate of Johns Hopkins University with a Bachelor's degree in Electrical Engineering and currently works at ARL HRED. He serves as one of the Army's top experts for digital human figure modeling in the field of workspace analysis and has been instrumental in developing HRED's digital library of soldier clothing and equipment models, as well as models of various Army vehicles, aircraft, and weapon platforms.

Andrea S. Krausman, Ph.D., is a research psychologist at ARL HRED. She received an M.S. in Industrial and Systems Engineering from Virginia Tech and is currently a Ph.D. candidate there. She has been involved in research investigating virtual reality as a training aid for dismounted soldiers, effects of physical exertion and chemical protective clothing on cognitive performance, and using tactile displays as an alternative communication medium for dismounted soldiers. Her current research efforts are aimed at investigating team collaboration in distributed environments.

Brent Lance, Ph.D., is a computer scientist at ARL HRED. Dr. Lance received a B.S. in Computer Engineering & Computer Science from the University of Southern California, and his Ph.D. in Computer Science from USC's Information

Sciences Institute. He did his postdoctoral work jointly between the ARL and USC's Institute for Creative Technologies. His current research focuses on the development of Army-relevant brain–computer interaction technologies.

Anna L. Mares received a B.S. in Electrical Engineering from the University of Texas at El Paso. In 2007, Ms. Mares received the prestigious ARL Award for Analysis for her analytical contributions in identifying significant issues involving the Patriot Air and Missile Defense system usage during Operation Iraqi Freedom. Her efforts resulted in recommendations for material and training solutions which increased friendly protection involving the Patriot air and missile defense system.

Kaleb McDowell, Ph.D., is Chief of the Translational Neuroscience Branch at ARL HRED. He received a B.S. in Operations Research and Industrial Engineering from Cornell University and a doctorate in Neuroscience and Cognitive Sciences from the University of Maryland. Dr. McDowell has been the organization's driving force in developing a research vision for translating basic neuroscience into military-relevant neurotechnologies that address the critical needs of our soldiers. For his efforts in this area, he received the 2011 ARL Award for Leadership.

James E. Melzer is with Rockwell Collins Optronics in Carlsbad, CA where he is Manager of Research and Technology. He has been designing and developing helmet-mounted displays and protective headgear systems for fighter and helicopter pilots and for ground soldiers for 25 years. Recently, he was the lead for the US Army's Future Force Warrior and the Air Warrior Integrated Headgear product teams, which focused on advanced hearing, vision and ballistic protection, as well as multispectral sensors, auditory and visual displays for the next generation.

Diane Kuhl Mitchell is an engineering psychologist with ARL HRED. In 2004 her crew size analysis of the non-line-of-sight cannon won the AMC Award for Meritorious Achievement in Systems Analysis. For 15 years Ms. Mitchell designed laboratory and field experiments investigating issues critical for the field artilleryman. In 1997 she earned her M.S. in Industrial Organizational Psychology and began building Improved Performance Research Integration Tool (IMPRINT) task-network models to analyze issues critical to soldier performance and its impact on system design.

Frank Morelli earned a B.A. in Psychology with a minor in Cognitive Neuroscience from Temple University and an M.A. in Cognitive Psychology from the University of Delaware. He is a research psychologist with research interests that include human cognition and perceptual processing. He served a tour in Iraq as an AMC FAST science advisor and is a recipient of the US Army Commander's Award for Civilian Service, the US Army Material Command (AMC) Civilian Campaign Medal, and the US Army Achievement Medal for Civilian Service.

Anthony W. Morris, Ph.D., is Chief of ARL HRED's AMCOM Field Element. He received his Ph.D. in Experimental Psychology from the University of Connecticut, specializing in Human Factors and Performance Learning within the framework of Ecological Psychology. He is currently working on projects that focus on pilot performance modeling for utility rotorcraft and other vehicle operations. His expertise covers issues of information detection and action control for navigation and instrumentation operation for human–machine interface design.

Linda Mullins research interests include the effects of encapsulation on dismounted warrior performance, the effects of auditory versus visual presentation of information on soldier performance, and the extent to which aviator behavior and performance are affected by aircrew life support and protective equipment. Most recently, Linda was a researcher in the Soldier Performance Division of ARL HRED.

Kelvin S. Oie, Ph.D., is a research kinesiologist with ARL HRED. Dr. Oie received his doctorate in Neuroscience and Cognitive Sciences from the University of Maryland where his research focused on human multisensory integration. At ARL, he has been integral to the development of the Army's research focus in neuroscience, and has managed two of the Army's leading basic and applied research programs in the field to advance our knowledge of and capabilities for studying human brain function in real-world environments.

Debbie Patton is a psychology technician and is currently pursuing a master's degree in Experimental Psychology from Towson University. She is internationally recognized for her work in stress and performance using state-of-the-art field practical psychological and physiological assessments. She is the recipient of the Commander's Award for Civilian Service and the Systems Analysis Award. She is a former chair of the Stress and Workload sub-TAG of the Department of Defense Human Factors Engineering Technical Advisory Group.

Beth Plott is a Division Manager at Alion Science and Technology. Ms. Plott received an M.S. in Civil and Environmental Engineering from Colorado State University and a B.S. in Chemical Engineering from the University of Colorado. In her current position, she is responsible for a number of programs in the areas of software development, modeling and simulation, HSI, and systems analysis. Ms. Plott has managed a variety of projects relating to the development and application of human performance and workload modeling tools.

Elizabeth S. Redden, Ph.D., is Chief of ARL HRED's Field Element located at the US Army Infantry Center in Fort Benning, GA. Her responsibilities include MANPRINT and human factors engineering for infantry systems. She has designed and executed many experiments and tests of the human factors design of infantry systems and of infantry situation awareness. She has also been a featured speaker at numerous national and international conferences on night vision and recently served as the US lead for the HUM Group Land Systems Technical Cooperative Panel. She has received the Superior Civilian Service Award, the Commander's Award for Civil Service and in 1995, she received the very first Department of the Army, MANPRINT Practitioner of the Year award.

Valerie J. Berg Rice, Ph.D., is the Chief of ARL HRED's Field Element at the US Army Medical Department Center and School, Fort Sam Houston, TX. Dr. Rice (Colonel, Retired) has Masters degrees in Health Care Administration and Occupational Therapy and a doctorate in Industrial Engineering and Operations Research. She is a board-certified ergonomist, a registered and licensed occupational therapist, and a fellow of the American Occupational Therapy Association. She has served on the Board of Directors for HFES.

Charneta L. Samms is an industrial engineer and lead of the Tools Development Team at ARL HRED. Ms. Samms earned a B.S. in Industrial Engineering from Morgan State University and a Masters of Industrial and Systems Engineering from Virginia Tech. Ms. Samms is currently responsible for the development, management and distribution of IMPRINT.

Pamela A. Savage-Knepshield, Ph.D., is a research psychologist and Chief of the Human Factors Integration Division at ARL HRED. She earned a doctorate in Psychology from Rutgers University and holds one patent assigned to AT&T Corp. In 2008, she received the Department of the Army's MANPRINT Practitioner of the Year award. In 2010, she returned from serving a tour in Iraq as an AMC FAST Science Advisor and is a recipient of the Superior Civilian Award and the Global War on Terrorism Civilian Service Medal.

Angelique A. Scharine, Ph.D., is a research psychologist and lead of the Auditory Research Team at ARL HRED. She has been involved in the development and acquisition of soldier headgear, communications headsets and hearing protection. She is currently an active participant of three Acoustical Society Association working groups developing standards related to the measurement of speech over communications systems and the effectiveness of hearing protectors. She received her doctorate in Psychology from Arizona State University.

Jennifer C. Swoboda is a research psychologist at ARL HRED. Much of her work has focused on the development of command and control human performance models using Command, Control, and Communications—Techniques for the Reliable Assessment of Concept Execution (C3TRACE), a PC-based modeling environment which is used to study the effects of information flow and quality on operator performance. More recently, she has been providing empirical data to substantiate modeling efforts focusing on physical impairment and soldier performance. Ms. Swoboda received a B.A. in Psychology from Washington

College and has completed graduate level classes at Towson University and Virginia Tech.

Richard A. Tauson is a research psychologist at ARL HRED in APG MD. He has worked in the development and evaluation of artillery, medical, and command and control systems, in addition to research effects of vehicle motion on human performance. Mr. Tauson has a B.S in Psychology from Pennsylvania State University and an M.A. from Towson State University.

Timothy L. White is a computer scientist at ARL HRED in APG, MD. He received his B.S. and M.S. in Computer Science from Jackson State University. Mr. White is pursuing his doctorate in Applied Experimental and Human Factors Psychology at the University of Central Florida. He has developed and supported a number of research efforts on the use of tactile display systems to improve the situational awareness and reduce the workload of mounted and dismounted soldiers.

Preface

Designing Soldier Systems focuses on contemporary human factors issues in the design of soldier systems and describes how they are being investigated and addressed by the US Army Research Laboratory in conjunction with some of our collaborators. Current research thrusts covered include: (1) *human systems integration* to optimize soldier-system performance through the use of soldier-centered design techniques and human performance modeling tools; (2) *human robotic interaction* to develop human factors design principles and technology that enhance soldier—robotic interaction through workload management and bio- and cognition-inspired designs for robots; (3) *social/cognitive network science* to improve cognitive performance for enhanced distributed collaboration and decision-making in complex network-enabled operations, drawing from cognitive science, computer science, and social network innovations; and (4) *neuroscience* to optimize soldier performance through system design that is consistent with brain function by taking into account its limitations and exploiting its potentials.

The book is organized in three main parts: "Understanding Human Performance with Complex Systems," "Overcoming Operational and Environmental Conditions;" and "Assessing and Designing Systems." Chapters cover many of the challenges that our users, like many others, encounter when using systems to accomplish their work, including information overload, blind faith in technology, and excessive physical and cognitive workload. Also covered are operations on the move, at night, and in extreme environments, as well as the challenges of learning and maintaining situation awareness in a combat/military environment. Lastly, the book discusses design tools and techniques that we have found effective during system design (human figure modeling, cognitive workload modeling, and user-centered design), as well as those that hold great promise to inform design and enhance future military operations (brain—computer interaction technologies and social network analysis).

The editors would like to express their sincere appreciation to the dedicated authors and reviewers who took a great deal of time out of their personal lives to make this book possible. We are also most appreciative of the staff at Ashgate Publishing for their assistance and patience during the editorial process and for providing us a venue to share our lessons learned as well as our thoughts on areas requiring further research.

Pam Savage-Knepshield
John Martin
John Lockett
Laurel Allender
Aberdeen Proving Ground, MD

List of Abbreviations

3DSSPP	Three-Dimensional Static Strength Prediction Program
AAR	after action review
AAS	Advanced Automation System
ABT	air-breathing threat
ACH	Advanced Combat Helmet
ADA	Air Defense Artillery
AEWE	Army Expeditionary Warrior Experiment
AFQT	Armed Forces Qualifying Test
AI	articulation index
AIT	Advanced Individual Training
AiTR	aided target recognition
AMC	Army Materiel Command
AMD	air and missile defense
ANR	active noise reduction
ARL	Army Research Laboratory
ASVAB	Armed Services Vocational Aptitude Battery
BAL	blood alcohol level
BCI	brain-computer interface
BCITs	brain-computer interaction technologies
BCS	Battle Command System
C2	command and control
C2V	Command and Control Vehicle
C3TRACE	Command, Control, and Communications—Techniques for the Reliable Assessment of Concept Execution
CAD	computer-aided design
CAT	Callsign Acquisition Test
CBA	Capabilities-Based Assessment
CBIRF	Chemical Biological Incident Response Force
CBRN	Chemical, Biological, Radiological, and Nuclear
CDTF	Chemical Defense Training Facility
C-IED	counter-IED
CMM	coordinate measuring machine
CNV	contingent negative variation
COA	course of action
CoHOST	Modeling of Human Operator System Tasks
CPASE	Cognitive Performance Assessment for Stress and Endurance
CSP	Common Spatial Patterns
CVC	combat vehicle crewman

Designing Soldier Systems

CW	cognitive workload
DAC	Department of the Army Civilian
DBBL	Dismounted Battlespace Battle Lab
DEL	Design Eye Line
deoxy-Hb	deoxygenated hemoglobin
DES	discrete educational software
DHCP	Dynamic Host Configuration Protocol
DISALT	Dismounted Infantryman Survivability and Lethality Test Bed
DoD	Department of Defense
DSB	Defense Science Board
ECG	electrocardiogram
EDA	Electrodermal activity
EEG	Electroencephalogram
EEPP	Energy Expenditure Prediction Program™
EMG	electromyography
ENVG	enhanced night vision goggle
EOD	Explosive Ordnance Disposal
EOG	electrooculography/electrooculogram
ERD	event related desynchronizations
ERN	error-related negativity
ERP	event related potentials
ErrP	error-related potential
ERS	event related synchronizations
FA	field artillery
FAA	Federal Aviation Administration
FAP	false-alarm prone
FAST	Field Assistance in Science and Technology
FBCB2	Force XXI Battlefield Command Brigade and Below
FCC	Fire Coordination Center
FECC	Fires and Effects Coordination Cell
FMC	fully mission capable
fMRI	functional magnetic resonance imaging
FMS	Flight Management System
fNIRs	functional near-infrared spectroscopy
FOBs	Forward Operating Bases
FoV	field of view
GPS	global positioning system
GS	General Schedule
GUI	graphical user interface
HBA	Human Behavior Architecture
HFE	human factors engineering
HFM	Human Factors and Medicine Panel; Human Figure Modeling
HFP	human factors practitioner
HITL	human-in-the-loop

HSI	human systems integration
HMD	helmet-mounted display
HMEE III	High Mobility Engineer Excavator Type III
HMI	human-machine interface
HMMWV	high mobility multipurpose wheeled vehicle
HMS	Handheld, Manpack, and Small Form Fit
HPMs	human performance models
HRED	Human Research and Engineering Directorate
HRI	human robot interaction
HRTF	Head Related Transfer Function
HSAs	human system analysts
HUD	heads-up-display
I2	image intensification
ICG	impedance cardiogram
IEDs	improvised explosive devices
IES	Immersive Environment Simulator
IID	Interaural Intensity Difference
IMPRINT	Improved Performance Research Integration Tool
IMT	individual movement techniques
IP	Internet Protocol
IPE	Individual Protective Equipment
IR	Infrared
ISC	Information Systems Command
ISO	International Organization for Standardization
ISR	Intelligence, Surveillance, and Reconnaissance
ITD	Interaural Time Delay
JFCOM	Joint Forces Command
JND	Just Noticeable Difference
JTF	Joint Task Force
JTRS	Joint Tactical Radio System
JURS2	Joint Urban Resolve 2
KD	Known Distance
KPPs	key performance parameters
LDA	Linear Discriminant Analysis
LSE	Logistics Support Element
MA&D	Micro Analysis and Design
MAACL-R	Multiple Affect Adjective Checklist-Revised
MANOVA	Multivariate Analysis of Variance
MANPRINT	Manpower and Personnel Integration
MAV	micro air vehicle
MCAM	medical cost-avoidance model
MEG	magnetoencephalography
MG	Major General
MGB	Medium Girder Bridge

MIDS	Multimodal Information Design Support Tool
MMH	manual material handling
MNVD	Monocular Night Vision Device
MOPP4	Mission Oriented Protective Posture
MOS	military occupational specialty
MOUT	Military Operations in Urban Terrain
MP	miss prone
MPV	Mine Protection Vehicle
MRT	Multiple Resource Theory/modified rhyme test
MWL	mental workload
NASA	National Aeronautics and Space Administration
NASA-TLX	National Aeronautics and Space Administration-Task Load Index
NBC	Nuclear, Biological, and Chemical
NRC	National Research Council
NIOSH	National Institute for Occupational Safety and Health
NVDs	night vision devices
NVGs	night vision goggles
OCU	operator control unit
ODS	Operation Desert Storm
ODT	Omni-Directional Treadmill
OEF	Operation Enduring Freedom
OIF	Operation Iraqi Freedom
OpDemo	Operational Demonstration
Ops	Operations
ORD	Operational Requirements Document
OSHA	Occupational Safety & Health Administration
OTW	out the window
oxy-Hb	oxygenated hemoglobin
PASGT	Personnel Armor System for Ground Troops
PCA	principal component(s) analysis
PEL	permissible exposure level
PEO	Program Executive Office
PET	Positive Omission Tomography
PETER	Performance Evaluation Tests for Environmental Research
PM	Project Manager
RAMS	Readiness Assessment Monitoring System
RDECOM	Research, Development and Engineering Command
REF	Rapid Equipping Force
RFIs	Requests for Information
RNLE	Revised NIOSH Lifting Equation
ROI	return-on-investment
RSOV	Ranger Special Operations Vehicle
RSTA	reconnaissance, surveillance, and target acquisition
RSVP	rapid serial visual presentation

RT3	Reconfigurable Table Top Trainer
RTOS	Reconfigurable Tactical Operations Simulator
RWL	Recommended Weight Limit
S&T	science and technology
SA	situation awareness; situational awareness
SARA	Standardized Stress and Readiness Assessment
SBSOD	Santa Barbara Sense of Direction
SD	Spatial Disorientation
SEs	system engineers
SME	subject matter expert
SNA	social network analysis
SNRs	signal-to-noise ratios
SoS	system of systems
SOSI	System of Systems Integration
SpA	spatial ability
SpO2	oxygen saturation in the blood
SRE	Specific Rating of Events
SSE	Squad Synthetic Environment
SSVEP	steady-state visually evoked potential
STATs	Science and Technology Assistance Teams
STI	speech transmission index
STRAP	System for Tactile Reception of Advanced Patterns
SUBJ	Subjective Stress Scale
SVM	Support Vector Machines
SVS	Soldier Visualization Station
SWAT	Subjective Workload Assessment Technique
SWIR	short-wavelength IR
T&E	test and evaluation
TacVis	tactile and visual alerts
TARDEC	Tank Automotive Research Development and Engineering Command
TBM	tactical ballistic missile
TCM	TRADOC Capability Manager
THINK ATO	Tactical Human Integration of Networked Knowledge Army Technology Objective
TOC	Tactical Operations Center
TRADOC	Training and Doctrine Command
TSAS	Tactile Situation Awareness System
TTPs	tactics, techniques, and procedures
UAV	unmanned aerial vehicle
UGV	unmanned ground vehicles
UI	user interface
UML	Unified Modeling Language
USAARMC	US Army Armor Center

VACP	visual, auditory, cognitive, and psychomotor
VBC	Victory Base Complex
VOG	videooculography
VWL	visual workload
WBV	whole body vibration

PART I
Understanding Human Performance
with Complex Systems

Chapter 1

Human Performance Challenges for the Future Force: Lessons from Patriot after the Second Gulf War

John K. Hawley and Anna L. Mares

Background

During the major combat operations phase of the Second Gulf War (Operation Iraqi Freedom (OIF), March and April 2003), US Army Patriot air and missile defense (AMD) units were involved in two fratricide incidents. In the first, a British GR-4 Tornado was misclassified as an anti-radiation missile and subsequently engaged and destroyed. The second fratricide incident involved a Navy F/A-18 Hornet that was misclassified as a tactical ballistic missile (TBM) and also engaged and destroyed. Three flight crewmembers lost their lives in these incidents. OIF involved a total of 11 Patriot engagements by US units. Of these 11, 9 resulted in successful TBM engagements; the other two (18 percent) were fratricides. Although significant in and of themselves, these fratricides opened the door for a unique look at the human performance problems introduced by increasing system and operational complexity in a major weapons system. The initial assessment was followed by a multi-year effort focused on remedying the problems identified during the initial incident investigation. Lessons and observations from the initial fratricide assessment and follow-on mitigation work set the stage for much of the discussion to follow. But first, we present a little background on the Patriot system.

Patriot is the Army's first-line AMD system. The system has been in the active force since the mid-1980s, but has been upgraded numerous times since first fielded. Initially, Patriot was intended as a defense against conventional air-breathing threats (ABTs). However, since Operation Desert Storm (ODS), the First Gulf War in the early 1990s, the system has been used primarily against TBMs. Future usage scenarios portray the system being used against a spectrum of air threats, including TBMs, conventional ABTs, cruise missiles, and unmanned aircraft systems. The range of potential air threats in the contemporary operating environment, along with major changes in the system's employment concept, have significantly increased the complexity of the air battle management[1] problem for

---

1   The term "air battle management" as used here is analogous to the term "battle command" as used in non-AMD Army organizations.

Patriot and other AMD systems. Arthur (2009) argues that systems tend to become more complex as they are enhanced to meet new usage requirements.

Figure 1.1 presents an overview of the Patriot AMD system and its contemporary operating environment. To begin, the system itself is technically and tactically complex. Patriot currently employs more than 3.5 million lines of software code in air battle management operations. Number of lines of software code is sometimes used as a rough proxy for system complexity. An additional layer of operational complexity is added by a requirement for Patriot air battle management crews to coordinate with other systems, composing what is termed an "integrated air and missile defense system." These associated systems include the Air Force's Airborne Warning and Control System, the Army's Joint Land Attack Cruise Missile Defense Elevated Netted Sensor and Terminal High Altitude Area Defense systems, the Navy's Aegis sea-based missile defense system, and various space-based National Assets. The current Patriot as part of an integrated AMD system of systems illustrates the complexity associated with contemporary concepts for joint, network-enabled operations. The requirement to coordinate with the associated systems also means that from an analytical perspective, Patriot must be approached as a system of systems.

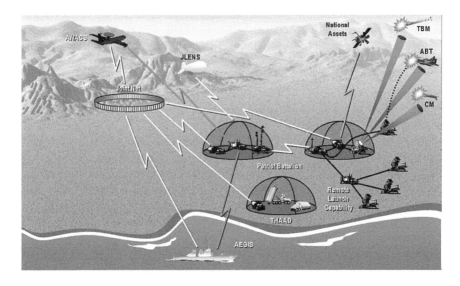

Figure 1.1   The Patriot AMD system and its operating environment

Given that Patriot is an existing system and has been in the Army's inventory since the early 1980s, what do lessons from Patriot tell us about system complexity and its impact on future Army systems and future military operations? We think that in some respects Patriot provides a glimpse into the future of military systems and operations. As Patriot has evolved over the past 25 years, it has acquired features and characteristics that are more typical of systems the Army will employ in the future than those in the current inventory. In drawing a parallel between current Patriot operations and future Army systems, we are not suggesting that future systems or their performance demands will be identical to Patriot. Rather, we assert that many emerging systems may result in an operator and crew performance environment similar to Patriot's. At a conceptual level, many of the performance demands placed on individual soldiers and crews will be similar to those encountered in Patriot.

When considering human performance topics in a system of systems setting similar to Patriot, one cannot avoid the issues of system complexity and the impact of complexity on system users. Complexity is a consistent theme in much of the discussion to follow. That said, a necessary first step is to define the term "complexity," or, more specifically, the "state of being complex." The dictionary definition of "complex" (Merriam-Webster 2009) refers to a system having many interconnected or related parts, or a system that has a complicated structure—not simple or straightforward. A second theme that emerges in an attempt to define complexity is the degree of orderliness or predictability of a system or process (Dekker 2005). An unpredictable, disorderly system or process, by definition, presents a high degree of complexity for users. Unpredictability is moderated by the amount of time available to effect control.

A definition that appears to encompass both of these definitional themes is given in Hollnagel and Woods 2005. Following these authors, system or process complexity is a function of (1) the number of parameters needed to fully define the system or process in space and time; and (2) the amount of information needed to fully comprehend the system or process and its operating environment. The impact of complexity on users revolves around the second portion of the Hollnagel and Woods definition: that is, the amount of information needed to comprehend the system and operating environment at any point in space and time. This requirement to fully comprehend the system and its operating environment is the key to maintaining effective human control and has significant implications for system design and user job preparation. The next part of this chapter begins our discussion of the impact of complexity on a system's users. Again, the context for this discussion is an assessment of actual events involving the Patriot AMD system during the major combat operations phase of OIF.

The Patriot Vigilance Project

A team from the US Army Research Laboratory (ARL) began looking into the OIF Patriot fratricides and the more general issue of Patriot human–system performance at the invitation of the then Fort Bliss, TX, Commanding General, Major General (MG) Michael A. Vane. Fort Bliss was the site of the Army's Air Defense Artillery (ADA) Center and School. In his own words, MG Vane was interested in operator vigilance and situational awareness (SA) as they relate to the performance of automated air defense battle management systems. (The generally accepted definition of SA is from Endsley, Bolte, and Jones (2003), who define it as the *perception* of elements in the environment, the *comprehension* of their meaning, and the *projection* of their status in the near future). MG Vane was particularly concerned by what he termed a "lack of vigilance" on the part of Patriot air battle management crews, along with an apparent "lack of cognizance" of what was being presented to them on situation displays, with a resulting "unwarranted trust in automation." The project team spent most of the summer and fall of 2004 reviewing the OIF fratricide incidents—reading documents, interviewing knowledgeable personnel in the Fort Bliss area, and observing Patriot training and operations. An initial report of findings was delivered to MG Vane in October 2004. ARL's assessment was not intended to be just another exercise in "Monday morning quarterbacking." Instead, the focus was to look into the deeper story behind the events leading to the OIF fratricides from a human performance perspective and to identify actionable solutions. MG Vane's reference to lack of vigilance on the part of Patriot operators led to the effort being called the "Patriot Vigilance project."

The ARL assessment team organized its presentation to MG Vane around two central themes, denoted (1) *undisciplined automation* during Patriot development; and (2) *automation misuse* on the part of Patriot crews. These themes and related contributors are shown graphically in Figure 1.2. A more detailed discussion of the findings of ARL's initial fratricide assessment is provided in Hawley and Mares (2006). What follows is a summary of material from that source intended to set the stage for the ensuing discussion. The first theme, or contributing factor, was termed "undisciplined automation." This is defined as the automation of functions by designers and subsequent implementation by users without due regard for the consequences for human performance (Parasuraman and Riley 1997). Undisciplined automation tends to define the operators' roles as by-products of the automation. Every function that can be automated is automated. Operators are left in the control process to monitor the engagement process and respond to system cues. In the case of Patriot, little explicit attention was paid during system design and subsequent testing to determine (1) what residual functions were left for the operators to perform, (2) whether operators actually could perform these functions, (3) how operators should be trained to perform properly, or (4) the impact of potential operator training deficiencies on the system's automated engagement decision-making reliability.

The downstream impact of undisciplined automation was exacerbated by two contributing or secondary factors: (1) unacknowledged system fallibilities, and (2) a fascination with and "blind faith" in technology. An unacknowledged system fallibility is a system deficiency that is known but not satisfactorily resolved. For example, a series of Patriot operational tests going back to the 1980s indicated that the system's automated engagement logic was subject to track misclassification problems—system fallibilities. The Patriot system classifies acquired tracks as

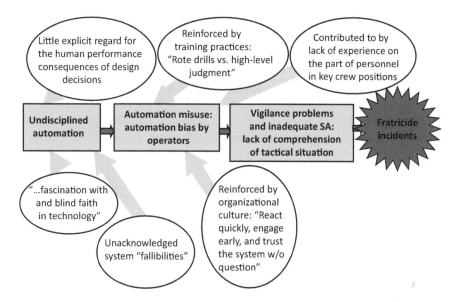

Figure 1.2   Human performance contributors to the Patriot system's fratricides during Operation Iraqi Freedom

conventional air-breathing, rotary wing, ballistic missile, cruise missile, anti-radiation missile, or other relevant category on the basis of flight profiles and other track characteristics such as point of origin and compliance with Airspace Control Orders. A misclassification occurs when the system-generated category designation does not match the track's actual status. These sources of automation unreliability were not satisfactorily addressed during system software upgrades, nor did information about them find its way into operator training, air battle management doctrine, crew procedures, or unit standard operating procedures. It should be noted that there are no technical quick fixes or easy solutions to such classifications problems. Research, experimentation, and operational experience suggest that any solution lies in the domain generally labeled situation-specific trust in automation (Hawley and Mares 2007). Operators and crews must have the technical and tactical expertise to determine when they can trust machine-generated solutions and when they should question such results. In the case of Patriot, system developers continued to pursue technology-centric solutions to automation reliability problems (e.g., increased use of artificial intelligence, non-cooperative target recognition, improved identification friend or foe query systems, etc.). When challenged concerning the system's track classification accuracy, the claim was repeatedly made that a technical fix for these problems was "just around the corner." But the basic problem remained unresolved: the system was not completely reliable in critical functional areas, most notably track classification and identification. To make matters worse, users were not informed regarding these reliability problems, or if they were informed, little if any effective responsive action was identified for them to take. Training enhancements and other crew-oriented solutions to the classification reliability problem were not considered.

In the aftermath of the First Gulf War (ODS), the air defense user community acquiesced in the developmental community's apparent lack of concern for problems with Patriot's track classification accuracy. Emboldened by Patriot's seeming success in engaging the Iraqi TBM threat during ODS, Patriot's organizational culture and command structure emphasized reacting quickly, engaging early, and trusting the system without question (Hawley and Mares 2006). The view that the system could be trusted without question persisted in spite of several "close calls" during ODS attributable to track classification problems and test results indicating that the system was not always accurate. This cultural norm was exacerbated by the air defense community's traditional training practices, which were criticized in the Army's post-fratricide review as emphasizing what were termed "rote drills" versus the exercise of high-level judgment. The Patriot user community approached training for air battle operations in much the same manner as less cognitively oriented tasks such as system movement and setup. The emphasis during training was on mastering routines (crew drills) rather than critical thinking and adaptive problem-solving (Hawley and Mares 2006). Inadequate individual and crew training were further impacted by the branch's traditional methods of assigning personnel to air battle management crews. Traditional personnel assignment practices tended to place inexperienced personnel in key air battle management crew positions. Moreover, routine personnel administration practices tended to rotate crew members out of battle staff positions and to other jobs rather quickly. The result was that tactical crews were generally formed from the unit's newest and least experienced personnel. System and operational complexity, inadequate training, and crew inexperience are a dangerous combination. As used here, the term "inadequate training" refers to training that is (1) too short to produce necessary levels of operator competence, (2) ill-focused, in the sense that training content does not address critical operator or crew skills, or (3) inappropriate, in that the instructional methods used are not suitable for the job's skill content.

Before the first missile round was fired during OIF, the stage was thus set for the second primary contributor, termed "automation misuse" (Parasuraman and Riley 1997). Automation misuse took the form of extensive automation bias on the part of Patriot crews. Automation bias is defined as unwarranted overreliance on automation, which has been demonstrated to result in failures of monitoring (vigilance problems) and accompanying decision biases (unwarranted and uncritical trust in automation—"let's do what the system recommends"). Recall that these are the very concerns expressed by MG Vane in his kickoff discussion with ARL's project staff.

One must be careful not to lay too much blame for these shortcomings at the feet of the Patriot crews involved in the OIF incidents. The roots of their apparent human performance shortcomings can be traced back to systemic problems resulting from decisions made years earlier by concept developers, software engineers, procedures developers, testers, trainers, and unit commanders. In hindsight, the most surprising aspect of ARL's fratricide assessment is that there really were no surprises. The OIF Patriot air battle management crews did what they had been trained to do, what Patriot's culture and command structure emphasized and reinforced, and what the automation literature suggested they likely would do in such circumstances.

Patriot is a very lethal system. It can be argued that the system was not properly managed during OIF. Driven by technological opportunity and mission expansion, the Patriot air battle management crew's role changed from traditional operators to supervisory controllers, whose primary role is to supervise the subordinate automated engagement routines controlling the system (Sheridan 1992). That is, the subordinate systems automatically close a control loop on the task or process, and the crew is expected to monitor these systems for correct performance and intermittently respond to system cues when necessary. Control is indirect through the automated engagement system as opposed to direct as in traditional manual control. The supervising crew is thus "on" the control loop versus "in" the loop, as with traditional manual control. The term "on-the-loop" versus "in-the-loop" is becoming standard to reflect this role change (USAF 2009: 14 and section 4.6). This change in terminology might appear minor, but it is significant for design and training.

The Story Behind the Story

ARL's report to MG Vane recommended two primary actionable items to address the problems identified during the fratricide incident assessment:

1. Reexamine air defense battle management automation concepts to emphasize *effective operator control*: look into ways to mitigate SA problems resulting from undisciplined automation of Patriot control functions.
2. Develop more effective air battle management teams: reexamine the level of *expertise* required to employ systems such as Patriot on the modern battlefield.

A later report on Patriot system performance during the Second Gulf War prepared by the Defense Science Board (DSB) reinforced ARL's recommendations concerning the importance of effective operator control and improved operator and crew proficiency (DSB 2004).

Considering ARL's first recommendation, it was clear when evaluating the Patriot fratricide incidents within the context of both known human performance vulnerabilities and classic design errors (as suggested in Woods and Dekker 2001) that the primary human performance problem was loss of effective control. In a system like Patriot, effective control means that operators and not the automated system are the ultimate decision-makers in engagement decisions. Decisions to shoot or not to shoot must be made by crews having (1) the technical potential for adequate SA, and (2) the expertise to understand the significance of the information available to them. This latter characteristic has sometimes been called "sensemaking ability" (Weick, Sutcliffe, and Obstfeld 2005).

All factors considered, the Patriot crews presented only the illusion of effective control. Hollnagel and Woods (2005) assert that being in control of a process means that operators know and understand what has happened and are reasonably able to predict what will happen next. Note that these characteristics correspond closely to the second and third stages of Endsley, Bolte and Jones's (2003) definition of SA, namely comprehension and prediction. Hollnagel and Woods (2005) also list three principal "enemies" of effective human control: (1) lack of time, (2) lack of knowledge of the system's status, and (3) lack of operator competence. ARL's assessment for MG Vane suggested that all three of these deficiencies contributed to the OIF fratricides. In retrospect, it might be said that the OIF Patriot operators simply lacked essential resources to maintain actual control, most notably time and competence. Both of these factors also can contribute to a lack of knowledge of the system's actual status. Operators either do not know where to seek essential information or they lack the time to do so.

ARL's second recommendation to MG Vane concerned reexamining the level of expertise required of Patriot operators and air battle management crews. In phrasing this recommendation, we deliberately chose to use the term "expertise" rather than "training." What is expertise and how is it different from the skills imparted through traditional training? In present usage, the term "expertise" refers to a capability for consistently superior performance on a specified set of representative tasks for a domain (Ericsson and Charness 1994). The term also implies the ability to perform adequately under conditions of uncertainty (Ericsson 2009). Expertise is a function of operator knowledge, skill, aptitudes, and job-relevant experience. Expertise-focused training concentrates on whole-job performance rather than individual task performance, as is often the case in routine, Task-Conditions-Standards focused Army training (Kahneman and Klein 2009). The ultimate performance target for battle management crews in a system like Patriot is recognitional decision-making, sometimes called "skilled intuition." Simon (1992) defined skilled intuition as a form of complex pattern recognition: The tactical situation provides cues, cues provide access to stored information (declarative knowledge and experiential), and the stored information suggests a response. Research has indicated that concentrating on the performance of individual tasks versus whole-job proficiency during training will not always result in the development of necessary levels of expertise as the term is used here (Schneider 1985).

Defining What Right Looks Like: The Human Factors Practitioner as Change Agent

After reviewing the initial fratricide assessment results, the Army Training and Doctrine Command (TRADOC) Capability Manager (TCM) for Patriot pushed for ARL's work to continue into a second phase. The assessment phase of the Patriot Vigilance project resulted in a technical report addressing the impact of automation on air defense battle management operations (Hawley, Mares, and Giammanco 2005). The TCM specifically requested that ARL expand this material and prepare two more detailed reports, one concerned with design for effective supervisory control (Hawley and Mares 2006) and one addressing training for automated battle management systems (Hawley, Mares, and Giammanco 2006). In the TCM's words, the intent of these reports was to inform the air defense community on "what right looks like" in each of these topic areas. The TCM wanted a guide to implementing ARL's recommendations.

Air Battle Management Initiatives: Reinforcing Effective Operator Oversight

In late summer 2005 after MG Vane had left Fort Bliss for another assignment, the ARL project staff briefed his replacement on the status and results of the Patriot Vigilance project and immediate follow-on work. Following this meeting, the new Commanding General, MG Robert P. Lennox, formally requested that ARL continue working with the air defense community on implementation. Follow-on implementation had two key aspects. The first aspect was to serve as the Manpower and Personnel Integration (MANPRINT—the Army's human system integration initiative) evaluator during an operational test of a major software upgrade for the Patriot system. This upgrade was developed to address several of the Patriot system's operational deficiencies that were considered to have contributed to the unacceptable fratricide rate.

In addition to wide-ranging software modifications, the Patriot operational test was expanded to address a number of changes consistent with ARL's first actionable item concerning a review of automation concepts in air defense battle management to emphasize effective operator control. The centerpiece of these changes was the integration of what was termed a Fire Coordination Center (FCC) into the Patriot battalion tactical operations center. The second aspect of follow-on implementation was to evaluate the FCC concept.

The FCC represented an attempt to reemphasize effective human control and also make air defense battle management more of a team-based, collaborative activity. Its introduction potentially represented a significant step forward in addressing the SA problems that contributed to Patriot's battle management deficiencies during OIF. First, however, it had to be demonstrated (1) that the FCC provided the incremental SA essential for more reliable engagement decision-making and (2) that the new configuration could do so in a timely manner.

Given the engagement timelines involved, decision cycle time is a significant issue in air defense battle management.

We should also note that the FCC represented a means to provide a more complete battle management work system—in a macroergonomic sense—for Patriot than had existed previously (Hendrick and Kleiner 2002). Hendrick and Kleiner define a work system as a system that "involves two or more persons interacting with some form of (1) hardware and/or software, (2) internal organizational environment, (3) external environment, and (4) organizational design" (2002: 399). Under this definition, Patriot's battle management work system includes all the battalion components illustrated in Figure 1.1 plus all of the external components to which the organization is linked. Based on our initial fratricide assessment, the ARL project team observed that Patriot's battle management problems had more to do with lack of a well-defined and properly functioning work system to accomplish the objectives of air battle management than it did with the design of individual equipment items per se ("microergonomics" as used by Hendrick and Kleiner).

From fall 2005 through summer 2006, during the New Equipment Training and collective training period for the operational test, the ARL team's observations regarding the progress of training for the test unit sounded an alarm. Training was not progressing satisfactorily. Training *events* were being completed, but individual and crew *performance objectives* were not being met. Many of the training issues identified during our follow-up to the initial fratricide inquiry were surfacing and were not being addressed adequately by training events in the test unit. We viewed these training deficiencies as a serious problem because inadequate test player training would compromise the validity of test results and undermine the basis for evaluating the value added of software changes or the FCC concept (Hawley 2007a).

As was feared, pretest training inadequacies partially compromised the validity of operational test results and made it difficult to evaluate the utility of the FCC concept. With respect to the FCC, the principal reasons for the pretest training inadequacies were that (1) operational procedures had not been properly developed and validated prior to the test, and, consequently, (2) validated training was not available prior to the test. Because of these limitations, we concluded that test results could not be used cleanly to answer either of the questions concerning the FCC posed previously.

In spite of the difficulties involving the FCC, the ARL team was able to make several defensible observations regarding Patriot training as experienced by the test unit. These observations, along with our persistent advocacy, led the ADA School to agree that the level of expertise required to employ the Patriot system properly with the software upgrade exceeded the current training standard. The earlier Army board of inquiry report on the fratricides had reached a similar conclusion, noting that "the system [Patriot] is too lethal to be placed in the hands of crews trained to such a limited standard" (Hawley and Mares 2006). This convergence of evidence supported the conclusion that modifications to current air defense training practices were required. Moreover, these modifications would require not simply more "traditional" training, but performance-oriented strategies focused on deliberate practice of key performance elements. Deliberate practice denotes a long period of active learning during which job performers refine and improve their skills under the supervision of an instructor or coach (Ericsson and Charness 1994). As a variety of research suggested, the problem was not so much one of inadequate training *equipment* as inadequate training *methods* and *time for training*. The ARL team advocated this position strongly in a second training report prepared in the aftermath of the training inadequacies leading up to the operational test (Hawley and Mares 2007). This report also laid out a blueprint for how air defense battle management training had to be modified to achieve the necessary ends.

Note that we had come full circle. By pursuing our objective of enhancing effective operator control from a design point of view, we had returned to Patriot's training problem. The two initiatives are inextricably linked, and the linkage is stronger than we had initially assumed. Well-designed materiel still requires well-trained soldiers for effective use. In that sense, design is not a substitute for training. Generalizing this observation beyond AMD, Fischerkeller, Hinkle, and Biddle (2002) concluded that battle outcomes across a range of engagement situations are determined primarily by the *interaction of skill and technology*. These authors based their conclusion on a range of historical and analytical evidence.

Training Initiatives

In addition to a general agreement that changes in training rigor and instructional strategies were warranted, the ADA School identified an additional training capability gap. This gap concerned the organic simulation capability available to air defense units. The School concluded that units might benefit from a capability to train Patriot air battle management crews that supplemented their organic embedded training capability and would better support performance-oriented instructional strategies focused on deliberate practice. The ARL team had observed that embedded training in the Patriot control vans was not a particularly conducive training environment for a deliberate practice regimen (Hawley, Mares, and Giammanco 2006). This observation is particularly true for the initial and mid-range portions of the training sequence wherein the skills required to take advantage of full-task training using the system's embedded capability should be developed.

To address this need, the School identified an existing device, the Reconfigurable Tactical Operations Simulator (RTOS), as potentially meeting the requirement for a standalone simulation capability to supplement Patriot units' organic embedded training capability. The RTOS is a part-task, less-than-full-fidelity Patriot simulator and has been used since the late 1980s to support air defense exercises as well as experimentation and analysis. However, it had not been used explicitly as a training device.

To begin exploring these issues, the School organized the RTOS Operational Demonstration (OpDemo). The OpDemo was structured as a joint project involving the School and an operational Patriot unit (the 5th Battalion, 52nd Air Defense Artillery Regiment (5–52 ADA)). Its objectives were to (1) demonstrate and evaluate modified instructional methods for use in Patriot unit training, and (2) assess the potential utility of an RTOS-like device to supplement unit training assets. ARL participated in a technical advisory capacity. In our view, the demonstration provided an opportunity to show off several of the modified training practices we had consistently advocated, to do so in a high-profile setting involving an operational Patriot unit, and to maintain the momentum for training change coming out of the operational test.

Results from the OpDemo indicated that (1) the RTOS, as an exemplar for a part-task, less-than-full-fidelity training device, had potential utility to support Patriot unit training, (2) the training strategy focusing on deliberate practice was effective for the trial modules used, and (3) the overall training package was well received by participants. Beyond these specific conclusions, positive results from the OpDemo gave the air defense community a green light to pursue further applications of an RTOS-like training device and modified instructional strategies. Demonstration results also helped to forge a consensus among decision-makers and opinion leaders that the exercise was a success, and that ARL's view of the way forward for training reform was both potentially useful and feasible. Such consensus was important to maintaining the momentum for training reform initiatives because it helped to offset the considerable resistance to changes in training equipment and methods that exists in some segments of the air defense community. As an added benefit, the training setup used during the demonstration represented a partial prototype for a solution to the training deficiencies that contributed to the Patriot fratricides and that showed up again during the run-up to the operational test. Hawley, Mares, Fallin, and Wallet (2007) provide a more complete description of the RTOS OpDemo.

Where Do We Stand Now?

After more than five years of working with the air defense community, the ARL team has a mixed record of success. We noted previously that the FCC remains a promising but not fully validated concept. The FCC has been implemented on a limited scale, but is not standardized across operational units. A formal procurement action to deploy a materiel replacement for the somewhat ad hoc FCC—termed the "AMD Integrated Battle Command System"—has recently been completed and development of that system is proceeding.

The ARL team's experiences in the area of training reform have been more positive. The RTOS OpDemo was generally considered a success. Partially based on these results, the ADA branch started actively looking into the utility and feasibility of deploying the RTOS or a comparable device to Patriot battalions, down to the battery level. An RTOS-like training device (denoted the Reconfigurable Table Top Trainer (RT3)) has replaced the aging Patriot Conduct of Fire Trainer for institutional training. At the time of this writing, more than 150 RT3 workstations are in use to support Patriot institutional training at the Army's new ADA School at Fort Sill, OK. Plans to deploy the RT3 to Patriot units are proceeding, albeit somewhat slower. The current goal is to place 15 RT3 workstations in each Patriot battalion. There has been little follow up on the ARL team's initiative to restructure air and missile defense battle management training using a deliberate practice instructional model.

It should be emphasized that the team considers this latter action (that is, changes in instructional methods and practices) to be the most important aspect of Patriot and broader AMD training reform. As illustrated in the previous paragraph, training reform efforts stemming from the Patriot Vigilance effort gained traction in the materiel arena, but the "soft" changes (e.g., restructuring training methods and practices) necessary to make effective use of this new materiel have been much harder to bring about. Deploying new training equipment without corresponding changes in training methods and practices is not likely to lead to desired operator and crew performance improvements. More than 50 years of training research has shown that instructional design issues generally trump issues pertaining to training equipment and simulator fidelity (Salas, Bowers, and Rhodenizer 1998). As these authors asserted in the title of their article, "It is not how much you have, but how you use it."

Challenges for the Future Force

Until now, we have been telling the story of the Patriot fratricides during OIF, describing ARL's assessment of human performance contributors to those events and recounting our efforts to implement a number of Patriot-specific fixes in the areas of air battle management operations and training. Now, in this second part of the chapter, we generalize these themes and discuss their relevance to emerging Army systems and operations.

Do lessons and observations from Patriot generalize to other Army systems and to future operations? Previously, we argued that they do, in two important ways. First, as the Patriot system has evolved over the past 25 years, it has acquired features and characteristics more typical of systems the rest of the Army will employ than those in the current inventory. Patriot provides an operator and crew performance setting similar to that being developed for Future Force battle command and related applications. Second, the system's technical and tactical complexities necessarily "flow through" directly to air battle management operators and crews. The issue of complexity flow-through to operators is controversial, and we will briefly discuss it before addressing specific challenges for the Future Force. Complexity flow-through to operators is a complicating factor in each the specific challenges.

The conventional wisdom in some system development circles is that technical and tactical complexity can be reduced to manageable levels (i.e., to match some "innate ability" of human operators) through "proper" system and interface design. The mantra of "design for simplicity" advocates that the need to cope with complexity can be reduced or eliminated by simplifying information presentation and making system use itself "intuitive." Hollnagel and Woods (2005) argue that such an approach is doomed to fail, because complexity is not reduced but merely hidden from users.

The problem with masking complexity from users can be summed up in a concept from control theory termed the "Law of Requisite Variety" (Hollnagel and Woods 2005: 40). Simply put, the Law of Requisite Variety states that the variety or variability of a process to be controlled must be matched or exceeded by the variety or variability of the controlling entity. Control is degraded if the controller has less variety than the controlled process itself. Simplifying information presentation to users can result in operator control variability being reduced, and that strategy lessens the potential for effective process control. It should be noted that the Law of Requisite Variety does not pertain to complexity in the use of the control system itself, merely information presentation using that system. Providing a user-friendly control system is beneficial because it reduces one source of complexity for operators—a control system that is hard to operate. However, control system operation versus information presentation on that system is not the same thing and should not be confused. The Law of Requisite Variety pertains to the latter and not to the former.

Citing the Law of Requisite Variety, Hollnagel and Woods (2005) argue that complexity for operators cannot be reduced to some arbitrary low level while still enabling the operators to maintain effective control. Rather, complexity must be managed by increasing response variability on the part of the controller. To do that, complexity must be made visible, and controllers must be provided with the resources to cope with that complexity. The resources that most impact an operator's ability to control a system are time, knowledge of the system's status, and competence. Time to effect control might be a problematic issue in many situations, but enhanced knowledge of a system's status and operator competence can be addressed through better system and display design (e.g., design to support SA as discussed in Endsley, Bolte, and Jones 2003) coupled with suitable operator training and follow-on job preparation.

Automation

The first major challenge facing developers and users of complex military systems is the effective use of automation. Sheridan defined automation as "the automatically controlled operation of an apparatus, a process, or system by mechanical or electrical devices that take the place of human organs of *observation, decision,* or *effort*" (1992: 3; emphasis added). In another view, Bainbridge (1987) characterized automation as *classical* versus *contemporary*. Classical automation is an older and more traditional variety and primarily involves allocating human *physical* functions to the machine. Contemporary automation is information-technology-based and also involves allocating human *sensory* and *cognitive* functions to the machine.

The popular concept of automation is that of a complex of machines performing their intended function with little or no human intervention, other than as a tender. Scenes from automated automotive assembly plants and other industrial facilities reinforce this notion. Sheridan (1992) remarked that there is an unfortunate tendency on the part of laypersons to view automation as an all-or-none phenomenon. A system is controlled either manually or automatically with nothing in between. He notes, however, that all-or-none control is the exception rather than the rule. Automated systems that do not leave some residual functions for human controllers are rare.

Extending automation into the cognitive (decision-making) domain coupled with the problem of unresolved residual human operator functions contributed greatly to the Army's problems with the Patriot system during OIF. Researchers such as David Woods (Hollnagel and Woods 2005, Woods and Hollnagel 2006) assert that when automation is extended into the sensory and cognitive domains, the so-called brittleness problem of automata often becomes an issue. The brittleness problem refers to the machine's inability to reliably handle unusual or ambiguous situations. Activities in the sensory and cognitive domains are less well defined, "fuzzy" if you like, than activities in the physical domain. Hence, such activities are more prone to the brittleness problem. There in an increased likelihood of what Singer (2009) refers to as "Oops moments," or mistakes on the part of the automated system. The Patriot fratricides during OIF are examples of such "Oops moments." They resulted from machine misclassifications that were not detected and countered by the air battle management supervisory controllers.

Another way of looking at the brittleness issue is termed the Catch 22 of human supervisory control (Reason 1990, Sheridan 1992, 2002). The logic of the Catch 22 situation is illustrated in the following sequence of statements:

1. Automation is introduced because it is thought to be able to do the job better than human operators: humans are becoming the weakest link in defense systems (Singer 2009).
2. Human operators have been left on the control loop to *monitor* that the automated system is performing correctly and to override the automation when it is *wrong*: the operators' role is changed to that of supervisory controllers.
3. The tacit assumption is that human operators can properly decide when the automated system's decisions should be overridden: humans are expected to compensate for machine unreliability—a consequence of the brittleness problem.

4. Humans suffer from a variety of physical and cognitive limitations that make it difficult to meet this expectation.
5. Undisciplined (sometimes called clumsy) automation can make the operators' situation more challenging.

The Catch 22 sequence illustrates the logical inconsistency in the way that many current systems are designed and later used. To begin, automation is justified because humans are becoming the weakest link in increasingly complex defense systems. But humans are left on the control loop to verify that the machine's performance is in accord with human intent. However, in many situations—including Patriot during OIF—this can be an unrealistic task demand. Humans make very poor monitors of complex processes (Davies and Parasuraman 1982). Moreover, the search for the disconfirming evidence necessary to override a machine decision or recommendation can require more resources (time, knowledge, and competence) than controllers can bring to bear on the situation. Norman (2007) states bluntly that any system operating between the extremes of full automation (no human involvement) and complete manual control (no machine involvement in at least the cognitive aspects of process or task control) occupies a "dangerous middle ground." In a similar vein, Woods and Hollnagel (2006: 125) cite what is termed "Robin Murphy's Law" to highlight the problem of being on this dangerous middle ground. The gist of Robin Murphy's Law is that *any* automated system will fall short of its anticipated level of autonomy, and this shortfall will lead to problems in achieving and maintaining effective human oversight. Maintaining effective human supervisory control of a complex process or task setting is a problematic issue.

A second aspect of the conundrum of human supervisory control is the residual functions problem. Recall Sheridan's earlier assertion that automated systems that do not leave some residual functions for human operators are rare (Sheridan 1992, 2002). During ARL's five-year Patriot Vigilance effort we frequently encountered two common misconceptions about automation that relate directly to the residual functions problem. The first misconception was that automation simply shifts tasks from operators to the machine. Hollnagel and Woods (2005) call this misconception the "substitution fallacy." We observed that belief in the substitution fallacy was widespread in the air defense community. Personnel in Patriot units accepted it; and people in the air defense command structure accepted it (Hawley and Mares 2006). We also observed that the substitution fallacy led to an uncritical use of automation in air defense operations. This uncritical use led to a general complacency about automation and any potential downside, i.e., "trust the system."

The second misconception, somewhat related to the first, was that automation simplifies training requirements. That is, if a function is automated, it is not necessary to worry about training personnel to carry it out. This second misconception contributed to the Patriot training situation characterized by the Army's OIF board of inquiry as emphasizing rote drills versus the exercise of high-level judgment. Simply put, Patriot training methods did not change to reflect a significantly changed job for air battle management crews.

It is axiomatic in the human factors literature that automation does not simply eliminate human operator tasks; rather, it changes the nature of the work that human operators do (Parasuraman and Riley 1997). Operators are recast as supervisory controllers. Similarly, it is also well documented in the human factors literature that automation does not reduce training requirements. Instead, it changes training requirements (in accord with changed job requirements) and often makes job preparation more demanding in terms of expertise and aptitude (Blumberg and Alber 1982, Boehm-Davis, Curry, Weiner, and Harrison 1983, OTA 1984, Zuboff 1988).

We have observed a similar pattern of development, training, and use in AMD systems other than Patriot, as well as in information-technology-dominated, non-AMD systems where automation is starting to be applied. Systems are developed following what has been termed a "Left-Over" philosophy of automation (Hollnagel and Woods 2005). Following the Left-Over approach, the technical parts of the system are designed to do as much as is feasible (usually from technical and efficiency points of view), while the rest is left for the operators to handle. The impact of these residual functions on operators and their ability to meet the associated demands typically are not evaluated until late in the system development process (e.g., at some initial test involving users). By then, the system's concept is set and it is often too late to do anything if serious human–automation integration problems are encountered. To some extent, the Patriot developmental community can be excused for having followed these practices. They were standard at the time Patriot development began, and no less an authority than Chapanis (1970) advocated the Left-Over approach and argued that it was reasonable.

If one accepts that automated support is required in some situations (such as battle management for AMD systems) and that automated systems to meet these requirements will be developed and deployed, what is the solution to the design problem? The consensus solution is generally to follow a design approach termed "human-centered automation" (for example, see Wickens et al. 1998, Endsley, Bolte, and Jones 2003, Hollnagel and Woods 2005, Woods and Hollnagel 2006). We will not go into the details of human-centered automation in this chapter other than to characterize its general philosophy. The gist of human-centered automation is that from the beginning of the system development process, developers must confront the substitution myth and challenge it. Technology can amplify human expertise, but not substitute for it. This caveat is particularly true for activities in the cognitive domain. The role of humans in system use and the larger socio-technical work system must be considered early on and evaluated in terms of its functional feasibility. We must emphasize, however, that none of this is straightforward or easy. Hollnagel and Woods (2005) remark, for example, that human-centered automation is an attractive but ill-defined concept that often raises more problems than it solves.

In a military setting, the traditional approach to system development also can present an obstacle to the development of effective automated systems. This obstacle pertains to what might be termed the "irreversible waterfall" from requirements definition through to deployment and field use. To illustrate this problem, briefly consider the case of the Federal Aviation Administration (FAA) Advanced Automation System (AAS) for air traffic control. Following the cancellation of the AAS in the mid-1990s, the FAA requested that the National Research Council Committee on Human Factors assess the reasons for the failure of the AAS initiative and provide the agency with recommendations for developing future automated air traffic control systems (Wickens et al. 1998). The Committee's ensuing recommendations to the FAA are relevant to the military automation problem as well as to the future of air traffic control.

To begin, the Committee supported the use of human-centered automation strategies in the development of future air traffic control systems. However, beyond human-centered automation, the Committee recommended an incremental approach to developing complex automated systems termed "build a little, test a little" (Wickens et al. 1998: 214). The panel noted that developers of complex automated systems have run into difficulties when they have followed a design approach that involved building a complete system according to complex specifications that must be amended as unforeseen technical problems develop, new technologies appear, and as more is learned about human performance on that system. They advocated instead a design approach based on early concept exploration prior to the development of system specifications. The Committee also noted that the successful deployment of automated systems requires that concept validation, prototype evaluation, and operational testing begin early in the developmental process. Hence, at no point in the developmental process has the design flowed down the "waterfall" in a way that prevents activities from being halted, reversed, or restarted.

To sum up, research and experience indicate that successful development of complex automated systems must (1) follow a human-centered approach, (2) be developed incrementally (build a little, test a little), and (3) be evaluated early and continuously. Unfortunately, this advice is not often followed in the development of Army systems. Routine system development practices usually result in Army systems being developed following the Left-Over approach, using detailed specifications developed upfront (the irreversible waterfall), and with most evaluation left until formal tests conducted prior to prescribed review points. Achieving effective human-automation integration is a technical challenge. But an even more daunting challenge to the development of future systems will be to modify traditional system development practices.

Training and Job Preparation

The second major challenge facing developers and users of complex military systems is training and job preparation. Recall from our previous discussion that the Army's fratricide board of inquiry, the Defense Science Board's assessment of Patriot system performance during OIF, and ARL's Patriot Vigilance assessment for MG Vane all cited inadequate training of Patriot air battle management crews as a major contributor to those incidents. Recall also that ARL's fratricide incident report called for a review of the level of *expertise* required to employ a system such as Patriot on the modern battlefield. Note also that the ARL team deliberately chose to use the term *expertise* instead of the Army's more standard term *training* when phrasing this recommendation. The skill development process underlying expertise is a function of extended practice following what has been termed a "deliberate practice instructional regimen:" focused repetition of key elements of job performance followed by feedback that is rapid and unequivocal (Ericsson 2009, Kahneman and Klein 2009).

Given the centrality of user expertise in the emerging warfighting environment, an obvious follow-on question is, "How is such expertise developed, and why is traditional Army training deficient in this respect?" Three training features are generally considered necessary for the development of expertise (Kozlowski 1998): (1) extensive deliberate practice defined as focused, job-relevant practice with expert feedback, (2) scenarios characterized by increasing variability and novelty that challenge routine skills, and (3) a focus on developing sensemaking skills that facilitate an operator's ability to recognize when to shift from automatic processing (rote drills) to critical thinking and problem solving. Adaptive expertise, or the ability to perform adequately under conditions of uncertainty, will develop as a natural consequence of the long-term application of this progressive instructional strategy. It should be emphasized, however, that all practice is not equal. Developing expertise requires a hands-on learning environment and hours and hours of practice under the supervision of a coach or mentor. How much time is necessary? Norman (1993) asserts that for any complex activity, a *minimum* of 5,000 hours of deliberate practice—2 years of full-time effort—are required to turn a beginner into a proficient operator. Proficient, in this context, describes a user who has developed the capability necessary to perform appropriately in a high-skill, knowledge-intensive job setting. Other research on the development of what are termed "high-performance" skills also supports this two-year rule (Schneider 1985).

We are not suggesting 5,000 hours as a hard-and-fast training duration benchmark for Patriot or any other Army system. Rather, we provide the 5,000-hour figure as reference point derived from the expertise literature. It should be noted, however, that training for FAA radar air traffic controllers and Israeli Patriot crews both involve more than 5,000 hours of training prior to full job qualification (Hawley, Mares, and Giammanco 2006). Training times for US Patriot crews vary, but are considerably shorter than the nominal 5,000-hour benchmark.

Beyond training times, there also is the issue of training methods. Crew-drill-oriented training coupled with uncritical reliance on automated support will not produce expertise as described above, which complex military systems exemplified by Patriot now require. Patriot training is not alone in this deficiency and should not be singled out for unique criticism. In a comment on standard Army methods for training complex cognitive skills, Shadrick and Lussier (2009) remark that the Army does not routinely apply deliberate practice training methods to instill cognitive skills. Why is deliberate practice so critical to the development of expertise? Recall our previous statement that the ultimate performance target for air battle management crews in a system like Patriot is recognitional decision-making, sometimes called skilled "intuition." Skilled intuition is a form of complex pattern recognition: the tactical situation provides cues; cues provide access to stored information; and the stored information suggests a response. Deliberate practice is crucial to developing these cue–response associations.

As has been stated, situational cues provide access to stored information, and what is retrieved is what matches best based on past experience. Humans interconnect specific information about individual things into larger and larger chunks. Chunks are further connected into associated groupings often called "schema." Building these knowledge structures from the bottom up through practice is basically what training and mentoring are all about. We want trainees to develop chains of actions in response to certain states of the real world. This is achieved through repetition of these cue–chunk/schema–response associations. If we want trainees to apply their knowledge in real-world situations, then elements of the real-world problem situation must trigger a pattern match to retrieve the appropriate knowledge and solution procedure. Deliberate practice focuses explicitly on these aspects of real-world problem situations (Shell et al. 2009).

The conventional solution to this pattern-matching requirement is to have trainees use their knowledge in situations similar to the real world. In this manner, critical aspects of these real-world situations are connected to relevant chunks or procedural condition elements and facilitate a later pattern match. This simple idea is the logic behind simulation-based training. Simulation-based training is intended to increase the likelihood that a given situation will lead to a pattern match and subsequent retrieval of appropriate facts or procedural knowledge.

An obvious limitation of this general approach is that with finite training time it is not possible to provide direct or simulated experience for every possible real-world situation. We need trainees to develop ways to retrieve relevant knowledge in situations that they might not have experienced. Developing the capability to retrieve knowledge through pattern-matching in situations that have not been encountered brings critical thinking and problem-solving into play. Critical thinking and problem-solving skills enable operators to identify or create pattern matches when none may explicitly exist. If existing sensory input has not triggered a pattern match, the operator must either keep looking for more information or transform sensory input into something that does trigger a pattern match. Critical thinking and problem-solving involve both continued search for new information and *restructuring* and *transforming* available information into different configurations that might lead to a pattern match. High-quality feedback from instructor–mentors is a critical aspect of developing the knowledge structures (chunks, schemas, and connections) underlying the restructuring and transforming processes. The Patriot training community's failure to move beyond the first stage of pattern-matching and address critical thinking and problem- solving skills is the essence of the Army fratricide board of inquiry's critical remark about emphasizing rote drills versus the exercise of high-level judgment.

Developing expertise in the manner described above requires *time* (Norman's minimum of 5,000 hours) and experienced instructor–mentors (expert job performers) to provide focused, high-quality, real-time feedback to trainees. Research has also shown that job-relevant skills developed following this approach are highly situated to a specific job situation (Bransford and Schwartz 2009). This finding likely explains why expertise developed for one job situation does not transfer directly to another. It also casts some doubt on the ability of generic training programs for critical thinking or problem-solving to lead directly to performance improvement in specific job domains. Generic training on these topics might make job-focused training more efficient, but will not prove to be a suitable substitute for training within a specific job context. Expertise is situated, and training to develop that expertise also must be situated.

These observations regarding the training implications of complex military systems are not new or unique to AMD. For example, in the first of two reports on training for future conflicts, the Defense Science Board warned of an increasing risk that training failures will negate hardware promise (DSB 2001). The Board's 2003 follow-on report further remarked that the future will require that more of our people do new and more complicated things, and "meeting this challenge amounts to a qualitative change in the demands placed on our people that cannot be supported by traditional training practices" (DSB 2003: 38). The background theory on the necessity for and development of expertise is straightforward; the challenge going forward is applying this theory in the contemporary Army training environment.

Related areas where change must occur are personnel administration and crew and unit staffing practices. Put bluntly, the Army's personnel management system must support the development of individual, crew, and unit competence rather than impede it. The DSB noted, for example, that personnel policies have a huge influence on unit readiness, and that personnel turbulence limits unit performance by playing "musical chairs" with unit manning (DSB 2003). The DSB further warned that current personnel practices will not deliver the higher levels of cohesion that warfare transformation will require. Reformed training practices overlaid on the Army's current personnel management practices might not produce the desired results.

Test and Evaluation

The third major challenge facing developers of complex systems is test and evaluation (T&E) concepts and practices. One of the primary observations coming out of the ARL team's experiences with the Patriot system during an extended test sequence is that operational testing of complex systems must be more comprehensive and rigorous, particularly where the impact of human performance on system reliability and effectiveness is concerned. The importance of comprehensive and rigorous operational testing involving the human component is illustrated in Figure 1.3. As shown in this figure, notional control risk increases when moving from the center of the circle to the margins. Based on our observations, most training and testing take place near the center of the circle, in an area labeled "Median of practice"— things we understand and routinely do. This is like testing a new aircraft by only flying it straight and level in clear weather. While this is cheap and easy to do, it does not reveal problems that could show up as maneuver complexity increases or flying conditions deteriorate. Moving out from the Median of Practice toward what is termed "Margins of practice," two additional zones are encountered. These are a zone of "Known but unresolved issues" and, finally, an area of "Surprises"— events that might be theorized but have not yet been experienced. These regions are the source of most of a system's performance problems.

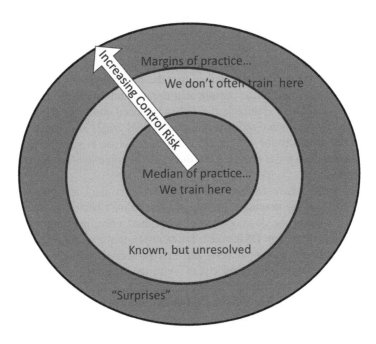

Figure 1.3   System control risk and testing and training practices

Training and testing become more expensive and technically difficult as one moves away from the Median of practice toward the Margins of practice. This is particularly true for complex systems like Patriot and other AMD systems. However, as the DSB stated in its post-OIF review of Patriot operations, training and simulations must support the ability to train in confusing and complex scenarios that contain unbriefed surprises (DSB 2004). In the ARL team's view, the DSB's training recommendation applies equally to testing practices. It can be taken as tantamount to a call to move away from the Median of practice during testing toward the fringes of the circle—the area of increasing control risk. For systems like Patriot, it is essential that developers and users understand the risks associated with potential usage situations. Only then can operators be properly prepared to exercise effective oversight of system operations (Hawley 2007a, Kahneman and Klein 2009). As the events of OIF illustrate, the combat environment can be expected to dish up situations on the Margins of practice. The Tornado fratricide resulting from a known track misclassification problem undeniably was such a situation.

Another problematic issue in contemporary operational testing is a strong hardware–software bias during test design and conduct. For example, during the Patriot operational test, ARL's MANPRINT support team encountered a strong bias on the part of testers to look at the system primarily from a hardware and software perspective. The impact of operator performance on overall system effectiveness was regarded as a secondary issue. Looking back on the situation during the Patriot test, the fact that MANPRINT is classified as a supportability issue and not a system performance issue during test design and conduct undoubtedly contributes to this hardware–software bias. Sidney Dekker argues that complex systems require an "overwhelming human contribution" for their effective operation. He goes on to state that "people are the only ones who can hold together the *patchwork of technologies* in their world; the only ones who can make it work in actual practice" (2002: 103; emphasis added). If a system's effectiveness depends on competent human performance, then testing of that system must be done with human operators who are appropriately trained and prepared to perform in their assigned roles.

Conversations with the Patriot Army Training and Doctrine Command (TRADOC) Capability Manager (TCM) during the operational test provided an opportunity to look deeper into the potential relationship between T&E practices and the later events of OIF. During a discussion of emerging test results, the TCM said something to the effect that "We've been doing human factors work on Patriot for more than 20 years. How did we miss all of the things you wrote about and we're now seeing? Certainly all of these things are not new." These questions and the implied challenge prompted the ARL project staff to ask the obvious follow-on questions, "How did we miss these high-driver issues? Or, did we note these things and nothing came of them?" These latter questions led to a review of available Patriot-related MANPRINT assessments and test reports going back to the start of the Army's formal MANPRINT initiative in the mid-1980s (Hawley 2007b). The team's review indicated that pre-OIF Patriot assessments focused on what might be termed "1472 issues" (control and display features, symbology concerns, and the like; the concerns addressed in MIL-STD-1472, Human Engineering) and user-jury-style, paper-and-pencil evaluations (the extensive use of after-action questionnaires and interviews during system evaluation).

We noted earlier that over the course of its 25-year operational life, Patriot evolved from a system that was quite literally just one step more advanced than its predecessor, the Hawk missile system, to the complex AMD system that exists today. As more technology and software capability were added to the Patriot system, the operators' roles evolved from traditional manual control to supervisory control, bringing with it everything that goes with that control mode. However, the MANPRINT assessments performed for progressive system upgrades were not substantially different from those performed early in the system's development. We also observed that several of the performance problems that became apparent during OIF were starting to show up in earlier test results. The bottom line with respect to the Patriot system's T&E program was that testing generally missed the elephants in the middle of the room. These were (1) an eroding potential for effective human control, and (2) the need for increased levels of operator expertise—not simply additional training of the "traditional" kind, but training with a focus on developing operator expertise. The DSB's post-OIF review of Patriot system performance focused quickly on these issues, as did ARL's Patriot Vigilance assessment. The Army's own fratricide investigation also alluded to inadequate training as a contributor to the incidents.

It should also be noted that problems similar to those encountered with Patriot during OIF were apparent in related domains going back to near the time the Patriot system was initially fielded. These domains include nuclear power operations and the 1979 Three Mile Island incident; Navy anti-air operations and the 1988 USS Vincennes–Iranian Airbus incident; and problems with the FAA's Advanced Automation System in the 1990s. All of these incidents were reported and discussed widely in the human factors literature, yet no mention of them is found in Patriot MANPRINT or test-related documents—despite the fact that Patriot's operational concept has many features in common with these systems. Is Patriot alone in these kinds of deficiencies? The available evidence suggests that it is not. A case can be made that the deficiencies we observed on Patriot are endemic to the Army's T&E enterprise (Hawley 2007a, Hawley and Mares 2009).

What changes to T&E concepts are suggested going forward? Based on our experiences, three avenues are suggested with respect to T&E of complex systems. First, MANPRINT testers—defined as test planners charged with structuring the human performance aspects of test and experimentation programs—must be proactive with respect to human performance problems identified in predecessor and parallel systems. For example, Three Mile Island, the Vincennes, and the FAA's AAS all were parallel to or even somewhat prior to the formulation of Patriot's test program. At a conceptual level, the problems that plagued these systems should have been expected to present a problem for Patriot. Upfront concept developers must address the issue of what can be done to avoid such problems. Later on, test and experimentation planners must explicitly look for them (Woods and Dekker 2001, Dekker 2002, Dekker 2006).

Second, there must be a greater emphasis on experimentation as opposed to formal T&E early on and over the course of system development. The focus of these experiments must be the "build a little, test a little" idea introduced earlier. They should be formative rather than summative in orientation—concerned with evolutionary improvement rather than a Go/No-Go evaluation. Systems undergoing formative evaluation are permitted to "fail," and we are developing an operational experience base with them. Another consideration is that early-on experiments permit problems to be identified and corrected before major resources commitments are made. It is often too late to do anything about problems identified during formal T&E. By then, the system's concept and much of its materiel is mostly set, and funding has been programmed.

Finally, formal T&E should occur only after the system has been shaken down during a series of earlier experimental exercises and an experienced test player cadre developed. Only then can the system be given a meaningful summative evaluation. At present, system evaluation is focused on formal T&E immediately prior to major system milestones. During these tests, system components often are immature, doctrine and procedures are not fully validated, and test players are frequently too inexperienced to provide meaningful human system integration results (Hawley 2007a). It is very difficult to draw meaningful test conclusions under such conditions. Experimentation and formal testing should be threaded together in the same fashion as recommended for formative and summative program evaluation (Alberts and Hayes 2002).

Lessons Must Be Learned

The fourth challenge facing developers and users of complex systems is incorporating the lessons of experience. That is, making system development and subsequent usage practices more evidence-based than is the situation at present. Change is a tough issue for any large, bureaucratic organization, and we are not so bold as to suggest that our perspicacity and insight into the change process are better than those qualities in others who work in this field. We will simply note from our experiences that the Army appears have a problem translating research, analysis, and test results into meaningful change. This is particularly true for the soft or non-materiel aspects of system development. We also observed that many of the Army's barriers to change have less to do with cost or technical feasibility than with culture, inflexible practices, and an often out-of-date regulatory structure. In other words, the barriers to change are mostly self-imposed. Difficulty in making soft change happen will present difficulties for personnel charged with developing the Future Force and realizing its potential. The noted defense analyst Anthony Cordesman has observed, for example, that "No country does better in making use of military technology than the US, but nor is any country so wasteful, unable to bring many projects to cost-effective deployment, and so prone to assume that technology can solve every problem" (Cordesman 2008: 53).

Former Chief of Staff of the Army, General Eric Shinseki, is credited with the observation that lessons are not learned until a resulting change occurs. If change does not occur, the lessons are merely "observed." For example, in the case of Patriot after OIF, a great many lessons were identified and slated for resolution. At the time of this writing, more than five years after the events of OIF, only a fraction of these lessons have been "learned" in the sense that a salutary change has occurred. Most issues remain unresolved. We remarked earlier on our own experiences with attempting to implement changes in air battle management organization, practices, and training for Patriot. We had some partial successes, mostly in the materiel arena, but soft changes have proven far more difficult to implement. This was in spite of considerable research and empirical evidence suggesting that these soft changes would prove beneficial.

Why are soft factors so difficult to change? Routine processes exist for modifying materiel—hardware and software—but similar mechanisms for bringing about non-materiel or soft changes (e.g., the doctrine, organization, training, leadership and education, personnel, and facilities process) are not as effectively implemented. We discussed earlier that ARL's test support team made a strong empirical case that the level of expertise required for upgraded Patriot operations "exceeds the current Army training standard." That phrase was picked up by the Department of Defense's Director of Operational Test and Evaluation and included in a later report to Congress (DOTE 2008). In spite of that evidence, the Army T&E community judged that they did not have the authority to require changes in Patriot institutional or unit training. The concerned Army commands could be advised of Patriot's training problems, but no formal response on their part was required.

Conclusion

This retrospective case study assessment of AMD operational practices along with the ARL team's successes, failures, and follow-on lessons was made possible because of a unique research opportunity that resulted from the Patriot fratricides during OIF. From the outset of the Patriot Vigilance effort, the ARL project staff was allowed relatively unfettered access to relevant documents, incident participants, technical experts, decision-makers, and opinion leaders from the AMD community. Our work would not have been possible without the support and participation of many members of that community. That said, we also want to emphasize that the preceding discussion is presented in the spirit of action research. Action research is characterized as self-directed research that any community of practice can do with the aim of improving its strategies, methods, and knowledge of the environment in which it practices (Schein 2004).

Throughout the chapter, we have argued that the Patriot system is more representative of systems the Army will acquire than of current systems. In that sense, Patriot's problems during OIF portend human performance issues that might accompany systems now on the drawing board. When the Patriot Vigilance project was initiated in the spring of 2004, the present lead author was serving as co-lead on the MANPRINT effort for the Army's Future Combat Systems program. He was struck by the parallels between current Patriot concepts and operations and what was planned for the FCS system of systems. It was also clear that even after more than 50 years of experience with Patriot and predecessor systems, major human performance problems were still apparent. The push of technology coupled with the pull of operational necessity—such as engaging TBMs—had led to the development of a class of systems that were different from the Army's norm. Further, the Army increasingly was having problems developing and properly supporting them once they were in the field. Moreover, the rest of the Army was now going down a similar technology-intensive path in system development and deployment. The ARL team came to believe there were lessons in the Patriot experience for the rest of the Army. We have attempted to articulate several of those lessons throughout the chapter.

There also are lessons in the Patriot experience for those whose primary interest is applied human factors in a systems acquisition setting (that is, MANPRINT). Previous discussions have alluded to several areas where improvements in applied human factors practice are suggested. However, we will highlight two areas where practice modification is potentially beneficial. First, it is necessary to take a sociotechnical systems perspective when doing applied human factors work on a complex system of systems like Patriot (Vicente 2006). We described earlier how many of Patriot's battle management modifications involved changes to the work system in which the system's hardware components were embedded. Our observations suggested that work-system-level, macroergonomic considerations are as, and possibly even more, important to system effectiveness than component-specific, microergonomic considerations.

Second, usability and operability assessments must attend to what is going on behind the operators' situation displays, so to speak. Technical and tactical complexity, cognitive workload, and other factors potentially impacting operator performance are driven mostly by the requirement to make sense of what is presented to operators on their displays (Klein 2009). Investigating sensemaking involves more than an assessment of display design and ease of use issues considered out of a valid operational context. Assessment in a dynamic, externally valid environment is essential.

Finally, applied human factors practitioners must bear in mind that in some sense they are acting as change agents. Although the ARL team did not begin with this intent in mind, we became an indirect change agent—a catalyst for change, working through others. We realized early in the Patriot Vigilance project that to be effective in helping to bring about beneficial changes, we could not stop with the initial incident assessment and recommendations. We had to become involved in implementing those recommendations. That brought us square into the change arena with all of the difficulties and frustrations associated with that role. It might be said we often became the face of and primary spokesman for soldier-performance-related change in the AMD community.

References

Alberts, D.S. and Hays, R.E. 2002. *Code of Best Practices for Experimentation*. DoD CCRP Publication Series. Washington, DC: US Government Printing Office.

Arthur, W.B. 2009. *The Nature of Technology: What it is and How it Evolves*. New York: Free Press.

Bainbridge, L. 1987. Ironies of automation, in *New Technology and Human Error*, edited by J. Rasmussen, K. Duncan and J. Leplat. New York: Wiley, 276–83.

Blumberg, J. and Alber, A. 1982. The human element: Its impact on the productivity of advanced batch manufacturing systems. *Journal of Manufacturing Systems*, 1, 43–52.

Boehm-Davis, D.A., Curry, R.E., Weiner, E.L. and Harrison, R.L. 1983. Human factors of flight-deck automation: Report on a NASA-industry workshop. *Ergonomics*, 26, 953–61.

Bransford, J.D. and Schwartz, D.L. 2009. It takes expertise to make expertise, in *Development of Professional Expertise: Toward Measurement of Expert Performance and Design of Optimal Learning Environments*, edited by K.A. Ericsson. Cambridge: Cambridge University Press, 432–48.

Chapanis, A. 1970. Human factors in systems engineering, in *Systems Psychology*, edited by K.B. DeGreene. New York: McGraw-Hill, 28–38.

Cordesman, A.H. 2008. *The Lessons of the Israeli–Lebanon War*. Washington, DC: Center for Strategic and International Studies.

Davies, D. and Parasuraman, R. 1982. *The Psychology of Vigilance*. New York: Academic Press.

Dekker, S. 2002. *The Field Guide to Human Error Investigations*. Burlington, VT: Ashgate.

——. 2005. *Ten Questions about Human Error: A New View of Human Factors and System Safety*. Mahwah, NJ: Erlbaum.

——. 2006. *The Field Guide to Understanding Human Error*. Burlington, VT: Ashgate.

DSB (Defense Science Board) 2001. *Training Superiority and Training Surprise*. (Final Report of the DSB Task Force on Training Superiority and Training Surprise). Washington, DC: Office of the Undersecretary of Defense for Acquisition, Technology, and Logistics.

——. 2003. *Training for Future Conflicts*. (Final Report of the DSB Task Force on Training for Future Conflicts). Washington, DC: Office of the Undersecretary of Defense for Acquisition, Technology, and Logistics.

——. (2004). *Patriot System Performance*. (Final Report of the DSB Task Force on Patriot System Performance). Washington, DC: Office of the Undersecretary of Defense for Acquisition, Technology, and Logistics.

DOTE (Director, Operational Test and Evaluation) 2008. *FY 2007 Assessment of the Ballistic Missile Defense System*. Washington, DC: Office of the Secretary of Defense.

Endsley, M.R., Bolte, B. and Jones, D.G. 2003. *Designing for Situation Awareness: An Approach to User-Centered Design.* New York: Taylor & Francis.

Ericsson, K.A. 2009. Enhancing the development of professional performance: Implications from the study of deliberate practice, in *Development of Professional Expertise: Toward Measurement of Expert Performance and Design of Optimal Learning Environments*, edited by K.A. Ericsson. Cambridge: Cambridge University Press, 405–31.

Ericsson, K.A. and Charness, N. 1994. Expert performance: Its structure and acquisition. *American Psychologist*, 49(8), 725–47.

Fischerkeller, M.P., Hinkle, W.P. and Biddle, S.D. 2002. The interaction of skill and technology in combat. *Military Operations Research*, 7(1), 39–56.

Hawley, J.K. 2007a. Training and testing: A complex and uneasy relationship. *ITEA Journal of Test and Evaluation*, 27(4), 34–40.

——. 2007b. *Looking Back on 20 Years of MANPRINT on Patriot: Observations and Lessons.* (Technical Report ARL-SR-158). Adelphi, MD: US Army Research Laboratory.

Hawley, J.K. and Mares, A.L. 2006. *Developing Effective Human Supervisory Control for Air and Missile Defense Systems.* (Technical Report ARL-TR-3742). Adelphi, MD: US Army Research Laboratory.

——. 2007. *Developing Effective Adaptive Missile Crews and Command and Control Teams for Air and Missile Defense Systems.* (Technical Report ARL-SR-149). Adelphi, MD: US Army Research Laboratory.

——. 2009. Training and testing revisited: A modest proposal for change. *ITEA Journal of Test and Evaluation*, 30(2), 251–57.

Hawley, J.K., Mares, A.L. and Giammanco, C.A. 2005. *The Human Side of Automation: Lessons for Air Defense Command and Control.* (Technical Report ARL-TR-3468). Adelphi, MD: US Army Research Laboratory.

Hawley, J.K., Mares, A.L. and Giammanco, C.A. 2006. *Training for Effective Human Supervisory Control of Air and Missile Defense Systems.* (Technical Report ARL-TR-3765). Adelphi, MD: US Army Research Laboratory.

Hawley, J.K., Mares, A.L., Fallin, J.I. and Wallet, C. 2007. *Reconfigurable Tactical Operations Simulator (RTOS) Operational Demonstration in 5–52 Air Defense Artillery.* (Technical Report ARL-SR-156). Adelphi, MD: US Army Research Laboratory.

Hendrick, H.W. and Kleiner, B.M. (eds) 2002. *Macroergonomics: Theory, Methods, and Applications.* Mahwah, NJ: Lawrence Erlbaum.

Hollnagel, E. and Woods, D.D. 2005. *Joint Cognitive Systems: Foundations of Cognitive Systems Engineering.* New York: Taylor & Francis.

Kahneman, D. and Klein, G. 2009. Conditions for intuitive expertise: A failure to disagree. *American Psychologist*, 64(6), 515–26.

Klein, G. 2009. *Streetlights and Shadows: Searching for the Keys to Adaptive Decision Making.* Cambridge, MA: MIT Press.

Kozlowski, S.W.J. 1998. Training and developing adaptive teams: Theory, principles, and research, in *Making Decisions under Stress: Implications for*

Individual and Team Training, edited by J. Cannon-Bowers and E. Salas. Washington, DC: American Psychological Association, 115–53.

Merriam-Webster Online Dictionary. [Online]. Available at: http://www.merriam-webster.com/ dictionary/complex (accessed: July 25, 2012).

Norman, D.A. 1993. *Things that Make Us Smart*. Cambridge, MA: Perseus Books.

——. 2007. *The Design of Future Things*. New York: Basic Books.

OTA (Office of Technology Assessment) 1984. *Computerized Manufacturing Automation: Employment, Education, and the Workplace*. (OTA-CIT-235). Washington, DC: US Government Printing Office.

Parasuraman, R. and Riley, V. 1997. Humans and automation: Use, misuse, disuse, abuse. *Human Factors*, 39(2), 230–52.

Reason, J. 1990. *Human Error*. Cambridge: Cambridge University Press.

Salas, E., Bowers, C.A. and Rhodenizer, L. 1998. It is not how much you have but how you use it: Toward a rational use of simulation to support aviation training. *International Journal of Aviation Psychology*, 8(3), 197–208.

Schein, E.H. 2004. *Organizational Culture and Leadership*. San Francisco, CA: Jossey-Bass.

Schneider, W. 1985. Training high-performance skills: Fallacies and guidelines. *Human Factors*, 27(3), 285–300.

Shadrick, S.B. and Lussier, J.W. 2009. Training complex cognitive skills: A theme-based approach to the development of battlefield skills, in *Development of Professional Expertise: Toward Measurement of Expert Performance and Design Of Optimal Learning Environments*, edited by K.A. Ericsson. Cambridge: Cambridge University Press, 286–311.

Shell, D.E., Brooks, D.W., Trainin, G., Wilson, K.M., Kauffman, D.F. and Herr, L.M. 2009. *The Unified Learning Model: How Motivational, Cognitive, and Neurobiological Sciences Inform Best Teaching Practices*. New York: Springer.

Sheridan, T.B. 1992. *Telerobotics, Automation, and Human Supervisory Control*. Cambridge, MA: MIT Press.

——. 2002. *Human and Automation: System Design and Research Issues*. New York: Wiley.

Simon, H.A. 1992. What is an explanation of behavior? *Psychological Science*, 3, 150–61.

Singer, P.W. 2009. *Wired for War: The Robotics Revolution and Conflict in the 21st Century*. New York: Penguin Press.

USAF (United States Air Force) 2009. *Unmanned Aircraft Systems Flight Plan, 2009–2047*. Washington, DC: Department of the Air Force.

Vicente, K. 2006. *The Human Factor: Revolutionizing the Way People Live with Technology*. New York: Routledge.

Weick, K.E., Sutcliffe, K.M. and Obstfeld, D. 2005. Organizing and the process of sensemaking. *Organizational Science*, 16, 409–21.

Wickens, C.D., Mavor, A.S., Parasuraman, R. and McGee, J.P. (eds) 1998. *The Future of Air Traffic Control: Human Operators and Automation*. Washington, DC: National Academy Press.

Woods, D.D. and Dekker, S. 2001. Anticipating the effects of technological change: A new era of dynamics for human factors. *Theoretical Issues in Ergonomic Science*, 1(3), 272–82.

Woods, D.D. and Hollnagel, E. 2006. *Joint Cognitive Systems: Patterns in Cognitive Systems Engineering*. New York: Taylor & Francis.

Zuboff, S. 1988. *In the Age of the Smart Machine: The Future of Work and Power*. New York: Basic Books.

Chapter 2

Who Needs an Operator When the Robot is Autonomous? The Challenges and Advantages of Robots as Team Members

Keryl A. Cosenzo and Michael J. Barnes

Unmanned air and ground systems (that is, robots) are an integral part of current and future military operations. Robotic control is currently a continuous and cognitively demanding task. More autonomous robotic systems with diverse roles, tasks, and operating requirements are being designed to exploit future battle spaces. These advancements will provide a tactical advantage against conventional and unconventional warfare. This advantage, however, may come at a cost. The prevailing expectation in the robotics community is that autonomy will enable robots to function with little or no human intervention. However, we propose that most of the contemplated systems will require either active human control or, at a minimum, human supervision with the possibility of intervention. All levels of the military command structure will use robotic assets. Use of the robot goes beyond moving the platform from point A to point B. The operator, who is responsible for the robots, will have a multitude of tasks to complete: route planning for the robot, monitoring the robot during the mission, monitoring and interpreting the sensor information received by the robot and communicating the sensor information with others. Control and use of these assets will no longer be the responsibility of a few specially trained operator(s), but the responsibility of many. In the most extreme case, the operator will be responsible for multiple systems while on the move and while under enemy fire. As a result, the addition of robotics can be considered a burden on the operator(s) if not integrated appropriately into the system. We will discuss the challenges of robotics on soldier performance and solutions to maximize the benefits of this game-changing technology.

Background

Military unmanned systems operations are continuous and cognitively demanding. In Operation Enduring Freedom and Iraqi Freedom, robots are being employed and have been effective in combat tasks that are "dull, dirty, and dangerous" (*Economist* 2007, Axe 2008). Remotely controlled robots play a vital role in current operations, including disarming roadside bombs (that is, improvised explosive devices (IEDs)) and performing ground-based and air-based patrol missions.

Small ground and air robotic platforms have been used successfully in current operations. Small unmanned air vehicles (UAVs) range in size from small hand-launchable (for example, RAVEN) to larger systems with a launching system (for example, SHADOW). UAVs are being used to support the conflicts in Iraq and Afghanistan. At the Platoon Leader and Division levels, they provide ground maneuver commanders with timely and critical combat information (Kappenman 2008). Army UAVs have flown over 375,000 hours and 130,000 sorties as of 2008. UAV autonomy ranges from teleoperation to semi-autonomy. Current UAV capabilities have evolved from an intelligence role to a tactical role in reconnaissance, surveillance, and target acquisition (RSTA); communications relay; convoy overwatch; and cooperative target engagement via human–robot teaming. Army UAVs provide immediate responsiveness with eyes on the target without waiting for an information request from the Division level. A more recent application for UAVs has been in the IED defeat mission, where there has been sensor-to-shooter handoff for IED emplacers. Linking the unmanned systems to the manned systems provides an overwatch on a target while a commander can determine the prosecution strategy.

Ground robotic systems, unmanned ground vehicles (UGV), are also being used in current operations. In the war against insurgents and IEDs, robots are one of the military's best weapons (Singer 2009). However, the use of UGVs is not as prolific as UAVs (Department of Defense 2009). There are currently about 6,000 UGVs in theater. Unlike UAVs, UGV autonomy is currently teleoperational only. UGVs are small platforms, such as the Packbot or MARCbot, and are used for reconnaissance for infantry and engineering units, convoy operations, and for IED defeat.

Future Force Concepts and Robotics

As described previously, robots have been an invaluable weapon in the current fight and will continue to be so in the future. Currently, robots are used by a few specialized operators and they are not fully integrated in the team. Future unmanned systems are envisioned as force multipliers by virtue of the fact that they are unmanned. It is postulated that robots will enable the same number of personnel to control a greater area of responsibility and take on more mission capacity, than if there were no robotic team members (Department of Defense 2009). The robots will conduct long, persistent reconnaissance and surveillance in all domains to include battlespace awareness, lethal and non-lethal force application, logistics, and force protection. This vision relies on more autonomy in navigation and mobility than currently used in theater. The goal is to have interchangeability across platforms, where the controller-to-platform ratio is 1-to-n, and a single controller can monitor and supervise multiple systems across domains as a collaborating team. The vision of the future for military robotics is that they will enable mission endurance over extended periods of time.

To achieve this vision, autonomy and supervisory control will be very important. In military operations, the soldier who is responsible for the robots will have a multitude of tasks to complete including route-planning for the robot, monitoring the robot (current and proposed route and mission plan) during the mission, monitoring and interpreting the sensor information received by the robot, and communicating the sensor information with others (Chen, Haas, and Barnes 2007). Embedded in these high-level tasks are smaller, more specific tasks, which include the following:

- assuring the platform is within expected or acceptable geographic and mission parameters;
- assuring the payload is best used to meet mission requirements;
- selecting best sensor(s) to use for a particular target/mission;
- determining most effective method to provide sensor coverage;
- determining if object is target of interest;
- determining if the target should be fired upon and whether call for fire should be issued;
- determining what information needs to be disseminated to others.

Due to the multitasking nature of the military environment, the soldier's tasks, robotic tasks, and crew safety can be compromised during high workload mission segments if autonomy and decision aids are not applied appropriately (Chen and Joyner 2009, Mitchell and Henthorn 2005). Our premise is that for this vision to be realized, the development of autonomy will not be sufficient. A paradigm shift is going to have to occur where new military tactics and teaming procedures will need to be developed for manned and unmanned teaming. The robot will need to become a team member, and the soldier will need to trust in that team member and the information and decision being made by that robotic entity.

Approach

Research on human robot interaction (HRI) has led to significant insights into the challenges of integrating unmanned systems into the human team. Robots are an asset, especially when the insertion of a human is not feasible or desirable (Casper and Murphy 2003, Riley, Murphy, and Endsley 2006). The effectiveness of HRI drives overall operational performance. The HRI community has researched issues such as levels of autonomy (for example, Wickens, Dixon, and Ambinder 2006), trust in automation (for example, Parasuraman, Sheridan, and Wickens 2000, Parasuraman and Riley 1997), individual differences (for example, Chen and Joyner 2009), multitasking constraints on robotic operations (for example, Cosenzo, Parasuraman, Novak, and Barnes 2006), and scaleability and interface concepts (for example, Redden, Carstens, and Pettitt 2010), to name a few. Research advancements are being made that will enable the realization of the goals to have robots as team members, assets not burdens, and force multipliers.

We will discuss HRI in terms of three dimensions: level of automation, specialized interfaces for supervisory control, and soldier–robot teaming. These dimensions are interrelated and a single system may fall on multiple points in this space. The main issues we will discuss will be type of automation and, in particular, the implications of the operator moving from manual to supervisory control, multitasking for workload issues, and types of interfaces and their impact on the previously mentioned dimensions. We will report data from field tests, simulations, and laboratory experiments in order to elucidate problems encountered and solutions we were able to verify.

Level of Autonomy

There are four general operational modes for unmanned systems: remote control, teleoperation, semi-autonomous, and fully autonomous (Huang, Messina, and Albus 2007). The difference between these modes is the level of autonomy on the unmanned system. Autonomy is the gradual property of the system related to the degree of intervention required by the operator. In remote control and teleoperation, the human operator is in direct control, either onsite (in the case of remote control) or remotely (in the case of teleoperation), of all the capabilities of the platform (such as sensing, perceiving, analyzing, planning, deciding, acting, and communicating). In semi-autonomous mode, the operator and the robot each controls some aspects of the robot's capabilities. In full autonomy, the unmanned system determines and executes all of the robot's capabilities; the operator's role is supervisory.

There are trade-offs to the varying levels of autonomy, especially in terms of situational awareness (SA). For robotic applications, SA is being aware of where the vehicle is, what it is doing, and how the activities are going to lead to a mission goal. With teleoperation or remote control, the operator has diminished sensory information and limited attentional resources to dedicate elsewhere, outside the robot control, which can contribute to poor SA outside of the robot's behavior. When autonomy is added (to compensate for this SA loss and to alleviate operator taskload) there can still be SA problems due to operator complacency. To further complicate the issue, it is plausible that robot control is only a subset of the tasks for which an individual is responsible.

Teleoperation

Because of their flexibility and low cost, manually controlled robots will remain an important option for current and future military operations. Also, for the foreseeable future, supervision of autonomous and near-autonomous systems will require human intervention for safety and tactical reasons (Chen, Haas, and Barnes 2007). The main issues we will address are those related to the ease of teleoperation while concurrently conducting combat tasks for mounted and dismounted operations.

In current military operations, the operator who is teleoperating a robot is holding a display and viewing a screen while driving the robot. An additional soldier is maintaining security and SA of the environment and may also be assisting the robot operator with monitoring the robot's video imagery. Thus, the task of teleoperating a robot requires two or more operators, one for robot mobility, and one or two more people for maintenance of SA of local environment and of the robot. This role structure is also evident in search-and-rescue operations (for example, Murphy 2004). Especially for the dismounted military missions, it is essential that soldiers are able to maintain SA while conducting the robotic missions. This will require heads-up scanning or tactical procedures that allow them to split their attention efficiently between their robotic tasks and their immediate environment. It is equally important that control devices be lightweight and accessible so as not to impede mobility.

Redden and her colleagues have investigated a number of interface options for dismounted operations during field tests at Fort Benning (Figure 2.1). They found that 3.5-inch displays were sufficient for small UGV performing simple navigation and target acquisition tasks (Redden et al. 2008). A goggle-mounted display with one eye uncovered was also a possible option; however, the display had problems with binocular rivalry and there were multiple operator complaints about it. Furthermore, during subsequent studies they found that equipping soldiers with monocular helmet-mounted displays resulted in poorer UGV navigation compared than hand-held devices (Redden et al. 2010).

For more complex navigation tasks, additional information such as a map display will be needed for UGV operations. Redden et al. (2010) examined three options to combine map and video control: (1) toggle between two 3.5-inch displays; (2) use a 6.5-inch split-screen display; or (3) use a 3.5-inch display with a tactile belt to replace the visual map display. They found that navigation with a 3.5-inch display with tactile augmentation performed as well as with the 6.5-inch split-screen display. This indicated that tactile belts were an efficient means of displaying map information while reducing display real estate. Toggling back and forth between the two 3.5-inch displays proved to be less effective. In a similar fashion, field tests at Fort Bliss showed that a tactile belt whose tactors produced a limited lexicon could be used effectively to direct soldiers controlling UGVs to targets hidden in desert terrain, when compared to the use of chat and voice communications (Hutchins et al. 2009).

These and other studies suggest that a mix of modalities, rather than depending solely on the visual modality will result in more flexible interface options for robot control (Haas and van Erp 2010). This is particularly true because soldiers already encounter a surfeit of visual and auditory noise during combat. Speech command and control is also being considered for robotics. Speech-based control

Figure 2.1   Soldier using the goggle-mounted display (Redden et al. 2008)

has been demonstrated to be a natural way for humans to interact with robots, with the advantage that it allows direct access to complex menu structures and discrete functions. However, the current speech-based algorithms are not perfect, and research has shown that they are inefficient when processing continuous commands, such as turning functions (Redden, Carstens, and Pettitt 2010).

As a whole, the research suggests that there is not an ideal solution for soldiers to maintain their mobility while conducting heads-up control of teleoperated robots. Tactical solutions, such as bounding (moving and stopping to control), proved to be popular with soldiers and were effective when compared to attempting robotic control during soldier movement. It is possible that a new generation of helmet-mounted displays (HMD) and speech-based controls will overcome these obstacles by improving soldiers' SA during movement. On a positive note, lightweight interfaces that are able to control multiple unmanned systems will be feasible in the near future.

Semi-autonomy

In the near future, the robotic operator will use semi-autonomous control to direct the remote UGV to its destination. Autonomy is not yet perfect and there will be times that require operator intervention (Alban et al. 2009). When intervention via teleoperation is required, the operator's ability to divide his/her attention to other tasks is compromised. Chen and colleagues (Chen and Terrence 2008, 2009, Chen and Joyner 2009) conducted a series of experiments examining the multitasking constraints on operator performance during robotic missions. Overall, the results have indicated that there are severe multitasking requirements for the robotic operator. For example, when the operator must maintain local security (in the case of a gunner) while conducting robotic missions, the operator evidence showed increased workload and loss of local security SA when robotic tasks were added to the mix. Operator performance decrements for the primary (non-robotic) local security task were almost linear as a function of robotic task difficulty (teleoperation—semi-autonomy). Further, Chen and Joyner (2009) showed that target-detection, via the robot's cameras, was better during teleoperation compared to the semi-autonomous robotic conditions. The secondary task of local security around the operator's own vehicle showed the most severe performance costs in terms of missed threats near the manned vehicle during teleoperation. This was most likely due to the increased attentional demands required for teleoperation in the robotic task, causing an attention deficit in the local security task.

Thus, there are situations in which the additional attentional focus required for manual control may actually be advantageous for robotic tasks. The problem is that for difficult tasks, research shows that manual control diminishes cognitive resources available for finding targets. The opposite can occur for automated conditions, where the lack of manual control causes operator complacency effects (Parasuraman, Cosenzo, and DeVisser 2009). Autonomous solutions must be used cautiously and applied appropriately to enable soldiers to maintain an optimal amount of attentional resources among tasks as a function of changes in environment or workload.

Supervisory Control

The research discussed so far has identified the challenges of robotic operations when the ratio of operator to robot is 1:1, and the level of autonomy is either teleoperation or semi-autonomy. The military vision is to achieve a ratio of 1:n (that is, more than one robot). To achieve this vision, higher robotic and supervisory control concepts beyond teleoperation and semi-autonomy will be needed. If full autonomy can be realized, then unmanned systems will become ubiquitous in future battle spaces. This will be not only because of their force multiplication potential but also because unmanned systems can operate in places manned vehicles cannot and, most importantly, because robotic systems increase soldier survivability. However, the development of full autonomy and a 1:n ratio (or beyond) will result in hundreds of unmanned vehicles crowding a battle space that is already crowded with manned vehicles during a major conflict. Such a multitude of systems will require that unmanned systems achieve a measure of autonomy and also an ability to communicate with human supervisors as both soldiers and unmanned systems operate in the fog of war (Chen, Barnes, and Harper-Sciarini 2010). To add further uncertainty, supervising automated systems creates its own problems. In particular, there are classical cases wherein automation has resulted in catastrophic accidents because the supervisor either over-relied on the automated system or ignored it when they should not have (Parasuraman, Sheridan, and Wickens 2000). These problems will manifest themselves and undoubtedly be exacerbated in future battle spaces, as multiple supervisors coordinate multiple unmanned systems (Barnes, Parasuraman, and Cosenzo 2006, Cosenzo, Parasuraman, and de Visser 2010).

Cosenzo and colleagues (including Cosenzo et al. 2006, Parasuraman, Cosenzo, and de Visser 2009) examined the impact of multiple unmanned system control (UAV and UGV) on operator performance. In this research, the unmanned systems were ranged in level of autonomy from semi-autonomous to full autonomy. In all cases, the operator needed to monitor the robots and provide limited intervention (begin mission, go to next waypoint, stop). Cosenzo et al. (2006) found that participants were good at integrating information received from the UAV (full autonomy) and UGV (semi-autonomous) during a simulated reconnaissance mission. However, the multitasking requirements of monitoring and controlling multiple systems decreased performance of the individual tasks. Participants typically took longer to respond to communications when they also had to identify many targets in the UAV video imagery, and when the semi-autonomous UGV required interventions.

Parasuraman and colleagues further expanded this paradigm and investigated the effects of imperfect automation on system performance with multiple semi-autonomous UGVs (de Visser and Parasuraman 2007, de Visser et al. 2008). In these experiments, the UGV was able to identify objects in the environment that required further discrimination by the operator (that is, enemy or friendly). De Visser and Parasuraman (2007) manipulated the reliability of the automation and number of UGVs to monitor or control. They found that the worst performance occurred in the high task-load (more UGVs to monitor/control) condition in the low reliability condition. Further, there were performance benefits of imperfect automation on performance for reliability levels above 70 percent. These results show that vehicle autonomy is not the full solution to achieving the 1:n vision for robotics. Even with vehicle autonomy, operator mission performance can suffer due to the multiple tasks a soldier must attend to during a mission.

Supervisory control will require calibrated trust, specialized displays, and communications between humans and unmanned systems, as well as coordination among unmanned systems. Trust is a research topic in its own right, and involves both extrinsic and intrinsic factors that influence it. Extrinsic factors are caused by design flaws or environmental factors that cause either over-reliance or under-reliance on the automated systems. For example, changing when decision cues are presented, multitasking requirements, and reliability level of the systems all influence whether the supervisor over-relies or under-relies on the automation (Chen, Barnes, and Harper-Sciarini 2010). The factors are not always straightforward. Based on a meta-analysis, Wickens and Dixon (2005) found that automated systems with reliability of less than .70 resulted in degraded performance. Chen and Terrence (2009) found that personality factors interacted with type of unreliability and trust for low reliability aids. Participants with a high perceived confidence tended to under-rely on false alarm-prone devices, whereas low-confidence participants over-relied on devices with high miss rates. It is interesting to note that there was no significant difference in their overall performance, only on what type of errors (over-reliance or under-reliance) each personality type manifested. This is important because it implies that trust is also related to intrinsic factors, depending on the personal characteristics of the supervisor and not solely on characteristics of the automation (for example, Lee and See 2004). This, in turn, suggests not only that supervisory control systems need to be designed carefully in terms of cue presentation, multitasking requirements, and reliability level but also that individual differences need to be considered for training regimens and decision support to ensure appropriate trust in automation.

Supervisory control in future military environments will require the operator to intervene immediately whenever unmanned systems encounter dangerous situations. Chen, Barnes, and Harper-Sciarini (2010) suggested a number of graphic techniques to present information efficiently for supervisory control of multiple unmanned systems. Figure 2.2 depicts an Air Force example of augmented reality with graphics, symbol, and UAV flight information overlaid on streaming video of terrain (Calhoun and Draper 2010).

Figure 2.2   Symbology generated with LandForm® SmartCam3D®, Rapid Imaging Software, Inc. (adapted from Calhoun and Draper 2006, with permission)

Other techniques to enable supervisory control include configural displays in which the process being displayed is shown as an emergent property of the display itself. For example, multiple UAVs can be shown on the same display with planning cost shown as distorted triangles that can be compared easily "at a glance" (Cummings and Bruni 2009, as reported in Chen, Barnes, and Harper-Sciarini 2010). The point of using emergent properties is to present intuitive change information showing the unfolding of whatever processes are being represented. Both types of displays are designed for rapid interventions, and in the latter case, for a quick comparison of the status of multiple unmanned systems.

Communication protocols need to be developed between the human supervisor and the unmanned systems to ensure not only that the supervisor is able to intervene but also that the unmanned system is able to report its status and serious problems encountered during its mission. This will require a two-way protocol that enables both the human and the UV to maintain a shared awareness of the battle situation. For example, Redden, Carstens, and Pettitt (2010) gathered information from soldier participants in order to create a vocabulary that both the robot and the human population could use efficiently. The results were used to develop intuitive command phrases. Simple rules, such as having the soldier always command with the verb first, resulted in performance gains including fewer speech-based errors and a greater number of correctly remembered commands by the operator. However, even natural language interfaces will not be sufficient for robotic control, as supervisory control is required of multiple autonomous systems. The next part of this chapter will cover issues for collaboration among unmanned systems, including coordination among multiple operators and coordination among multiple unmanned systems.

It is foreseeable that groups of collaborative unmanned systems will exist that have a higher-order cognitive ability beyond going to a designated waypoint. Groups of systems can have a set of rules that they follow and use to dictate group behavior and dynamics. One such example is a robotic swarm. Robotic swarms are arrays of small, relatively simple robots capable of autonomous travel and operation (Haas et al. 2009). With swarms, the soldier must be kept cognizant of swarm operations through an interface that allows him or her to monitor status and/ or institute corrective actions. Haas and colleagues are developing algorithms and devices that allow soldiers to efficiently interact with a robotic swarm conducting a reconnaissance mission within convoy operations. Specifically, the researchers are working to provide metacognition algorithms that enable swarm members to efficiently monitor changes in swarm status as the swarm executes its mission. A second concept for supervisory control of multiple robots being investigated is RoboLeader (Figure 2.3) by Chen and colleagues (Chen and Barnes 2010, Chen, Barnes, and Qu 2010). RoboLeader is a robotic surrogate for the human operator and an intelligent agent that was developed to interpret the operator's intent. RoboLeader can issue detailed command signals to a team of robots of lower capabilities, enhancing the overall human–robot teaming performance. With the RoboLeader capabilities, dependence on operator instructions is reduced, and the level of autonomy in operation of the unmanned systems can be improved through the implementation of algorithms such as real-time path-planning, cooperative control, and multi-objective decision of tactical strategies. RoboLeader can help the operators with their route-planning tasks, and it is being expanded to deal, more specifically, with dynamic re-tasking requirements for persistent surveillance of a simulated urban environment, based on various battlefield developments (Chen, Barnes, and Qu 2010).

Soldier–Robot Teaming and Future Systems

How soldiers interact with heterogeneous robotic systems and how they interact with other soldier controller/supervisors will be of paramount importance in future battles. The n:n (that is, many soldiers to many robots) paradigm creates its own set of issues wherein multiple aerial systems are functioning with multiple manned and unmanned ground systems. Human factors issues include teaming relationships, interface design for coordination, and span of control with heterogeneous systems.

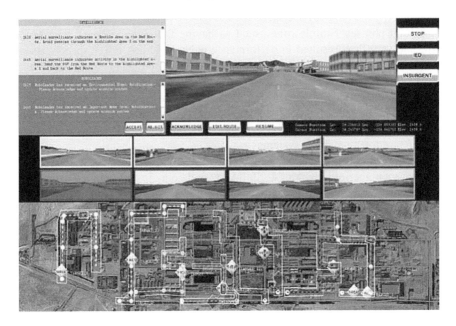

Figure 2.3   RoboLeader interface

Teaming includes teams of human supervisors, teams of autonomous robots, and teams of human and interacting robots.

Early research indicated that human teams operate unmanned systems more efficiently using a ratio of many operators to one system (Murphy 2004). This is the result of the superior SA that a multi-person team affords in comparison to a single human trying to control and maintain SA during difficult missions (Barnes and Evans 2010). For example, Jentsch, Evans, and Ososky (2010) examined human–robot team performance conducting a surveillance mission with mobile robotic systems in a 1/36th-scale toy city representing an Iraqi urban environment. The experiments assumed semi-autonomous vehicles so that teleoperation (or driving performance) was not a variable. Their main finding was that a single operator had significantly poorer target identification performance in comparison to a two-person team when operating from a single UGV; the single-operator performance was even worse when attempting to find targets with two semi-autonomous UGVs. Jentsch, Evans, and Ososky (2010) reported additional studies investigating heterogeneous systems requiring teamwork between operators of UGVs and UAVs attempting to find targets under a bridge that neither system could locate by itself. The initial experiment found that shared awareness was increased by nearly twofold if the operators shared their imagery, compared to teaming with chat communications. Curiously, allowing the operators to share control of the UVs resulted in a sharp decrease in the number of targets acquired, indicating that information and not control is the important teaming variable.

During further experiments, researchers reported an increase in targets found if an additional supervisor was available to coordinate missions. The role of the additional supervisor was particularly effective when communications were not perfectly reliable between the UAV and UGV operators, showing the importance of coordination in uncertain environments.

The design solutions presented so far include increasing the number of operators to improve SA and general surveillance performance. However, one goal of the unmanned systems revolution is force multiplication, which is an unlikely result if the number of soldiers required for supervision continually increases. We have discussed intelligent agents and swarm solutions; Jentsch, Evans and Ososky (2010) also examined the efficacy of robot-to-robot communications, decreasing the amount of interaction required by humans. In this paradigm, the soldier set goals for route reconnaissance, and robot teams would attempt to find the best routes by communicating back and forth without operator intervention. For easy routing missions, there was no advantage for robot-to-robot teams. In fact, if communications were unreliable, having humans communicate with robots directly improved performance because operators were able to adjust route plans more rapidly if they were directly involved in coordination.

In a second study, operator workload was increased by making the routes the robots were required to transverse more difficult. Under these conditions, robot-to-robot communications did improve performance compared to requiring direct human control of both robots. This finding agrees with most of the results we observed previously involving automating systems. For relatively easy tasks, engaging humans in the robotic tasks improves their SA (Kaber and Endsley 2004), but as task difficulty increases, the operator loses cognitive capacity and automation improves performance, suggesting once again the benefits of adaptive systems (Parasuraman, Cosenzo, and De Visser 2009). The n:n paradigm is a particularly daunting problem that has prompted possible solutions such as fan-out metrics and call centers. So far, however, the examples addressed and the research evaluations have been applied to relatively simple domains (Goodrich 2010, Lewis, and Wang 2010). Until simulations are more realistic in terms of the number and complexity of systems, and we are able to conduct field trials with many interacting heterogeneous systems, the full scope of the problems involved and the possible solutions will remain elusive.

Conclusions

Practical applications of robotics to the current military environment have shown that UAVs and UGVs are an integral part of soldier operations. Robots provide timely and critical combat information for soldiers and support soldiers in the continuous and demanding tasks of RSTA and IED defeat. Currently, robotic technologies are used by a select group of trained soldiers. In the future, the role that robots will play in military operations will be even greater and will be used by the majority of soldiers across the battlefield. The soldier may be the robot operator or just a recipient of the information from the robot team member, or both, depending on specific conditions and mission needs.

The defense community's expectation for robotics is that robots will function with little or no human intervention. However, we propose and have shown that this will not be the case. Most of the contemplated systems will require either active human control or, at a minimum, supervision with the possibility of intervention. The task demand of controlling a robot via teleoperation consumes all an individual's resources, making the completion of secondary tasks almost impossible. The addition of semi-autonomy for robot mobility alleviates this burden to a degree. To fully realize the vision of the seamless integration of robots into human teams, and where a single controller can monitor multiple systems across domains as collaborating team, higher levels of autonomy will need to be employed with novel interface and supervisory control techniques. As was previously stated, this will require a paradigm shift where new military tactics and teaming procedures need to be developed. The robots will become team members. The soldier will need to trust in that team member, and in the information provided and decision being made by that entity. Then, the robots will have become true team members working alongside and with soldiers.

References

Alban, J., Cosenzo, K., Johnson, T., Hutchins, S., Metcalfe, J. and Capstick, E. 2009. Robotics collaboration Army technology objective capstone soldier experiment: Unmanned system mobility. *Proceedings of the AUVSI Conference*, Washington, DC, August 2009. CD-ROM.

Axe, D. 2008. *Warbots*. Ann Arbor, MI: Nimble Books.

Barnes, M.J. and Evans, A.W. 2010. Soldier–robot teams in future battlefields: An overview, in *Human-Robot Interactions in Future Military Operations*, edited by M.J. Barnes and F.G. Jentsch. Farnham: Ashgate, 9–30.

Barnes, M., Parasuraman, R. and Cosenzo K. 2006. Adaptive automation for military robotic systems, in *Uninhabited Military Vehicles: Human Factors Issues in Augmenting the Force* (NATO Publication No. RTOTR-HFM-078), 420–40.

Calhoun, G.L. and Draper, M.H. 2010. Unmanned aerial vehicles: Enhancing video display utility with synthetic vision technology, in *Human-Robot Interactions in Future Military Operations*, edited by M.J. Barnes and F.G. Jentsch. Farnham: Ashgate, 229–48.

Casper, J. and Murphy, R.R. 2003. Human–robot interactions during the robot-assisted urban search-and-rescue response at the World Trade Center. *IEEE Transactions on Systems Man and Cybernetics Part B-Cybernetics*, 33(3), 367–85.

Chen, J.Y.C. and Barnes, M.J. 2010. Supervisory control of robots using roboleader. *Proceedings of the HFES 54th Annual Meeting*, San Francisco, CA, October 2010, 54(19), 1483–1487.

Chen, J.Y.C. and Joyner, C.T. 2009. Concurrent performance of gunner's and robotics operator's tasks in a multi-tasking environment, *Military Psychology*, 21(1), 98–113.

Chen, J.Y.C. and Terrence, P.I. 2009. Effects of imperfect automation and individual differences on concurrent performance of military and robotics tasks in a simulated multi-tasking environment. *Ergonomics*, 52(8), 907–920.

Chen, J.Y.C., Barnes, M.J., and Harper-Sciarini, M. 2010. Supervisory control of multiple robots: Human–performance issues and user–interface design. *IEEE Transactions on Systems, Man, and Cybernetics, Part C: Applications and Reviews*, 41(4), 435–54.

Chen, J.Y.C., Barnes, M.J. and Qu, Z. 2010. RoboLeader: An agent for supervisory control of multiple robots. *Proceedings of the 5th International ACM/IEEE Conference on Human-Robot Interaction*, Nara, Japan, March 2010, 81–82.

Chen, J.Y.C., Haas, E.C. and Barnes, M.J. 2007. Human performance issues and user interface design for teleoperated robots. *IEEE Transactions on Systems, Man, and Cybernetics—Part C: Applications and Reviews*, 37(6), 1231–1245.

Cosenzo, K., Parasuraman, R. and de Visser, E. 2010. Automation strategies for facilitating human interaction with military unmanned vehicles, in *Human–Robot Interactions in Future Military Operations*, edited by M.J. Barnes and F.G. Jentsch. Farnham: Ashgate, 103–124.

Cosenzo, K., Parasuraman, R., Novak, A. and Barnes, M. 2006. *Adaptive Automation for Robotic Military Systems*. (Technical Report ARL-TR-3808). Aberdeen Proving Ground, MD: US Army Research Laboratory.

Department of Defense 2009. FY2009-2034 unmanned systems integrated roadmap. Available at: http://www.acq.osd.mil/psa/docs/UMSIntegratedRoadmap2009. pdf (accessed: July 19, 2012).

de Visser, E. and Parasuraman, R. 2007. Effects of imperfect automation and task load on human supervision of multiple uninhabited vehicles. *Proceedings of the Annual Meeting of the HFES*, Baltimore, MD, October 2007, 1081–1085.

de Visser, E., Horvath, D., Parasuraman, R. and Cosenzo, K. 2008. *Adaptive Automation for Human Interaction with Uninhabited Vehicles*. Paper presented at the Midyear Meeting of Division 21 of the American Psychological Association. Fairfax, VA: George Mason University.

Economist, The 2007. *Robot Wars*. Available at: http://economist.com/printedition/ PrinterFriendly. cfm?story_id=9249201 (accessed: July 19, 2012).

Goodrich, M. 2010. On maximizing fan-out: Towards controlling multiple unmanned vehicles, in *Human–Robot Interactions in Future Military Operations*, edited by M.J. Barnes and F.G. Jentsch. Farnham: Ashgate, 375–96.

Haas, E. and van Erp, J. 2010. Multimodal research for human–robot interactions, in *Human-Robot Interactions in Future Military Operations*, edited by M.J. Barnes and F.G. Jentsch. Farnham, UK: Ashgate, 271–92.

Haas, E., Fields, M., Hill, S. and Stachowiak, C. 2009. *Extreme Scalability: Designing Interfaces and Algorithms for Soldier–Robotic Swarm Interaction.* (Technical Report ARL-TR-4800). Aberdeen Proving Ground, MD: US Army Research Laboratory.

Huang, H., Messina, E. and Albus, J. 2007. *Autonomy Levels for Unmanned Systems (ALFUS) Framework.* NIST Special Publication, 1011–II–1.0.

Hutchins, S., Cosenzo, K., McDermott, P., Feng, T., Barnes, M. and Gacy, M. 2009. An investigation of the tactile communications channel for robotic control. *Proceedings of the HFES Annual Meeting*, San Antonio, TX, October 2009, 53(4), 182–86.

Jentsch, F.G., Evans, A.W., III and Ososky, S. 2010. Model world: Military research conducted using a scale MOUT facility, in *Human–Robot Interactions in Future Military Operations*, edited by M.J. Barnes and F.G. Jentsch. Farnham: Ashgate, 419–34.

Kaber, D.B. and Endsley, M. 2004. The effects of level of automation and adaptive automation on human performance, situation awareness and workload in a dynamic control task. *Theoretical Issues in Ergonomics Science*, 5, 113–53.

Kappenman, J. 2008. Army unmanned aircraft systems: Decisive in battle. *JFQ*, issue 49 (Second Quarter), 20–23. [Online]. Available at: http://www.dtic.mil/cgi-bin/GetTRDoc?AD=ADA516600 (accessed: July 24, 2012).

Lee, J.D. and See, K.A. 2004. Trust in automation: Designing for appropriate reliance. *Human Factors*, 46, 50–80.

Lewis, M. and Wang, J. 2010. Coordination and automation for controlling robot teams, in *Human–Robot Interactions in Future Military Operations*, edited by M.J. Barnes and F.G. Jentsch. Farnham, UK: Ashgate, 397–418.

Mitchell, D. and Henthorn, T. 2005. *Soldier Workload Analysis of the Mounted Combat System (MCS) Platoon's Use of Unmanned Assets.* (Technical Report ARL-TR—3476). Aberdeen Proving Ground, MD: US Army Research Laboratory.

Murphy, R.R. 2004. Human–robot interaction in rescue robots. *IEEE Transactions on Systems, Man, and Cybernetics—Part C: Application sand Reviews*, 32(2), 138–53.

Parasuraman, R. and Riley, V. 1997. Humans and automation: Use, misuse, disuse, abuse. *Human Factors*, 39, 230–53.

Parasuraman, R, Cosenzo, K. and DeVisser, E. 2009. Adaptive automation for human supervision of multiple unmanned vehicles: Change detection, situation awareness, and workload. *Military Psychology*, 21, 270–97.

Parasuraman, R., Sheridan, T.B. and Wickens, C.D. 2000. A model of types and levels of human interaction with automation. *IEEE Transactions on Systems, Man, and Cybernetics—Part A: Systems and* Humans, 30, 286–97.

Redden, E.S., Carstens, C.B. and Pettitt, R.A. 2010. *Intuitive Speech-Based Robotic Control.* (Technical Report ARL-TR-5175). Aberdeen Proving Ground, MD: US Army Research Laboratory.

Redden, E.S., Pettitt, R.A., Carstens, C.B. and Elliott, L.R. 2008. *Scalability of Robotic Displays: Display Size Investigation*. (Technical Report ARL-TR-4456). Aberdeen Proving Ground, MD: US Army Research Laboratory.

Redden, E.S., Pettitt, R.A., Carstens, C.B. and Elliott, L.R. 2010. *Scaling Robotic Displays: Displays and Techniques for Dismounted Movement with Robots*. (Technical Report ARL-TR-5174). Aberdeen Proving Ground, MD: US Army Research Laboratory.

Riley, J.M., Murphy, R. and Endsley, M.R. 2006. Situation awareness in control of unmanned ground vehicles, in *Advances in Human Performance and Cognitive Engineering Research: Human Factors of Remotely Piloted Vehicle*, edited by N. Cooke, H. Pringle, H. Pedersen and O. Conner. Amsterdam: Elsevier, 359–71.

Singer, P.W. 2009. *Wired for War*. New York: Penguin Press.

Wickens, C.D. and Dixon, S.R. 2005. *Is There a Magic Number 7 (to the Minus 1)? The Benefits of Imperfect Diagnostic Automation: A Synthesis of the Literature*. (AHFD-05-01/MAAD-05-01). Urbana-Champaign, IL: University of Illinois.

Wickens, C.D., Dixon, S.R. and Ambinder, M. 2006. Workload and automation reliability in unmanned air vehicles, in *Human Factors of Remotely Operated Vehicles*, edited by N.J. Cooke, H.L. Pringle, H.K. Pederson and O. Connor. Amsterdam: Elsevier, 209–22.

Chapter 3

Effects of Operators' Spatial Ability on Their Performance of Target Detection and Robotics Tasks

Jessie Y.C. Chen

This chapter summarizes five human-in-the-loop simulation experiments that investigated the effects of operators' spatial ability (SpA) on their performance of target detection and robotics tasks. In the first experiment, the operator's task was to conduct route reconnaissance missions using one or three heterogeneous robots to detect enemy targets along a designated route. In the second, third, and fourth experiments, the operators performed three tasks concurrently (that is, gunnery targeting, robotics, and communication tasks), and the types of robotics tasks were manipulated (that is, monitoring the streaming video from an autonomous unmanned ground vehicle (UGV), managing a semi-autonomous UGV, and teleoperating a UGV). In the third experiment, an aided target recognition (AiTR) capability (delivered via visual and/or tactile cueing) was implemented to assist the operators with their primary (gunnery) task. In the fourth experiment, the reliability levels of the AiTR system were manipulated (i.e., false-alarm prone (FAP) vs miss prone (MP)). Finally, in the fifth experiment, a military reconnaissance environment was simulated to examine the performance of UGV operators who were instructed to utilize streaming video from an unmanned aerial vehicle (UAV) to navigate his/ her ground robot to the locations of the targets. Details of the experiments can be found in Chen, Durlach et al. 2008, Chen and Joyner 2009, Chen and Terrence 2008, Chen and Terrence 2009, and Chen 2010).

Operator Spatial Ability

SpA has been found to be a significant factor in virtual environment navigation (Chen, Czerwinski, and Macredie 2000, Stanney and Salvendy 1995), learning to use a medical teleoperation device (Eyal and Tendick 2001), a target search task (Chen, Durlach et al. 2008, Chen 2010, Chen and Barnes 2012), and robotics task performance (Cassenti et al. 2009, Lathan and Tracey 2002, Menchaca-Brandan et al. 2007). For example, Lathan and Tracey (2002) demonstrated that people with higher SpA performed better in a teleoperation task through a maze, finishing their tasks faster and with fewer errors. Recent studies conducted by the US Air Force (Chappelle et al. 2010a, 2010b) indicated that SpA was one of the main abilities that should be taken into account when selecting personnel to pilot UAVs and operate UAV sensors. In another recent study by the US Army Research Laboratory, Cassenti et al. (2009) demonstrated that robotics operators with higher SpA (measured by a mental rotation test) performed robot navigation tasks significantly better than those with lower SpA. Our previous studies (Chen, Durlach et al. 2008, Chen 2010, Chen and Barnes 2012) also found SpA to be a good predictor of the operator's target detection task performance and scanning effectiveness.

SpA may also influence the efficacy of spatial cueing displays. Stanney and Salvendy (1995) found that high SpA individuals outperformed those with low SpA on tasks that required visuospatial representations to be mentally constructed. Although spatial tactile alerts have been found to be effective in various tasking environments such as aviation and driving (Ho, Tan and Spence 2005, Sklar and Sarter 1999), it remains unclear to what extent the impact of SpA has on the integration of spatial information gleaned from spatial tactile displays. On the other hand, the interconnections of sensory modalities at the level of spatial perception suggest that the effects of SpA on effectiveness of spatial visual displays may translate into differential effects of multisensory spatial displays across SpA levels (Spence 2002).

Spatial Ability Measures

The cube comparison test and hidden patterns test (Ekstrom, French, and Harman 1976), as well as the spatial orientation test, were used to assess participants' SpA in the five experiments reported in this chapter. The cube comparison test requires participants to compare, in 3 minutes, 21 pairs of 6-sided cubes and determine whether the rotated cubes are the same or different (only 3 sides of each cube are shown). The hidden patterns test a measure's flexibility of closure and involve identifying specific patterns or shapes embedded within distracting information. The spatial orientation test, modeled after the cardinal direction test (Gugerty and Brooks 2004), is a computerized test consisting of a brief training segment and 32 test questions. Both accuracy and response time were automatically captured by the program. Participants were designated as high or low SpA based on their composite score of the spatial tests (median split). In the fifth experiment, a survey on perceived sense of direction, the Santa Barbara Sense of Direction (SBSOD) scale, was used to assess participants' perceived abilities on navigation and wayfinding tasks. Hegarty et al. (2002) reported that this self-reported sense of direction is correlated with some spatial task performance (for example, imagining oneself taking a different perspective in the environment and learning the spatial layout of the environment). The following parts of this chapter summarize the five experiments in which we investigated human–robot interaction and the effects of operator SpA on his/her performance.

Experiment 1: Human-Robot Interaction in the Context of Simulated Route Reconnaissance Missions (Chen et al. 2008)

In this experiment, the operator's task was to conduct route reconnaissance missions using one or three robots to detect enemy targets along a designated route (Figure 3.1). There were four experimental conditions: semi-autonomous UAV (UAV condition), semi-autonomous UGV (UGV condition), teleoperated UGV (teleop condition), and mixed (1 UAV + 1 UGV + 1 teleop). Each participant conducted four missions, which corresponded to each experimental condition (that is, three missions with a different type of robotic asset and a final mission with all three robotic assets at their disposal). For the UAV and UGV conditions, operators assigned a set of waypoints, and then the robot(s) traveled the route automatically unless the operator intervened to alter its behavior. All vehicles were equipped with camera sensors, which could be panned and zoomed and could send streaming video back to the operator control station. As the robot traveled, the operator manipulated the sensors searching for targets. The teleoperated (teleop) robot was a ground vehicle that required the operator to remotely drive the UGV while simultaneously manipulating its sensors to search for targets. In contrast, the operator was only required to manipulate the sensors on the two semi-autonomous robots (UAV and UGV).

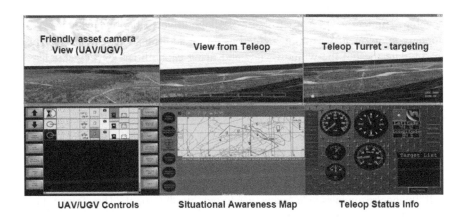

Figure 3.1   User interface of the simulation test bed in Experiment 1

The results suggest that providing robotic operators with additional assets may not be beneficial for enhancing target-detection performance. Essentially, participants failed to discover more targets with three robots compared with operating the UAV or the UGV alone. Participants with higher SpA were found to perform better in both speed and accuracy across platform types (p's < .05). Their superior performance was especially evident when they used the semi-autonomous UAV. More specifically, SpA was found to be positively correlated with the number of targets detected in the UAV, UGV, and the mixed conditions, all p's $< .05$. The correlations between SpA and mission completion times were all negative (that is, participants with higher SpA completed their missions faster); however, only the correlation in the UAV condition was significant, $p < .05$, whereas the correlation in the Teleop condition was marginally significant, $p = .056$. In summary, a difference in SpA rather than the number of robots was the determining factor in operator performance.

Experiment 2: Concurrent Performance of Gunner and Robotic Operator Tasks in a Multitasking Environment (Chen and Joyner 2009)

The second experiment examined the workload and performance of the combined position of gunner and robotics operator (Figure 3.2). The purpose was to determine whether participants could perform their primary gunner task while concurrently conducting their robotics tasks. The experiment was a within-subject design, and the experimental conditions included a gunner baseline (*Gunner Baseline* condition) and three concurrent task conditions in which participants simultaneously performed gunnery tasks and one of the following robotic tasks: monitor a UGV via the video feed (*Monitor* condition), manage a semi-autonomous UGV (*Auto* condition), and teleoperate a UGV (*Teleop* condition). The robotic operator control unit (OCU) used in this experiment was developed under the US Army Robotics Collaborative Technology Alliance. The gunnery component was implemented using an additional screen and controls to simulate the out-the-window view and firing capabilities. Participants used designated buttons on the joystick to manipulate the views, zoom in and out, and fire rounds.

In the experimental trials, participants' tasks were to use their robot to locate targets (that is, enemy soldiers) in the remote environment and also find/engage targets in their immediate environment. For the Monitor task, participants simply monitored the robot's video feed on the OCU and verbally reported the detection of targets. The Auto task required the participants to look for targets by monitoring the video feed as the robot traveled autonomously and examining still images/reconnaissance scans generated by the robot. The Teleop task required the participants to manually drive the robot (using a joystick) along a predetermined route while using the OCU to detect randomly placed targets at each scanning checkpoint. For both the Auto and Teleop tasks, upon detecting a target, participants needed to place the target on the map, label the target, and then send a spot report. In addition to the gunnery and the robotics tasks, the participants also simultaneously performed a communication task (that is, answering simple reasoning questions, repeating short statements, or responding to call signs).

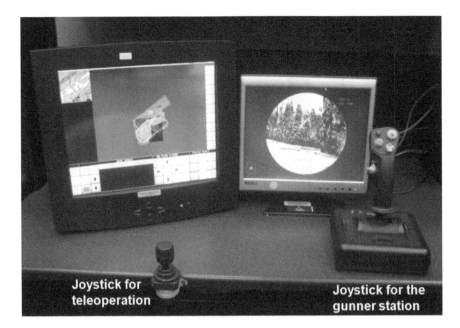

Figure 3.2   OCU (left) and gunnery stations (gunner's out-the-window view) (right)

Results showed that a gunner's target detection performance degraded significantly when he or she had to perform a concurrent task (monitor, manage, or teleoperate a UGV), $p < .001$. The gunner's performance was significantly different in each of the three concurrent task conditions, with the Monitor condition being the highest (that is, the least degraded) and the Teleop condition being the lowest (p's $< .05$). These results suggest that if it is necessary for the gunner to concurrently access information from robotic assets, the robotic tasks should be limited to activities such as monitoring. If excessive manipulation of the warfighter–machine interface is required, as in the Auto and Teleop conditions, their gunnery task performance will be significantly affected. Participants with higher SpA had significantly better gunnery task performance (that is, they detected more targets) than did those with lower SpA, $p < .005$ (Figure 3.3). However, a significant difference in performance on the robotics targeting task was not found between those with high and low SpA, suggesting that SpA helped operators reorient themselves to locate targets for their primary gunnery task when switching from their robotic tasks.

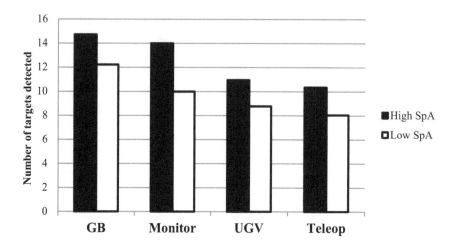

Figure 3.3   Effects of SpA on gunnery target detection performance

Experiment 3: Effects of Tactile Cueing on Concurrent Performance of Military and Robotics Tasks in a Simulated Multitasking Environment (Chen and Terrence 2008)

This experiment examined if and how tactile cueing, which delivered simulated AiTR capabilities (cues to a potential target position relative to the gunner), enhanced a gunner's performance in a military multitasking environment similar to the one in Experiment 2. The overall design of the study was a 2 × 3 repeated-measures design. The independent variables were robotics task type (auto vs teleop) and AiTR type (baseline (BL) no alerts vs tactile alerts only (Tac) vs tactile + visual alerts (TacVis)). In each experimental condition, participants were required to simultaneously perform the gunnery, robotics, and communication tasks, as described in Experiment 2.

Results showed that the gunner's target detection performance improved significantly when his or her task was assisted by AiTR (from about 52 percent in the BL condition to about 84 percent in the Tac and TacVis conditions) ($p < .001$). Consistent with the findings of Experiment 2, participants' SpA was found to be an accurate predictor of their gunnery performance ($p < .05$). Those with higher SpA consistently outperformed those with lower SpA throughout the scenarios (illustration A in Figure 3.4). Additionally, the data show that AiTR was more beneficial for enhancing performance on the concurrent robotics task for those with lower SpA than for those with higher SpA ($p < .05$) (illustration B in Figure 3.4). It appears that when AiTR was available to assist those operators with low SpA, the performance of their concurrent task was improved to a similar level as those with higher SpA. These results are consistent with findings previously reported (Young and Stanton 2007) and may have important implications for system design and personnel selection for future military programs.

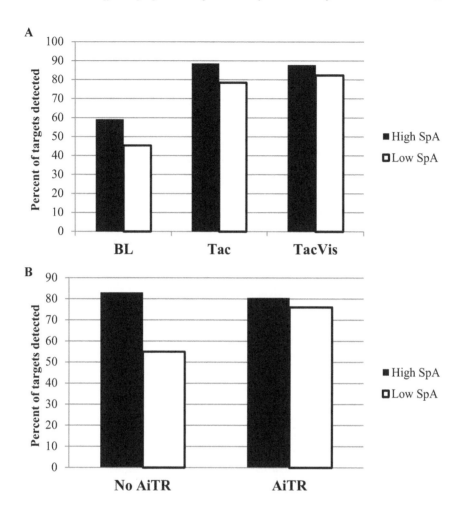

Figure 3.4   Effects of SpA on target detection performance for A: Gunnery; B: Robotics

According to a usability survey on preference of AiTR display, 65 percent of the participants responded that they either relied predominantly or entirely on the tactile AiTR display. Only 15 percent responded that they either relied predominantly or entirely on the visual AiTR display. AiTR preference was also significantly correlated with SpA (p's < .05). Perhaps those with higher SpA can more easily make use of the spatial tactile signals in the dual task-setting and therefore have a stronger preference for something that makes the gunner task easier to complete. Individuals with lower SpA, on the other hand, may not have utilized the spatial tactile cues to their full extent and therefore preferred the visual

AiTR display. However, this display preference may have caused degraded target detection performance on the gunnery task due to more visual attention being devoted to the visual AiTR display than the simulated environment. In contrast, those who had a higher SpA relied on the directional information of the tactile display to help them with the visually demanding tasks, resulting in more effective performance.

Experiment 4: Effects of Imperfect Automation on Concurrent Performance of Military and Robotics Tasks in a Simulated Multitasking Environment (Chen and Terrence 2008)

This experiment followed the paradigm of Experiment 3 and examined if and how tactile/visual cueing with imperfect reliability affected the operator's performance. The AiTR was either MP or FAP; the reliability level of both MP and FAP was 60 percent. The experiment was a 2 × 3 mixed design. The between-subject variable was AiTR type (FAP vs MP) and the within-subject variable was robotics task type (Monitor vs Auto vs Teleop).

Results showed that the operators' gunnery task performance in detecting hostile targets was significantly better in the Monitor condition than in the other two robotics task conditions ($p < .05$), consistent with the findings of Experiment 2. The workload associated with the monitor condition was significantly lower than the other robotics conditions, which was also consistent with Experiment 2. Taken together, these results suggest that the operator had more visual and mental resources for the gunnery task when the robotics task was simply monitoring the video feed, compared with the other two robotics conditions. Also consistent with the results of experiments 2 and 3, participants' SpA was found to be an accurate predictor of their gunnery performance ($p < .05$). The results also show that low SpA participants preferred visual cueing over tactile cueing, whereas high SpA participants favored tactile cueing over visual cueing ($p < .01$), which confirmed the trend that was observed in Experiment 3. Figure 3.5 shows the data from both experiments examined in the same analysis, $F(1,35) = 12.1$, $p = .001$. The implication of this cueing preference finding for user interface design is discussed in the following part of this chapter.

Experiment 5: UAV-Guided Navigation for Ground Robot Operations (Chen 2010)

The goal of Experiment 5 was to investigate UAV-guided navigation for ground robot teleoperations. We used a different paradigm and simulation environment, indicating that our SpA findings were not specific to a particular paradigm or measurement procedure. The overall design of the study was a 2 × 2 × 4 mixed design. The between-subject variable was Lighting condition in the simulated environment (day vs night vision). The within-subject variables were Target type (stationary vs moving target); and UAV type (no UAV vs micro air vehicle (MAV) vs large UAV—fixed view vs large UAV—orbiting view). A first-person-shooter

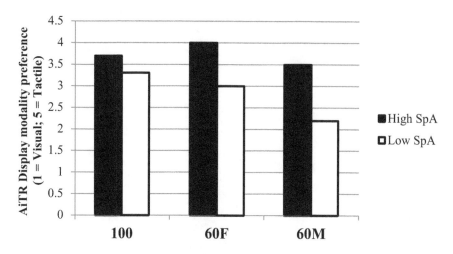

Figure 3.5   **Effects of SpA on operators' preference of cueing display modality (100 = 100 percent reliability (Experiment 30, 60F = 60 percent FAP (Experiment 4), 60M = 60 percent MP (Experiment 4))**

computer game, Half Life2®, was used to provide the simulation for the MAV and the UGV (Figure 3.6). The terrain database of the US Army McKenna Military Operations in Urban Terrain (MOUT) of Fort Benning, GA, was used as the simulation environment for this experiment. The first-person-shooter perspective of Half Life2 was used to simulate the view from the UGV. Participants used voice commands (for example, forward, backward, turn left, turn right, scan for targets, engage target) to control the UGV's navigation. Half Life2 also provides a spectator's view (the altitude of which can be manipulated by the user), which was used to simulate the view from the MAV. Participants used a joystick to control the movement of the MAV. Another set of simulations was used to provide the large UAV views (Figure 3.6). The large UAV with a fixed view was simulated as hovering above the MOUT at 100 meters, and the orbiting UAV was simulated as orbiting the MOUT at 15 miles per hour at the same altitude. Participants were able to see the entire MOUT site from the larger UAV video. However, the MAV was limited to lower-altitude flight (roughly the height of a three-story building) and thus did not provide as good a bird's-eye view of the environment as the larger UAVs.

Figure 3.6   UGV screen (left) and UAV screen (right)

In each scenario, there was one primary target (an enemy vehicle, which was a sport utility vehicle) and five secondary targets (stationary enemy soldiers) in the MOUT environment. In half of the scenarios the vehicle traveled slowly (about 10 miles per hour) in the MOUT environment following a predesignated route. In the other scenarios, it was stationary. The night-vision condition was rendered by adjusting the color setting of the computer monitors to render scenes as though seen through night-vision goggles. The dependent measures included the number of targets detected using the robotic assets, the amount of time it took the participants to find the targets, and the participants' perceived workload.

In addition to SpA, the effect of an operator's self-assessed sense of direction on his/her navigation performance was examined. According to Baldwin and Reagan (2009), those with a good sense of direction tend to rely more heavily on visuospatial working memory (for example, using cardinal directions and map-like representations) when navigating in virtual environments, whereas those with a poor sense of direction tend to rely more on verbal working memory (for example, using landmarks as references). Based on these findings, it is reasonable to expect that the lighting conditions would have a greater impact on those with a poor sense of direction, as objects in the environments would not be as useful to serve as landmarks in the night condition due to the loss of color cues at night, compared to the day condition.

The results showed significantly superior performance in map-marking accuracy by those participants with higher SpA, especially when they used the larger UAVs. Those with lower SpA, in contrast, did not appear to take advantage of the larger UAVs (compared to the other two conditions) as much as their higher-SpA counterparts (Figure 3.7). Additionally, there was a significant difference in other map-marking related performances (that is, target locations and landmark locations) between those with superior and poor SpA.

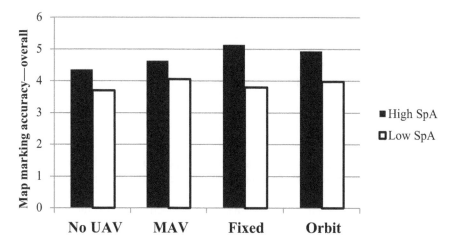

Figure 3.7   Effects of SpA on mapping performance (map-marking accuracy)

Additionally, participants' self-assessed sense of direction rating (based on the SBSOD scale) was an accurate predictor of both their map-marking performance (target locations) and target search time (including both primary and secondary targets). According to Hegarty et al. (2002), the SBSOD scale was more related to self-orientation within the environment than distance-estimation and map-drawing. However, the data from this study indicated that participants' SBSOD scores were not only predictive of their navigation-related measures (for example, target search time) but also their map-marking related performance. This finding is consistent with Prestopnik and Roskos-Ewoldsen (2000), that participants' speed of wayfinding performance was predicted by their self-assessed sense of direction rating.

The data also showed that the navigation performance of those with poorer sense of direction was especially affected by the Lighting factor. They took about the same amount of time to find targets as those with a better sense of direction in the Day condition; however, they took a significantly longer time to search for targets in the Night condition. This finding is consistent with what was reported in Baldwin and Reagan (2009), that those with a poor sense of direction tend to rely more on verbal working memory (for example, using landmarks as references) when performing navigational tasks. Since objects in the Night environment cannot serve as landmarks as effectively as in Day due to the loss of color cues, those with a poorer sense of direction cannot effectively utilize their navigational strategies involving landmarks. On the other hand, those with a better sense of direction tend to rely on visuospatial strategies (for example, using cardinal directions and map-like representations) when performing navigational tasks. Therefore, their navigation performance in the Night condition would not be severely affected, since color cues play a less important role in their navigational strategies.

Male participants performed significantly better than females in map-marking accuracy of target locations. This is consistent with previous findings of gender differences in spatial knowledge acquisition (Hurts 2006, Lawton and Morrin 1999, Prestopnik and Roskos-Ewoldsen 2000, Waller, Knapp, and Hunt 2001). Male participants also had a higher average SBSOD score than females (76 vs 64). However, this difference failed to reach statistical significance.

Conclusions

The data from these five experiments repeatedly show superior task performance by participants with higher SpA, and these results are consistent with past research (Lathan and Tracey 2002, Vincow 1998). Thomas and Wickens (2004) showed that there were individual differences in scanning effectiveness and the associated target detection performance. Although Thomas and Wickens did not examine the characteristics of those participants who had more effective scanning strategies, the findings of these five studies indicate that SpA may be an important factor for determining scanning effectiveness. The findings also support the recommendation by Lathan and Tracey and researchers from the US Air Force (Chapelle et al. 2010a, 2010b) that military missions can benefit from selecting personnel with higher SpA to operate robotic devices. However, especially during times of war, personnel selection is not always possible. Therefore, we will emphasize design and training solutions suggested by our results as well.

Decision aids might help mitigate differences. For example, the data show that AiTR was more beneficial for enhancing performance on robotics tasks that are concurrently conducted with gunnery tasks for those with lower SpA than for those with higher SpA. When AiTR was available to assist operators with low SpA, the performance of their concurrent task was improved to a similar level to those with higher SpA. These results are consistent with other findings showing that vehicle automation helps reduce the performance gap between experts and novices (Young and Stanton 2007). These results may have important implications for system design and personnel selection for future military programs.

The results also show that low-SpA participants preferred visual cueing over tactile cueing, and high-SpA participants favored tactile cueing over visual cueing. Perhaps those with higher SpA can more easily employ the spatial tactile signals in a dual task setting (that is, gunner and robotics) and therefore have a stronger preference for something that makes the gunner task easier to complete. Individuals with lower SpA, in contrast, may have not utilized the spatial tactile cues to their full extent and therefore continued to prefer the visual AiTR display. It is likely that in this study, those who preferred visual AiTR displays might be more iconic in their mental representations (Kozhevnikov, Hegarty, and Mayer 2002). However, this preference may have caused degraded target detection performance because more visual attention was devoted to the visual AiTR display than to the simulated environment. In contrast, those who were more spatial (that is, higher SpA) relied on the directional information provided by the tactile display to help them with the visually demanding tasks, resulting in more effective performance. These findings may have important implications for personnel selection, systems design, and training development. To better enhance task performance for low-SpA individuals, one design recommendation is that the visual cueing display should be more integrated with the visual scene. For example, augmented reality (that is, visual overlays) is a potential technique to embed directional information onto a video (Calhoun and Draper 2006).

Finally, past studies have shown that training on navigational strategies and spatial tasks can be beneficial for those with poor SpA (Gugerty and Brooks 2004, Lawton and Morrin 1999, Rodes, Brooks, and Gugerty 2005). Baldwin and Reagan (2009) also suggested that user interface designs should take operators' preference of navigational strategies into account and reduce extraneous demands that use the same working memory processes. A customizable user interface would be an ideal solution for users with various levels of SpA.

References

Baldwin, C.L. and Reagan, I. 2009. Individual differences in route-learning strategy and associated working memory resources. *Human Factors*, 51(3), 368–77.

Calhoun, G. and Draper, M. 2006. Multi-sensory interfaces for remotely operated vehicles, in *Human Factors of Remotely Operated Vehicles*, edited by N.J. Cooke, H.L. Pringle, H.K. Pedersen, O. Connor and E. Salas. Oxford: Elsevier, 149–63.

Cassenti, D., Kelley, T., Swoboda, J. and Patton, D. 2009. The effects of communication style on robot navigation performance. *Proceedings of the HFES 53rd Annual Meeting*, San Antonio, TX, October 2009, 53(4), 359–63.

Chappelle, W.L., Novy, P.L., Randall, B. and McDonald K. 2010a. Critical psychological attributes of US Air Force (USAF) predator and reaper sensor operators according to subject matter experts. *Aviation, Space, and Environmental Medicine*, 81, 253.

Chappelle, W.L., McMillan, K.K., Novy, P.L. and McDonald K. 2010b. Psychological profile of USAF unmanned aerial systems predator and reaper pilots. *Aviation, Space, and Environmental Medicine*, 81, 339.

Chen, J.Y.C. 2010. UAV-guided navigation for ground robot teleoperation in a military reconnaissance environment. *Ergonomics,* 53(8), 940–950.

Chen, J.Y.C. and Joyner, C.T. 2009. Concurrent performance of gunner's and robotic operator's tasks in a multi-tasking environment. *Military Psychology*, 21(1), 98–113.

Chen, J.Y.C. and Terrence, P.I. 2008. Effects of tactile cueing on concurrent performance of military and robotics tasks in a simulated multi-tasking environment. *Ergonomics*, 51(8), 1137–1152.

———. 2009. Effects of imperfect automation on concurrent performance of military and robotics tasks in a simulated multi-tasking environment. *Ergonomics*, 52(8), 907–920.

Chen, J.Y.C. and Barnes, M.J. 2012. Supervisory control of multiple robots: Effects of imperfect automation and individual differences. *Human Factors,* 54(2), 157–174.

Chen, C., Czerwinski, M. and Macredie, R. 2000. Individual differences in virtual environments—introduction and overview. *Journal of the American Society for Information Science*, 51(6), 499–507. [Online]. Available at: http://onlinelibrary.wiley.com/doi/10.1002/(SICI)1097-4571 (2000)51:6%3C499::AID-ASI2%3E3.0.CO;2-K/full (accessed: July 24, 2012).

Chen, J.Y.C., Durlach, P.J., Sloan, J.A. and Bowens, L.D. 2008. Human robot interaction in the context of simulated route reconnaissance missions. *Military Psychology*, 20(3), 135–49.

Ekstrom, R.B., French, J.W. and Harman, H.H. 1976. *Kit of Factor-Referenced Cognitive Tests*. Princeton, NJ: Educational Testing Service.

Eyal, R. and Tendick, F. 2001. Spatial ability and learning the use of an angled laparoscope in a virtual environment, in *Medicine Meets Virtual Reality 02/10, Digital Upgrades: Applying Moore's Law to Health*, edited by J.D. Westwood, H.M. Hoffman, R.A. Robb and D. Stredney. Amsterdam: IOS Press, 146–52.

Gugerty, L. and Brooks, J. 2004. Reference-frame misalignment and cardinal direction judgments: Group differences and strategies. *Journal of Experimental Psychology: Applied*, 10, 75–88.

Hegarty, M., Richardson, A., Montello, D., Lovelace, K. and Subbiah, I. 2002. Development of a self-report measure of environmental spatial ability. *Intelligence*, 30(5), 425–47.

Ho, C., Tan, H. and Spence, C. 2005. Using spatial vibrotactile cues to direct visual attention in driving scenes. *Transportation Research Part F: Traffic Psychology and Behaviour*, 8, 397–412.

Hurts, C. 2006. Effects of spatial intelligence and gender on wayfinding strategy and performance. *Proceedings of the HFES 50th Annual Meeting*, San Francisco, CA, October 2006, 1533–1537.

Kozhevnikov, M., Hegarty, M. and Mayer, R. 2002. Revising the visualizer–verbalizer dimension: Evidence for two types of visualizers. *Cognition and Instruction*, 20, 47–77.

Lathan, C. and Tracey, M. 2002. The effects of operator spatial perception and sensory feedback on human–robot teleoperation performance. *Presence*, 11, 368–77.

Lawton, C.A. and Morrin, K.A. 1999. Gender differences in pointing accuracy in computer-simulated 3D mazes. *Sex Roles*, 40(1), 73–92.

Menchaca-Brandan, M., Liu, A., Oman, C. and Natapoff. A. 2007. Influence of perspective-taking and mental rotation abilities in space teleoperation. *Proceedings of HRI '2007*, Washington D.C., March 2007, 271–78.

Prestopnik, J. and Roskos-Ewoldsen, B. 2000. The relations among wayfinding strategy use, sense of direction, sex, familiarity, and wayfinding ability. *Journal of Environmental Psychology*, 20(2), 177–91.

Rodes, W., Brooks, J. and Gugerty, L. 2005. *Using Verbal Protocol Analysis and Cognitive Modeling to Understand Strategies Used for Cardinal Direction Judgments*. Paper presented at the Human Factors of UAVs Workshop, Mesa, AZ, May 25–26, 2005.

Sklar, A. and Sarter, N. 1999. Good vibrations: Tactile feedback in support of attention allocation and human–automation coordination in event-driven domains. *Human Factors*, 41, 543–52.

Spence, C. 2002. Multisensory attention and tactile information-processing. *Behavioural Brain Research*, 135, 57–64.

Stanney, K. and Salvendy, G. 1995. Information visualization: Assisting low spatial individuals with information access tasks through the use of visual mediators. *Ergonomics*, 38(6), 1184–1198.

Thomas, L. and Wickens, C. 2004. Eye-tracking and individual differences in off-normal event detection when flying with a synthetic vision system display. *Proceedings of the HFES 48th Annual Meeting*, New Orleans, LA, September 2004, 48(1), 223–27.

Vincow, M. 1998. *Frame of reference and navigation through document visualizations*. Unpublished dissertation. University of Illinois.

Waller, D., Knapp, D. and Hunt, E. 2001. Spatial representations of virtual mazes: The role of visual fidelity and individual differences. *Human Factors*, 43(1), 147–58.

Young, M. and Stanton, N. 2007. What's skill got to do with it? Vehicle automation and driver mental workload. *Ergonomics*, 50, 1324–1339.

Chapter 4

Reducing Workload:
A Multisensory Approach

Linda R. Elliott and Elizabeth S. Redden

Advanced information distribution systems promise immediate connections from sensors to shooters. Emerging concepts in net-centric and asymmetrical warfare rely on the capability for sensors and warfighters to "push" information as well as pull information from centralized sources. These advancements enable information creation and distribution that challenge operators to perform at new levels of performance. However, the soldier warfighter can be overwhelmed by all the available information. Under these circumstances, poor decision-making, slower response times, and generally poor performance can result because the individual is too focused on processing information rather than performing tasks (Wickens 2002, 2008). Our challenge, as human factors researchers and designers, is to ensure future information displays are designed to optimize information perception, interpretation, and decision-making. One approach to enhanced displays in high-workload situations is to distribute incoming information through different sensory channels, as proposed in Multiple Resource Theory (MRT) (Wickens 2002) and Prenav theory (van Erp 2007). There is no doubt that visual displays enable easy and powerful perception, awareness, and comprehension. At the same time, presentation of certain types of information using different sensory modalities is predicted to guide attention, reduce workload, and support performance. In this chapter, we describe experiments and meta-analyses drawn from these tenets that predict easier workload and more intuitive processing of information.

Introduction

Current Western models of warfare, with their emphasis on network-centric systems and asymmetric warfare, are based upon the assumption that advanced information distribution capabilities will reach the lower echelons of command (Alberts, Garstka, and Stein 1999). More than ever, smaller, localized battles are fought in dynamic, uncertain, and unpredictable circumstances and information updates must be rapid and precise. The notion of network-centric warfare is contrasted with platform-centric warfare by its definition: "an information superiority-enabled concept of operations that generates increased combat power by networking sensors, decision-makers, and shooters to achieve shared awareness, increased speed of command, higher tempo of operations, greater lethality, increased survivability, and a degree of self-synchronization" (Alberts, Garstka, and Stein 1999). Network-centric warfare represents rapid dissemination and understanding of complex and dynamic information (Smith 2006). Within this strategy, heightened battlefield awareness and a faster operational tempo (that is, speed of operations) are assumed to result from greater capability to request, receive, interpret, and share information. It is clear that advancements in networked systems will challenge operators to achieve new levels of performance.

However, information alone is not sufficient for information superiority. The warfighter can be overwhelmed with quality information and multiple demands for attention (Mitchell et al. 2004). Under these circumstances, poor decision-making, slower response times, and generally poor performance can result because the individual is too focused on processing information rather than performing tasks (Wickens 2002, 2008). Designers of soldier systems strive to reduce this cognitive load through information fusion, battlefield visualization, and decision support systems that allow some level of system autonomy in operations. Even so, the need persists to better understand and enhance operator perception and understanding of information. Our challenge, as human factors researchers and designers, is to ensure that information is presented to the decision- maker in a manner that will most likely ease and enhance information perception, interpretation, and decision-making.

Soldier Workload

One need only scan the array of emerging soldier technologies to realize that cognitive task demands for soldiers are increasing in volume and complexity. Soldiers on the battlefield now have unprecedented information flow from numerous sources. For example, the emerging technologies assessed for infantry soldier combat teams during the Army Expeditionary Warrior Experiment (AEWE) included aerial and ground vehicles with sensor arrays, small stationary sensors, more robust communication capabilities, and improved visual capabilities (for example, weapon sights, binoculars, night vision, targeting aids, and so on) (Scalsky, Meshesha, and Struken 2009). Even the ground soldier ensemble assessed during the AEWE included technologies with high cognitive task demands. It consisted of a "laser rangefinder, visual displays, integrated load carrying equipment, protective clothing, helmet, speaker, microphone, computer, navigation [system], radio, and controls." This system is expected to allow soldiers to know their own locations and the locations of friendly and enemy assets, and facilitate rapid and informed battlefield decision-making. Strong relationships between communications, battlefield awareness, and overall force effectiveness are envisioned. The report states that force effectiveness is likely enhanced by emerging command, control, intelligence, surveillance, and reconnaissance technologies supported by a functional digital network. In fact, the technology to support command and control was expected to enhance several facets of performance, such as the following:

- positive control of forces beyond line of sight;
- complex/noncontiguous operations;
- increased flexibility in execution;
- rapid development and dissemination of orders;
- sustained momentum of operations and improved maneuver;
- increased tactical options available to the small dismounted force;
- improved planning and rehearsal;
- more effective distributions of combat power;
- enhanced survivability and lethality of the small dismounted force.

The emerging technologies, as assessed in carefully scripted mission scenarios, promise many improvements. The AEWE report concludes that improved force effectiveness will result from "(1) the synergy of Intelligence, Surveillance, and Reconnaissance assets working collaboratively to detect and identify enemy activity, (2) a dismounted Battle Command System (BCS) to share data and provide a common operational picture and control operations, and (3) accurate friendly position location information to aid in precision maneuver, synchronization and coordination of operations, and fratricide avoidance" (Scalsky et al. 2009).

From these plans and expectations for future soldier systems, we foresee higher levels of cognitive demands for information processing. The soldier is already equipped with an increasing array of technology-enabled capabilities and expectations for additional benefits are high. Certainly, training and rehearsals will mitigate the cognitive load, as soldiers gain experience and expertise. In addition, efforts are focused to making the systems lighter and more effectively mounted for the dismounted soldier. But it is clear that cognitive demands are increasing and it is critical to design systems that are easy to use, as well as easy to carry. Because most of these systems provide visual information (for example, maps, diagrams, photographs, text, and so on), there will often be multiple demands for focal visual attention. This increased demand for focal visual attention diminishes the capacity for task-sharing and attention allocation, especially in the context of unexpected changes and events (Wickens 2002).

While cognitive demands can be inferred from emerging technology, systematic analyses of these demands have already demonstrated high levels of workload. Human factors studies of soldier roles have shown significant overloading of the visual and auditory channels for positions such as Abrams tank commanders and drivers (Mitchell 2009), and ground/air robot controllers (Mitchell et al. 2004, Mitchell 2005, Mitchell and Brennan 2009a, 2009b, Pomranky and Wojciechowski 2007). For example, models of task performance and workload (Improved Performance Research Integration Tool (IMPRINT)) predicted workload of robot controllers to be high when the operator is moving as part of the weapons squad.

This workload is particularly high if the operator is also processing incoming information from communications or when manually teleoperating the robot while moving. Such reliance on visual and auditory information forces serial processing of tasks, resulting in delays and/or errors (Mitchell 2005).

A consequence of increased information distribution is the impact on soldier attention. While focused attention may be necessary for performance of some tasks, there will be more interruptions and thus more demands to switch attention and/ or monitor multiple sources of information. The higher-level function of attention management is associated with the ability of the operator to accomplish multiple tasks to "divide" attention, to some extent, and switch attention when necessary. It has been established that individuals cannot perform two information-processing tasks as quickly as one, though some people are better at it than others (Braune and Wickens 1986). Divided attention, in its pure sense, describes how people attend to or perceive two or more channels of perceptual information at once (Wickens 1992). However, this is usually accomplished through effective attention-switching. Demands for divided attention may refer to competing demands on the same perceptual channel, such as dichotic listening, where operators monitor two or more sources of audio information. Similarly, multiple demands for visual attention occur where visual attention must switch from the environment to one or more visual displays. Demands for divided attention can also refer to demands arising from two or more perceptual channels, such as when a driver listens to verbal directions over the radio while also focusing visual attention on the road, the terrain, and, perhaps, also a map of the area.

Management of attention becomes most difficult when cognitive demands are associated with two or more independent tasks that require both information input (for example, visual terrain, visual displays, audio instructions) and output (speech response). An example would be when a driver or a dismounted soldier on patrol is also navigating, searching the environment, and responding to radio requests for information. While this situation necessitates divided attention, the attention is not only between perceptual channels but also between competing tasks with independent goals (Wickens 1992). In contrast, both audio and visual information may be processed more easily when there is a common goal, such as when a driver is navigating and listening to navigation information, or when a gunner is visually searching for a target while listening to targeting information or relaying target information to others. In these situations both channels of information have a shared goal (for example, navigation, targeting) and so they are not nearly as challenging as when the driver has additional independent tasks such as driving while (1) listening and responding to tactical communications, or (2) visually tracking separate map-based information.

Soldiers typically experience several independent demands for attention. For example, dismounted soldiers on patrol may be navigating urban territory while listening to instructions or responding to requests for information, while constantly noting the locations and actions of nearby pedestrians and vehicles. Similarly, mounted soldiers will also experience multiple demands for attention.

In fact, the three basic (independent) tasks of soldiers are to "move, shoot, and communicate." Add to this additional demands associated with new technology (for example, unmanned ground and aerial vehicles, command and control information systems, and so on), and one can expect cognitive task demands to increase, and perhaps even overwhelm. Limitations on multitask behavior have been noted in numerous studies and situations, where operators were not able to effectively divide attention between required tasks, or dynamically prioritize and allocate attentional resources to competing threads of activity (Spink, Cole, and Waller 2009, Wickens 1991, 2008). Management of interruptions becomes a challenge and can deteriorate due to prospective memory failure (for example, inability to remember the previous task (Ellise and Kvavilashvili 2000)), ineffective prioritization (Hopp et al. 2005), or poor timing (Bailey and Konstan 2005, Bailey and Iqbal 2008). At some point, even experienced and expert operators will become overwhelmed.

MRT: Reducing Workload through Sensory Offloads

One approach that has reduced cognitive workload in high-workload situations, MRT, is based on advantages proposed to arise from multisensory display of information (Wickens 2002). According to Wickens:

1. People have several semi-independent cognitive resources.
2. Some resources can be used near-simultaneously without detriment to performance, while others cannot.
3. Tasks requiring the use of different resources can sometimes be effectively performed together.
4. Competition for the same resource can produce interference.
5. Dissimilar cognitive resources exist to process information from different sensory modalities (for example, visual, audio, tactile, olfactory, kinesthetic, and so on).

MRT defines these capacities and contingencies while predicting the degree to which information from a particular sensory channel can be effectively offloaded to another channel, given situational task demands. In conditions of high information flow, it has been found that information-processing is affected by the sensory channels used to perceive information. For example, in conditions of high visual workload, offloading information and alerts to the auditory channel has been found to significantly reduce overall workload and/or enhance operator performance for diverse military operations as well as in the laboratory (Bolia 2004, Burke et al. 2006, Wickens 2002, Wickens et al. 2003). Similarly, the addition of tactile cues has proven effective in realistic situations as well as in laboratory experiments (Chiasson, McGrath, and Rupert 2002, Elliott, Coovert, and Redden 2009, van Erp 2007, van Erp and Werkhoven 2006).

According to MRT, the effectiveness of offloading depends on many factors including (1) overall workload (for example, offloading is more likely to ease high workload), (2) visual attention requirements (for example, offloading is more likely effective when there are high requirements for both focal and ambient vision), (3) sensory cue overlap for common goals (for example, visual scanning and tactile/audio direction cues effectively combine to support a common task such as target detection or navigation), and (4) the degree to which separate tasks demand attention. Wickens (2002) provides extensive discussion and examples of these factors. MRT has driven many investigations to reduce workload and improve decision-making and performance through use of multisensory devices that convey task information through multiple or alternative sensory channels instead of the visual channel (Wickens 2002, 2008).

Display design based on MRT principles begins with an analysis of cognitive task demands associated with a particular job situation. Workload in Army positions is often assessed using IMPRINT models of task performance and workload that are based on task demand analyses as recommended by Wickens (Keller, 2002, Mitchell 2005, 2008, 2009, Mitchell et al. 2004, Mitchell and Brennan 2009a, 2009b). Jobs are described in a mission context by identifying task demands of different resources for each task. For example, driving a car includes task demands that include visual (ambient and focal), auditory, cognitive, and manual responses. While ratings could be generated for this task in general, they are usually formulated within a mission context. It should be noted that specific task demands will depend on situational demands (for example, night-driving, challenging terrain, multiple demands for attention) and experience of the operator (degree of automaticity of the task). Given a mission situation for a particular soldier role, all task demands are identified and rated with regard to demand for auditory, visual, cognitive, and psychomotor processing. Rating scales have been developed for each type of demand (Keller 2002, McCracken and Aldrich 1984). Seven-point scales range from 0.0 (no activity) to 7.0 for visual, auditory, cognitive, and motor demands. These scales were developed through paired comparison frequencies validated through inter-rater reliabilities (Keller 2002) and an example is provided in Table 4.1 (Mitchell 2009). Scales were also developed for auditory, cognitive, speech, fine motor, gross motor, and tactile.

To further identify when and where sensory offloads will benefit, after-task demands are identified, they are analyzed for conflict. Conflict can arise from several sources and one source of conflict is among perceptual modalities. The operator will experience more conflict if information streams from several sources in the same sensory modality at the same time. The operator should experience less conflict when information is distributed through more than one sensory modality. A practical example of this is the experience that a driver is better able to listen to communications and drive, as opposed to watching two visual displays simultaneously or listening to two people talking at the same time. There is still interference, but interference is less when distributed among different perceptual channels. This aspect of MRT supports the use of multisensory displays.

Table 4.1   Visual workload rating scales in IMPRINT

Descriptors	Scale Value	
	Visually unaided (naked eye)	Visually aided (night vision goggles)
Visually scan/search monitor (continuous/serial inspection)	6.0	7.0
Visually discriminate (detect visual differences)	5.0	7.0
Visually read (symbol)	5.0	–
Visually track/follow (maintain orientation)	4.4	5.4
Visually locate/align (selective orientation)	4.0	5.0
Visually inspect/check (discrete inspection/static condition)	3.0	5.0
Visually register/detect (detect occurrence of image)	3.0	5.0

Using the IMPRINT model, one can generate predictions of task performance and compare different task combinations based on the type and amount of interference between the tasks. A good overview of this approach is found in Chapter 19 in this volume, in which Mitchell and Samms provide a thorough review of IMPRINT theory and process.

To summarize, MRT proposes several ways to reduce workload. One means is through offloading to another sensory channel, given situation characteristics such as high multitask demands and minimization of task conflicts. To that end, numerous studies have supported MRT predictions for numerous task situations. These studies and results are well described in other sources (Wickens 2002, 2008). Also reflecting the effectiveness of combining visual and audio cues in a multisensory display, a meta-analysis of 24 studies showed visual–audio multisensory displays were more effective than all-visual displays (Burke et al. 2006). The accumulation of findings provides strong support for reducing visual workload by offloading to audio channels. However, the predictions generalize to tactile channels as well. In the next part of this chapter, we find further theory-based support for multisensory displays, with special consideration of the growing interest in tactile information channels and the potential for enhanced intuitive displays for procedural tasks.

Prenav, Reducing Workload through Intuitive Multisensory Response

MRT provides theory-based support for distribution of information across sensory channels. Still, distribution alone is not sufficient. Information that is offloaded must still be presented in a manner that is easily comprehended. For example, there is increasing consideration and use of tactile displays for attention management, direction and spatial orientation information, and short communications (van Erp 2007). Focusing on intuitive design has resulted in tactile displays that are effective for driving (van Erp and van Veen 2004; van Erp et al. 2006).

To emphasize this point, a cognitive model has been offered to describe and predict performance as a function of intuitive presentations. While MRT is based on models of attentional resources in high workload situations, the Prenav model of sensory channel information (van Erp 2007, van Erp and Werkhoven 2006) more specifically includes the tactile modality and has a strong emphasis on performance that can be automated, such as steering, navigation, and control systems. Prenav explains how tactile cues can affect attention, cognition, and performance, with particular regard for the effects of practice, automaticity, and intuitive response. Figure 4.1 illustrates the model components and processes.

According to van Erp and Werkhoven (2006), Prenav (Figure 4.1) is based on an integration of models for navigation and workload: Sheridan's model for supervisory (vehicle) control (1992), Wickens' information processing model (1984, 1992), Veltman and Jansens' workload framework (2004), and Rasmussen's framework on skill-, rule-, and knowledge-based behavior (1982, 1983). Prenav consists of two loops: the information processing loop and the workload loop. The information-processing loop, highlighted, describes cognitive deliberation and has the usual links from sensation to perception, perception to decision, and decision to action, and feedback from the environment or display. The perception and decision steps are represented in Prenav as the "cognitive ladder." The five arrows denote different modalities that can provide sensory input (for example, visual, auditory, tactile) and that can be processed as sensations in parallel, which is consistent with MRT propositions.

Prenav also addresses the contribution of intuitive cues by proposing two cognitive "short cuts," as represented by the thinner arrows from sensation to action and from perception to action, to better account for automatic or skill-based performance. These short cuts are related to concepts of automaticity (Shiffrin and Schneider 1984), skill-based learning (Rasmussen 1983), and recognition-based decision-making (for example, Klein 2008). These short cuts are proposed to reduce cognitive overload, thus preserving performance quality. The short cut from sensation to action accounts for reflexive or highly trained responses that require no cognition per se, such as maintaining balance while walking, orienting to a sudden loud noise, or suddenly braking at an unexpected obstacle. This short-cut is associated with control tasks such as in driving, where the driver automatically stays on the road based on visual, kinesthetic, and force-feedback cues. The short-cut from perception to action also describes highly automated reactions, where the sensation is processed enough to evoke recognition, such as slowing for a stop sign ahead. The action does not require effortful process, but instead is more automatic and recognition-based, for tasks such as steering. The more effortful

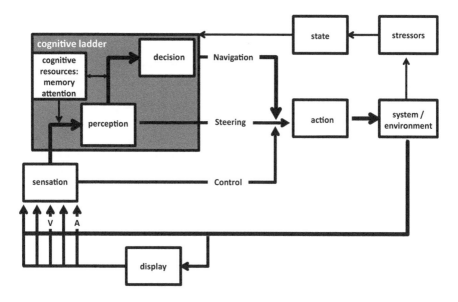

Figure 4.1   Prenav model for human behavior in platform navigation and control (van Erp and Werkhoven 2006)

process represented in the cognitive ladder pertains to navigation tasks such as waypoint selection, map interpretation, and rerouting.

Prenav also accounts for individual differences and environmental stressors as they impact the operator state, which affects the more deliberative cognitive ladder, but not the sensation–action loop. Thus, factors such as operator expertise, workload, or environmental constraints will impact the more deliberative processes of cognition. As workload increases, two types of bottlenecks are described: sensory overload and cognitive overload. Tactile displays are proposed to potentially reduce both types of overload, through use of the shortcuts from sensation and perception that bypass the cognitive ladder. Tactile cues are proposed to be more reflexive, particularly when presenting direction or spatial orientation information. In this manner, tactile cues are expected to reduce workload when used for direction and orientation.

Multisensory Tactile Displays

While earlier multisensory systems focused on integrating audio and visual displays, recent research has shown many effective applications of tactile displays. With the exception of tactile devices developed to aid persons with sensory impairments (for example, blind, deaf), tactile displays function primarily in multisensory situations. Much support has amassed for the use of the tactile channel as an additional means of information processing. Experiments and demonstrations have been conducted across a wide range of settings, from laboratory tasks to high-fidelity simulations and real-world environments. Operators have successfully perceived and interpreted vibrotactile cues in averse, demanding, and distracting situations such as combat vehicles (for example, Carlander and Eriksson 2006, Krausman and White 2008), aircrew cockpits (for example, Rupert 2000a, 2000b, McGrath et al. 2004), high-speed watercraft (Dobbins and Samways 2002), underwater environments (Self et al. 2007), and during strenuous movements (Pettitt, Redden and Carstens 2006, Redden et al. 2006). Also, tactile alerts have effectively prompted task-switching in high workload conditions (Hameed et al. 2006, Ho, Nikolic, and Sarter 2001, Hopp et al. 2005). Still, while many tactile displays have been found effective, it is not difficult to imagine a tactile display that is counterintuitive. As tactile displays are increasingly used for communication of more complex and multiple concepts, it will become evident that tactile and multisensory systems in general must be designed for rapid and easy comprehension.

Within the Prenav model, the tactile sense is described as highly intuitive and associated with fast reaction times. In fact, studies have found that tactile cues can be faster than other sensory channels when used for alarms or direction cues. For example, faster vehicle braking responses occurred in reaction to tactile rear-end collision warnings, compared to when visual cues were used (Scott and Gray 2008). In another context, operators fired more quickly in outdoor target practice using targets to the left, right, and center when using torso-mounted tactile direction cues than with visual cues (Gilson, Redden, and Elliott 2007). Tactile cues added to visual cues also yielded faster reaction times in simulation-based studies of operator decision-making and performance (for example, Calhoun et al. 2004, 2005, Forster et al. 2002). Van Erp and his colleagues found that adding tactile cues improved performance in visually demanding situations such as navigating a car through unfamiliar urban terrain (van Erp et al. 2004) and maintaining helicopter altitude in low visibility (van Erp et al. 2003). Similarly, Chiasson, McGrath, and Rupert (2002) reported faster navigation in air (high-altitude parachute environment), ground, and under water using tactile direction cues.

Tactile cues have been associated with improvements on other kinds of performance, in addition to reaction time. Users have successfully perceived and responded to tactile cues in various body locations. For example, torso-mounted tactors were found effective for alerts, directions, and orientation cues—for tasks such as decision-making, navigation, targeting, situational awareness, and spatial orientation (for example, Chiasson, McGrath, and Rupert 2002, Cholewiak and McGrath 2006, Elliott et al. 2007, van Erp 2007). Other locations that have been successfully used include the *wrist* (Dobbins and Samways 2002, Brewster and King 2005, Calhoun et al. 2004, Ho, Nikolic, and Sarter 2001), *fingers* (Hameed et al. 2006, Lathan and Tracey 2002), *palm of hand* (Tang, Beebe, and Kramer 1997), *arm* (Bloomfield and Badler, 2007), *abdomen* (Ho, Reed, and Spence 2007, Moorhead, Holmes, and Furnell 2004), *shoulder* (Hopp et al. 2005), and *back* (Hameed et al. 2007, Jones, Lockyer, and Piateski 2006, Lindeman et al. 2003).

Some ARL Soldier-based Experiments of Multisensory Tactile Displays

While tactile cues have been associated with better or faster performance, we cannot assume that all tactile cues will be intuitive. However, if we consider both MRT and Prenav predictions, we should expect that given a high workload multitask and multisensory situation, offloading from a saturated sensory channel (for example, visual, audio) to a tactile channel will reduce workload and improve performance, assuming the tactile display is intuitive. Several HRED experiments conducted with soldiers support these expectations. Krausman and her colleagues have demonstrated the value of tactile alerts to guide attention in simulations of tank leader communications and within simulated and actual vehicles with high levels of vibrations (Krausman, Elliott, and Pettitt 2005, Krausman, Pettitt, and Elliott 2007). Additional studies further support the ability of soldiers to detect not only single alerts but also patterns of multiple tactors to represent different communications (Pettitt, Redden, and Carstens 2006). It is particularly promising that these patterns can be perceived during strenuous movements (Redden et al. 2006).

Because tactile alerts have been demonstrated to effectively and easily portray spatial orientation (van Erp 2007, van Erp et al. 2007), it is reasonable to assume that a torso belt would convey direction information that is also immediately understood. Three experiments investigated the efficacy and suitability of a torso-mounted tactile belt for soldier navigation (Elliott et al. 2006, 2007, 2010). In Experiment 1, researchers used a tactile navigation system developed by The Netherlands Organisation for Applied Scientific Research (Toegepast-Natuurwetenschappelijk Onderzoek) for US Army soldier land navigation. Land navigation occurred at the Fort Benning, GA, training site for soldier navigation—a challenging terrain that requires careful attention to one's surroundings for threat from terrain and natural wildlife. Participants navigated three waypoints along 600 m through heavily wooded terrain, using (1) map and compass, (2) standard alphanumeric handheld global positioning system (GPS) device, and (3) the tactile GPS system, while also responding to radio requests for information. In Experiment 2, researchers used challenging terrain similar to Experiment 1, with the added challenge of night operations, during intermittent thunderstorms. Soldiers navigated three waypoints while also searching for live and silhouette targets using (1) a handheld GPS device, (2) a head-mounted map-based GPS, and (3) the tactile GPS system. For one leg, soldiers were told to move as quickly as possible. In another leg, soldiers were told to search the area surrounding them for targets. A third leg required soldiers to navigate around a natural obstacle in their path. Experiment 3 had participants navigate with (1) a commercial GPS arrow display, (2) the tactile GPS system, and (3) both together. Given this series of results, more extensively discussed elsewhere (Elliott et al. 2006, 2007), it was concluded that tactile navigation displays can be used in strenuous outdoor environments and can outperform visual displays under conditions of high cognitive and visual workload. In addition, soldier comments were highly in favor of this capability, stating that the system was "hands-free, eyes-free, and mind-free."

A torso-mounted tactile belt was also demonstrated as useful for robot control operations (Redden et al. 2009). In this experiment, three types of robot controller navigation map display configurations were evaluated for effects on beyond-line-of-sight robotic navigation tasks. First was a larger split-screen visual display that presented both a map display and a camera-based driving display on a 6.5-inch screen. Two smaller alternatives were also evaluated. One alternative was a 3.5-inch display that allowed the operator to toggle back and forth between the driving and map displays. The third option added a torso-mounted tactile display to the toggle-based display in order to provide direction information simultaneously with the camera display and thus reduce the need to toggle as frequently to the map display. Each display option was evaluated based on objective performance data, expert-based observations, and scaled subjective soldier questionnaire items. Findings indicated that operators' navigation performance with the multimodal 3.5-inch toggle display was as effective as their performance with the 6.5-inch split-screen display. Operator performance was significantly lower with the 3.5-inch toggle display without the tactile display.

Meta-analyses of Visual, Auditory, and Tactile Displays

Theories such as MRT and Prenav predict when and how multisensory displays can aid information processing and performance. Empirical studies serve to validate or refine theory. At this time there are many empirical studies that have (1) demonstrated usefulness of tactile displays, (2) compared usefulness of visual versus audio and/or tactile displays, (3) demonstrated usefulness of multisensory displays, or (4) compared unimodal versus multisensory displays. The variety of these studies ranges from very basic to very applied, very simple to very complex, and they span different types of tasks. In order to organize and interpret these findings, a multi-year effort was made to review and organize the literature and apply meta-analytic techniques to identify trends (Coovert et al. 2007, 2008).

The first step was a literature review and creation of a bibliographic database. Details regarding search strategy and criteria for inclusion are specified elsewhere (Coovert et al. 2007). Over 600 studies were initially reviewed and coded along numerous criteria that determined appropriateness for meta-analysis categories, using a widely used meta-analysis software package (Borenstein et al. 2005). The literature review culminated in an organized set of guidelines for display design and research (Coovert et al. 2006).

Preliminary Comparisons

As studies accumulated, they were organized by broad categories with regard to whether displays were audio, visual, tactile, or multisensory. Multisensory categories included all combinations (that is, audio and visual, audio and tactile, visual and tactile, and all three). Analyses compared multisensory to single-sensory displays.

Results indicate that in general, adding an additional modality to visual feedback improves performance. Both visual–auditory and visual–tactile feedback provided advantages in reducing reaction times and improving performance scores; however, they were not as effective in reducing error rates. Effects were moderated by task type, workload, and number of tasks. Visual-auditory feedback was most effective when a single task was being performed and under normal workload conditions (Burke et al. 2006). Visual–tactile feedback was more effective than visual displays when multiple tasks were performed and workload was high. When looking at different criteria, visual–tactile display was particularly effective at reducing reaction time and increasing performance (Prewett et al. 2006). Both types of multimodal feedback were effective for target acquisition tasks, but varied in effectiveness for other task types.

Additional Comparisons

Preliminary meta-analytic comparisons suggested that multisensory displays in general (for example, audio, tactile augments) can outperform visual displays. However, these comparisons were made on broad categories that combined studies using a variety of displays and context. Follow-on comparisons were structured to explore more specific comparisons. To accomplish this, a subset of the meta-analytic database was chosen for more detailed analyses focusing on studies of visual and tactile displays.

First, studies were organized by type of comparison. Preliminary comparisons did not distinguish whether the tactile cues were added to a visual display or were used to substitute for visual cues. This distinction is important, with regard to the amount of information being processed and the need for division of attention. For example, while audio cues have been found to enhance visual displays, it is usually because the audio information is either managing attention or offloading information that was portrayed visually. When one adds new audio information to a visual display that is already demanding, results can change. A common example of this nature is the use of cell phones while driving. Few people would argue that cell phone chatter would enhance driving performance. Therefore, the following comparisons have core distinctions. For the follow-on comparisons, our research group distinguished three types of comparisons: (1) adding a tactile cue to a baseline condition, where the tactile cue is providing additional information (for example, adding a tactile alert to a condition where there is no alert), (2) comparing a tactile cue with a visual cue that provides the same information (for example, comparing a tactile alert to a visual alert condition), and (3) comparing a visual cue condition with a multisensory (visual and tactile) condition. Then, we further categorized within these comparisons with regard to type of information, which ranged from simple alerts and single-direction cues to more complex tactile patterns representing spatial orientation or short communications. This organization is portrayed in Table 4.2 along with the studies comprising each comparison. The results are listed in Table 4.3. Studies included in these analyses are listed in the references, marked with an asterisk (*).

Table 4.2   Studies comprising meta-analyses of visual and tactile cues by information type

Information Type	Studies
Alerts	Hameed et al. 2006, Ho, Nikolic, and Sarter 2001, Hopp et al. 2005, Moorhead, Holmes, and Furnell 2004, Scott and Gray 2008
Direction	Bloomfield and Badler 2007, Chen and Terrence 2008, Davis 2007, Dorneich et al. 2006, Elliott et al. 2006, Glumm, Kehring, and White 2006, Hameed et al. 2007, Lindemann et al. 2003, 2005, Murray, Katzky, and Khosla 2003, Tan et al. 2003
Spatial Orientation	Aretz et al. 2006, Chiasson, McGrath, and Rupert 2002, Diamond et al. 2002, McGrath et al. 2004, Raj, Kass, and Perry 2000, Small et al. 2006, Tripp et al. 2007, van Erp, Meppelink, and van Veen 2002, van Erp, Veltman, and van Veen 2003, van Erp and Werkhoven 2006
Visual Versus Visual, and Tactile Comparisons	
Alerts	Calhoun et al. 2002, 2004, 2005, Forster et al. 2002, Hopp et al. 2005, Krausman, Pettitt, and Eliott 2007, Moorhead, Holmes, and Furnell 2004, Sklar and Sarter 1999
Direction	Bloomfield and Balder 2007, Elliott et al. 2007, Lathan and Tracey 2002, Tan et al. 2003, van Erp, Meppelink, and van Veen 2002
Spatial Orientation	Chiasson, McGrath, and Rupert 2002, Eriksson et al. 2006, McKinley and Tripp 2007, van Erp et al. 2007
Visual Versus Tactile	
Alerts	Calhoun et al. 2002, Forster et al. 2002, Hameed et al. 2007, Krausman, Elliott, and Pettitt 2005, Scott and Gray 2008, Sklar and Sarter 1999, van Erp et al. 2004
Communication	Brewster and King 2005, Ho, Nikolic, and Sarter 2001, Pettitt, Redden, and Carstens 2006
Direction	Bloomfield and Badler 2007, Cholewiak and McGrath 2006, Davis 2006, Elliott et al. 2006, 2007, Glumm, Kehring, and White 2006, Hameed et al. 2006, Lindeman et al. 2003, Savick et al. 2008, van Erp, Meppelink, and van Veen 2002, van Erp et al. 2004
Spatial Orientation	McKinley and Tripp 2007, Small et al. 2006

Table 4.3   **Results of visual versus tactile display comparisons: Hedge's *g* (weighted mean difference), standard error of *g*, and fail-safe *N* (number of non-significant studies required to change significance result)**

Results From Meta-analytic Comparisons Of Visual Versus Tactile Cues			
	Type of Comparison		
Task demand	**Adding tactile cue to baseline (B versus T)**	**Substituting tactile cue for a visual cue (V versus T)**	**Adding tactile cue to visual cue (V versus VT)**
Alerts	**1.60 (0.36) *FS* = 89	**1.81 (0.45) *FS* = 99	**1.08 (0.24) *FS* = 124
Direction	**0.91 (0.21) *FS* = 279	−0.20 (0.29)	*0.64 (0.26) *FS* = 24
Spatial Orientation	**1.13 (0.24) *FS* = 215	0.02 (0.70)	*0.93 (0.40) *FS* = 15
Communication	na	**2.29 (0.61) *FS* = 63	x *FS* = na
Total	** 1.15 (0.21) *FS* = 1709	0.95 (0.22)	**0.89 (0.18) *FS* = 431

*   *p < 0.05*

** *p < 0.01*

Baseline versus Tactile Comparison

This comparison category included studies that assessed the effectiveness of tactile cues when added to a baseline condition. For these comparisons, almost all studies favored the tactile condition. Most studies in this category were based on simulation and virtual reality-based situations. However, Elliott and colleagues (2006) and Dorneich and colleagues (2006) studies were based on actual land navigation (and secondary tasks) in the field. Performance measures varied, but most reflected some type of timed performance (for example, reaction time, time to navigate, time to visually search, number of targets or icons found within a time limit, and so on). Hedge's *g*, which represents the overall effect size across all studies, was 1.15 (standard error (*SE*) 0.21) and was significant at $p < 0.01$, in favor of the tactile cue condition. There was a consistent strong effect favoring the tactile condition across most studies and thus a high likelihood that tactile cues can improve performance when added as an alert, direction, or orienting cue. Analyses were then run with subgroups divided by type of information. All subgroup analyses were significantly in favor of the tactile condition, for alerts ($g = 1.60$, $SE = 0.36$, $p < 0.01$), direction cues ($g = 0.91$, $SE = 0.21$, $p < 0.01$), and spatial orientation cues ($g = 1.13$, $SE = 0.24$, $p < 0.01$).

Visual versus Tactile Comparison

This category included studies that compared tactile cues to visual cues presenting the same information for the same purpose. For studies in this review, visual and tactile alerts or direction cues have been compared for a variety of tasks, including simple reaction time or visual search tasks (for example, Forster et al. 2002) simulated driving and/or targeting tasks (Davis 2006, Glumm, Kehring, and White 2006, Savick et al. 2008, Scott and Gray 2008, van Erp, Meppelink, and van Veen 2002, van Erp et al. 2004), complex cockpit, unmanned aerial vehicle (UAV), or command simulations (Calhoun et al. 2002, Hameed et al. 2006, 2007, Ho, Nikolic, and Sarter 2001, Krausman, Pettitt, and Elliott 2007, McKinley and Tripp 2007, Sklar and Sarter 1999, Small et al. 2006), communicating information (Brewster and King 2005, Redden and Elliott 2007), localization from dense multitactor displays (Cholewiak and McGrath 2006), land navigation in the field (Elliott et al. 2006, 2007), and orienting in virtual environments (Bloomfield and Badler 2007).

Overall Hedge's *g* was 0.95, *SE* = 0.22; however, Hedge's *g* was not significant for this comparison, primarily due to the conservative test of the random effects model. Although the effect of replacing visual cues with tactile cues was large ($g = 0.95$, $SE = 0.22$), significant variation ($Q_{total} = 288.89$, $p < 0.01$) prevented a statistically significant result. Using a less conservative fixed-effects model would find this difference significant. The larger variation among studies was verified by a moderator analysis for type of task, which was significant. Tactile cues were more consistently effective for alerts and communication as opposed to direction and spatial orientation (Table 4.3). Most studies compared visual and tactile direction cues, albeit for different purposes (for example, navigation, targeting). For this category, Hedge's *g* was -0.203 ($SE = 0.294$, $p = 0.50$), indicating strong differences among studies with regard to visual versus tactile effect, and thus support for presence of a moderator variable, such as task purpose or type of tactor system. There were only two studies that compared visual and tactile cues for spatial orientation and neither was significant (Hedge's $g = 0.02$, $SE = 0.70$, $p = $ n.s.).

Visual versus Visual–Tactile Comparison

Visual versus visual–tactile (V vs. VT) studies compared a multisensory presentation of visual and tactile cues representing the same information, to conditions consisting of only visual cues. This category appears similar to the baseline (B) vs. tactile (T) comparison, but is distinguished by the multimodal and redundant nature of the VT condition. Studies ranged from simple reaction time or tracking tasks (Forster et al. 2002, Moorhead, Holmes, and Furnell 2004) to more complex situations such as communications (Bloomfield and Badler 2007), driving (van Erp, Meppelink, and van Veen 2002), cockpit or command and control simulations (Calhoun et al. 2002, 2004, 2005, Hopp et al. 2005, Krausman,

Pettitt, and Elliott 2007, Sklar and Sarter 1999), land navigation (Elliott et al. 2007), teleoperation in virtual reality environments (Lathan and Tracey 2002), more advanced tactile arrays for spatial orientation in flight (Chiasson, McGrath, and Rupert 2002, Eriksson et al. 2006, van Erp et al. 2007), and localization of a dense array of tactors (Tan et al. 2003).

Nearly all studies demonstrated a positive effect. Overall Hedge's g was 0.89, $SE = 0.18$, $p < 0.01$, in favor of multimodal cueing over visual-only cueing when analyzed across all studies. Results for studies using tactile cues as alerts were mostly in favor of tactile cues, with $g = 0.412$ and $p = 0.08$. For alerts, Hedge's g was 1.08 ($SE = 0.24$, $p < 0.01$), also in favor of the multimodal presentation. For studies in which multimodal cue patterns were used to convey direction information, the multimodal presentation was again favored, with Hedge's $g = 0.64$ ($SE = 0.26$, $p < 0.05$). Studies that compared visual versus visual–tactile cues for orientation had an overall Hedge's g of 0.93 ($SE = 0.40$, $p < 0.01$), also in favor of the multimodal presentation.

To summarize, when added to a baseline task or to existing visual cues, the addition of tactile cues enhanced task performance. Using tactile to replace visual cues produced mixed results that were moderated by cue information complexity. Tactile alerts were effective when replacing visual alerts, tactile communication cues were effective when replacing visual cues, but tactile direction cues varied widely in their performance, resulting in overall nonsignificance.

Discussion

It is apparent that technology that was previously the domain of commanders and specialists will continue to spread and become common on the battlefield. Soldiers and leaders will face increasing challenges to process more information rapidly, and use and maintain high technology assets. In this chapter, we discussed the potential of multisensory displays to provide information that is more rapidly and easily understood. We first discussed two theoretical frameworks (MRT and Prenav) that generate predictions regarding how and when multisensory displays would be more effective. We paid particular attention to displays with tactile cues, as almost all tactile displays are multisensory in context, such that there were many experiments that added tactile information to baseline multisensory conditions. In addition, many tactile applications have recently proven effective in operational situations.

There are several reasons to expect the addition of tactile information to multisensory context would enhance performance and lower workload. MRT emphasizes the role of sensory channels as somewhat separate in information processing resources, such that offloading information to a separate channel is expected to alleviate workload from a channel that is saturated. Thus, the emphasis is on alleviating information overload of a particular sensory channel, usually that of focal visual attention. Good examples of tactile cue effectiveness in high visual workload operations include tactile cues for navigation and spatial orientation for pilots (Eriksson et al. 2006, Raj, Kass, and Perry 2000, van Erp 2007), robot control operators (Redden et al. 2009), speedboat drivers and underwater divers (Chiasson, McGrath, and Rupert 2002), command and control decision-makers with multiple visual displays (Krausman, Elliott, and Pettitt 2005, Krausman, Pettitt, and Elliott 2007) and dismounted soldiers (Elliott et al. 2006, 2007, Pettitt, Redden, and Carstens 2006). In these situations, the common factor is near-information overload on the visual channel. Finally, operational conditions that degrade the visual or auditory channels (for example, weather, and noise) or that require stealth (for example, light security, radio silence) also support the use of tactile displays.

For simple tactile applications, improved attention management is usually the goal. Tactile cues have successfully been used for alerts, particularly for driving (Bloomfield and Badler 2007) and decision-making (Krausman, Pettitt, and Elliott 2007, Sklar and Sarter 1999). Tactile cues have also been successful in more complex applications to indicate direction or spatial orientation (van Erp 2007). This is consistent with Prenav theory, which emphasizes the intuitive characteristics of each sensory input. Tactile cue displays have been demonstrated to be very effective in portraying information such as direction (for example, targeting, navigation) and spatial orientation (for example, pilots, and astronauts).

There is also growing interest in tactile cue displays for more complex communications representing verbal concepts. MRT predicts effectiveness of this approach given high workload through other sensory channels. Prenav places more emphasis on ease of comprehension, such that the tactile cue displays would need to be trained to automaticity. Triangulation of both MRT and Prenav theories generates predictions of higher effectiveness when (1) there is already very high workload, (2) there is need for attention management cues (for example, alerts), (3) there is need for direction or spatial orientation information, or (4) tactile communication patterns are easily learned and recognized. Many questions remain, however, with regard to whether information gained through different senses should be consistent, redundant, and/or pertaining to the same task.

For the most part, meta-analyses of experiment-based data support predictions of effectiveness. Preliminary comparisons of unimodal and multimodal information displays favored multisensory displays. Multisensory categories included all combinations (that is, audio and visual, audio and tactile, visual and tactile, and all three). Analyses compared multisensory to single-sensory displays. Results indicate that in general, adding an additional modality to visual feedback improves performance. Both visual–auditory and visual–tactile feedback provided advantages in reducing reaction times and improving performance scores; however, they were not as effective in reducing error rates. Effects were moderated by task type, workload and number of tasks. Visual–auditory feedback was most effective when a single task was being performed and under normal workload conditions (Burke et al. 2006). Visual–tactile feedback was more effective than visual displays when multiple tasks were performed and workload was high. When looking at different criteria, the visual/tactile display was particularly effective at reducing reaction time and increasing performance (Prewett et al. 2006). Both types of multimodal feedback were effective for target acquisition tasks, but varied in effectiveness for other task types.

Further meta-analyses were performed to better understand the impact of adding tactile cues to visual information. These analyses distinguished comparisons where tactile cues offered additional information versus comparisons where tactile cues were redundant to visual information. Further, comparisons were structured to account for different types of information that was presented by tactile cues: whether the information was simply to manage attention (that is, alerts), direction information, or spatial orientation, or substitute for verbal information. Results indicated that the type of comparison did matter. When added to a baseline task or to existing visual cues, the addition of tactile cues enhanced task performance. Using tactile cues to replace visual cues produced mixed results that were moderated by cue information complexity. Tactile alerts were effective when replacing visual alerts, tactile communication cues were effective when replacing visual cues, but tactile direction cues varied widely in their performance, resulting in overall nonsignificance. Further experiments and analyses are needed to address the wide variation in task demands associated with direction cues.

Future Research

Although the meta-analyses provide a solid foundation upon which to build, there remains significant variation in study comparisons. Further research in this area should clarify moderating factors and refine guiding principles to determine when, where, why, and how to best employ alternate cues to support operator performance in demanding or complex environments. For example, while our focus was on performance in terms of efficiency (for example, time to complete) or accuracy, research should also look at the effects on other levels of performance complexity, from simple reaction times to complex decision-making under uncertainty, and particularly, on aspects of workload. Workload was ultimately excluded from our analyses because very few studies manipulated or measured subjective workload in a systematic manner. Future studies need to further test and refine theory-driven predictions with regard to overall workload and demands for attention management.

In addition, research should include examination of individual differences, particularly with regard to attention management skills, task experience, and automaticity. Attention management and other cognitive skills can be assessed via PC-based cognitive batteries such as the Automated Neuropsychological Assessment Metrics (Reeves et al. 2001) and the Synwin task (Elsmore 1994). Similarly, the role of expertise and automaticity needs further exploration. To date, experiments have focused on short-term performance, in which tasks are performed by novices or are associated with short periods of training.

It is increasingly apparent that "ease of comprehension" is a critical factor in the design of any perceptual cue, be it visual, audio, or tactile. This leads to questions such as: To what degree will initial differences among cue displays last? Can such differences be overcome or reversed with extended practice? Research comparing alternate and multiple modalities of cue information should be performed in task situations where these factors can be effectively manipulated and controlled, these designs would further predict and model the impact of levels of task demand and cue complexity.

Finally, we must consider effects across different task demand situations, from stationary control operators to expert first responders in a highly mobile emergency environment. Principles cannot be applied regardless of situation. Instead, designers must always consider in detail the specific cognitive and situation demands in order to best determine where bottlenecks occur and how they can be alleviated. Results demonstrated positive evidence for the effectiveness of multisensory and, particularly, tactile cues, but benefits are not definitive. The variation among studies indicates the need for careful consideration of task demands and cue characteristics by interface designers.

References[1]

Alberts, D., Garstka, J. and Stein, F. 1999. *Network Centric Warfare: Developing and Leveraging Information Superiority*. 2nd Edition. Vienna, VA: CCRP Press.

Aretz, D.T., Andre, T.S., Self, B.P. and Brenaman, C.A. 2006. *Effect of Tactile Feedback on Unmanned Aerial Vehicle Landings. Proceedings of the Interservice/Industry Training, Simulation, and Education Conference*, Orlando, FL, December 2006.*

Bailey, B. and Iqbal, S. 2008. Understanding changes in mental workload during execution of goal-directed tasks and its application for interruption management. *ACM Transactions on Computer–Human Interaction*, 14(4), Article 21: 1–27.

Bailey, B. and Konstan, J. 2005. On the need for attention–aware systems: Measuring effects of interruption on task performance, error rate, and affective state. *Computers in Human Behavior*, 22(4), 685–708.

Bloomfield, A. and Badler, N.I. 2007. *Collision Avoidance Using Vibrotactile arrays. Proceedings of the IEEE Virtual Reality Conference*, Charlotte, NC, March 2007, 163–70.*

Bolia, R. 2004. Special Issue: Spatial audio displays for military aviation. *International Journal of Aviation Psychology*, 14(3), 233–38.

Borenstein, M., Hedges, L., Higgins, L. and Rothstein, H. 2005. *Comprehensive Meta-Analysis, Version 2.0: A Computer Program for Research Synthesis*. Englewood, NJ: Biostat.

---

1   References included in the follow-on meta-analysis are marked with an asterisk (*).

Braune, R. and Wickens, C. 1986. Time-sharing revisited: Test of a componential model for the assessment of individual differences. *Ergonomics*, 29(11), 1 399–1414.

Brewster, S. and King, A. 2005. *The Design and Evaluation of a Vibrotactile Progress Bar. Proceedings of the First Joint Eurohaptics Conference and Symposium on Haptic Interfaces for Virtual Environment and Teleoperator Systems*, Pisa Italy, March 2005. Available at: http://www.computer.org/csdl/proceedings/whc/2005/2310/00/23100499-abs.html (accessed: July 25, 2012).*

Burke, J.L., Prewett, M., Gray, A., Yang, L., Stilson, F.R.L., Redden, E., Elliott, L.R. and Coovert, M.D. 2006. *Comparing the Effects of Visual–Auditory and Visual–Tactile Feedback on User Performance: A Meta-Analysis. Proceedings of the International Conference on Multimodal Interfaces*, Banff, Alberta, Canada, November 2006, 108–17.

Calhoun, G., Draper, M., Ruff, H. and Fontejon, J. 2002. *Utility of a Tactile Display for Cueing Faults. Proceedings of the Human Factors and Ergonomic Society 46th Annual Meeting*, Baltimore, MD, September–October 2002, 2118–2122.*

Calhoun, G.L., Fontejon, J., Draper, M., Ruff, H. and Guilfoos, B. 2004. *Tactile vs. Aural Redundant Alert Cues for UAV Control Applications. Proceedings of the HFES 48th Annual Meeting*, New Orleans, LA, September 2004, 137–41.*

Calhoun, G.L., Ruff, H.A., Draper, M.H. and Guilfoos, B.J. 2005. *Tactile and Aural Alerts in High Auditory Load UAV Control Environments. Proceedings of the Human Factors and Ergonomic Society 49th Annual Meeting*, Orlando, FL, September 1, 2005, 145–49.*

Carlander, O. and Eriksson, L. 2006. *Uni- and Bimodal Threat Cueing with Vibrotactile and 3-D Audio Technologies in a Combat Vehicle. Proceedings of the Human Factors and Ergonomics 50th Annual Meeting*. San Francisco, CA, October 2006, 1552–1556.

Chen, J.Y. and Terrence, P.I. 2008. Effects of tactile cueing on concurrent performance of military and robotics tasks in a simulated multitasking environment. *Ergonomics*, 51(8), 1137–1152.*

Chiasson, J., McGrath, B. and Rupert, A. 2002. *Enhanced Situation Awareness in Sea, Air, and Land Environment: Proceedings of NATO RTO Human Factors & Medicine Panel Symposium on Spatial Disorientation in Military Vehicles: Causes, Consequences and Cures*. No. TRO-MP-086, 1–10. La Coruña, Spain. [Online]. Available at: http://www.faa.gov /library/ online_libraries/aerospace_medicine/sd/media/MP-086-32.pdf (accessed: July 19, 2012).*

Cholewiak, R. and McGrath, C. 2006. *Vibrotactile Targeting in Multimodal Systems: Accuracy and Interaction. Proceedings of the 14th Symposium on Haptic Interfaces for Virtual Environment and Teleoperator Systems*, Arlington, VA, March 2006, 413–20.*

Coovert, M., Gray, A., Tolentino, A., Jagusztyn, N., Stilson, F., Klein, R., Willis, T., Rossi, M., Redden, E. and Elliott, L. 2006. *Guiding Principles for Tactile Technology: Implications from Theory and Empirical Findings. Proceedings of the Human Factors and Ergonomics 50th Annual Meeting*, San Francisco, CA, October 2006, 1682–1686.

Coovert, M.D., Gray, A., Elliott, L. and Redden, E. 2007. *Development of a Framework for Multimodal Research: Creation of the Bibliographic Database*. (Technical Report ARL-TR-4068). Aberdeen Proving Ground, MD: US Army Research Laboratory.

Coovert, M.D., Walvoord, A.A., Elliott, L.R. and Redden, E.S. 2008. A tool for the accumulation and evaluation of multimodal research. *IEEE Transactions on Systems, Man, and Cybernetics, Part C: Applications and Reviews*, 38(6), 850–55.

Davis, B. 2006. *Effect of Tactical Navigation Display Modality on Navigation Performance, Situation Awareness, and Mental Workload. Proceedings of the Human Factors and Ergonomics 50th Annual Meeting*, San Francisco, CA, October 2006, 2089–2093.*

———. 2007. *Effects of Visual, Auditory, and Tactile Navigation Cues on Navigation Performance, Situation Awareness, and Mental Workload*. (Technical Report ARL-TR-4022). Aberdeen Proving Ground, MD: US Army Research Laboratory.*

Diamond, D.D., Kass, S.J., Andrasik, F., Raj, A.K. and Rupert, A.H. 2002. Vibrotactile cueing as a master caution system for visual monitoring. *Human Factors & Aerospace Safety*, 2(4), 339–54.*

Dobbins, T. and Samways, S. 2002. *The Use of Tactile Navigation Cues in High-Speed Craft Operations: Proceedings of the RINA Conference on High Speed Craft: Technology and Operation*, London, November 2002, 13–20.

Dorneich, M., Ververs, P., Whitlow, S. and Mathan, S. 2006. *Evaluation of a Tactile Navigation Cueing System and Real-Time Assessment of Cognitive State. Proceedings of the HFES 50th Annual Meeting*, San Francisco, CA, November 2006, 2600–2604.*

Elliott, L.R., Coovert, M.D. and Redden, E.S. 2009. *Overview of Meta-analyses Investigating Vibrotactile Versus Visual Display Options: Proceedings of the 13th International Conference on Human–Computer Interaction, Part II: Novem Interaction Methods and Techniques*, San Diego, CA, July 2009, 435–43.

Elliott, L.R., Duistermaat, M., Redden, E.S. and van Erp, J. 2007. *Multimodal Guidance for Land Navigation*. (Technical Report ARL-TR-4295). Aberdeen Proving Ground: US Army Research Laboratory.*

Elliott, L.R., Redden, E., Pettitt, R., Carstens, C., van Erp, J. and Duistermaat, M.E.S. 2006. *Tactile Guidance for Land Navigation*. (Technical Report ARL-TR-3814). Aberdeen Proving Ground: US Army Research Laboratory.*

Elliott, L.R., van Erp, J.B.F., Redden, E.S. and Duistermaat, M. 2010. Field-based Validation of a Tactile Navigation Device. *IEEE Transactions on Haptics*, 3(2), 78–87.

Ellise, J. and Kvavilashvili, L. 2000. Prospective memory in 2000: Past, present and future directions. *Applied Cognitive Psychology*, 14(7), 1–9.

Elsmore, T.F. 1994. SYNWORK1: A PC-based tool for assessment of performance in a simulated work environment. *Behavior Research Methods, Instruments and Computers 1994*, 26, 421–26.

Eriksson, L., van Erp, J., Carlander, O., Levin, B., van Veen, H. and Veltman, H. 2006. Vibrotactile and Visual Threat Cueing with High G Threat Intercept in Dynamic Flight Simulation. *Proceedings of the HFES 50th Annual Meeting*, San Francisco, CA, October 2006, 50(16), 1547–1551.*

Forster, B., Cavina-Pratesi, C., Aglioti, S.M. and Berlucchi, G. 2002. Redundant target effect and intersensory facilitation from visual–tactile interactions in simple reaction time. *Experimental Brain Research*, 143(4), 480–87.*

Gilson, R.D., Redden, E.S. and Elliott, L.R. (eds) 2007. *Remote Tactile Displays for Future Soldiers*. (Technical Report ARL-SR-0152). Aberdeen Proving Ground, MD: US Army Research Laboratory.

Glumm, M., Kehring, K. and White, T.L. 2006. *Effects of Tactile, Visual, and Auditory Cues about Threat Location on Target Acquisition and Attention to Visual and Auditory Communications*. (Technical Report ARL-TR-3863). Aberdeen Proving Ground, MD: US Army Research Laboratory.*

Hameed, S., Ferris, T., Jayaraman, S. and Sarter, N. 2006. *Supporting Interruption Management Through Informative Tactile and Peripheral Visual Cues. Proceedings of the HFES 50th Annual Meeting*, San Francisco, CA, October 2006, 50(3), 376–80.*

Hameed, S., Jayaraman, S., Ballard, M. and Sarter, N. 2007. *Guiding Visual Attention by Exploiting Crossmodal Spatial Links: An Application in Air Traffic Control. Proceedings of the HFES 51st Annual Meeting*, Baltimore, MD, October 2007, 51(4), 220–24.

Ho, C., Nikolic, M. and Sarter, N. 2001. *Supporting Timesharing and Interruption Management Through Multimodal Information Presentation. Proceedings of the HFES 45th Annual Meeting*, Vol. 1, Minneapolis, MN, October 2001, 45(4), 341–45.*

Ho, C., Reed, N. and Spence, C. 2007. Multisensory in-car warning signals for collision avoidance. *Human Factors*, 49, 1107–1114.*

Hopp, P.J., Smith, C.A.P., Clegg, B.A. and Heggestad, E.D. 2005. Interruption management: The use of attention-directing tactile cues. *Human Factors*, 47(1), 1–11.*

Jones, L.A., Lockyer, B. and Piateski, E. 2006. Tactile display and vibrotactile pattern recognition on the torso. *Advanced Robotics*, 20(12), 1359–1374.

Keller, J. 2002. *Human Performance Modeling for Discrete-Event Simulation: Workload: Proceedings of the Winter Simulation Conference*, Berlin, 8–11 December 2002, 1, 157–62.

Klein, G. 2008. Naturalistic decision making. *Human Factors*, 50(3), 456–61.

Krausman, A. and White, T. 2008. *Detection and Localization of Vibrotactile Signals in Moving Vehicles*. (Technical Report ARL-TR-4463). Aberdeen Proving Ground, MD: US Army Research Laboratory.

Krausman, A.S., Elliott, L.R. and Pettitt, R.A. 2005. *Effects of Visual, Auditory, and Tactile Alerts on Platoon Leader Performance and Decision Making.* (Technical Report ARL-TR-3633). Aberdeen Proving Ground, MD: US Army Research Laboratory.*

Krausman, A.S., Pettitt, R.A. and Elliott, L.R. 2007. *Effects of Redundant Alerts on Platoon Leader Performance and Decision Making.* (Technical Report ARL-TR-3999). Aberdeen Proving Ground, MD: US Army Research Laboratory.*

Lathan, C. and Tracey, M. 2002. The effects of operator spatial perception and sensory feedback on human–robot teleoperation performance. *Presence: Teleoperation and Virtual Environments,* 11(4), 368–77.*

Lindeman, R.W., Sibert, J.L., Mendez-Mendez, E., Patil, S. and Phifer, D. 2005. Effectiveness of Directional Vibrotactile Cueing on a Building–Clearing Task. *Proceedings of the SIGCHI Conference on Human Factors in Computing Systems,* Portland, OR, April 2005, 271–80.*

Lindeman, R.W., Yanagida, Y., Sibert, J.L. and Lavine, R. 2003. *Effective Vibrotactile Cueing in a Visual Search Task. Proceedings of INTERACT 2003,* Zurich, September 2003, 89–96.*

McCracken, J.H. and Aldrich, T.B. 1984. *Analyses of Selected LHX Mission Functions: Implications for Operator Workload and System Automation Goals.* (Technical Report DTIC ADA232330). Fort Rucker, AL: Anacapa Sciences Inc.

McGrath, B.J, Estrada, A. Braithwaite, M.G., Raj, A.K. and Rupert, A.H. 2004. *Tactile Situation Awareness System Flight Demonstration Final Report.* (USAARL Technical Report 2004–10). Fort Detrick, MD: US Army Aeromedical Research Lab.*

McKinley R.A. and Tripp, L.D. Jr. 2007. Multisensory cueing to improve UAV operator performance during landing. *Aviation Space and Environmental Medicine,* 78(3), 338.*

Mitchell, D. 2005. *Soldier Workload Analysis of the Mounted Combat System (MCS) Platoon's Use of Unmanned Assets.* (Technical Report ARL-TR-3476). Aberdeen Proving Ground, MD: US Army Research Laboratory.

Mitchell, D.K. 2000. *Mental Workload and ARL Workload Modeling Tools.* (Technical Report ARL-TR-161). Aberdeen Proving Ground, MD: US Army Research Laboratory.

——. 2008. *Predicted Impact of An Autonomous Navigation System (ANS) and Crew-Aided Behaviors (CABS) On Soldier Workload And Performance.* (Technical Report ARL-TR-4342). Aberdeen Proving Ground, MD: US Army Research Laboratory.

——. 2009. *Workload Analysis of the Crew of the Abrams V2 SEP: Phase I Baseline IMPRINT Model.* (Technical Report ARL-TR-5028). Aberdeen Proving Ground, MD: US Army Research Laboratory.

Mitchell, D.K. and Brennan, G. 2009a. *Infantry Squad Using the Common Controller to Control an ARV-A (L) Soldier Workload Analysis.* (Technical Report ARL-TR-5029). Aberdeen Proving Ground, MD: US Army Research Laboratory.

——. 2009b. *Infantry Squad Using the Common Controller to Control a Class 1 Unmanned Aerial Vehicle System (UAVS) Soldier Workload Analysis.* (Technical Report ARL-TR-5012). Aberdeen Proving Ground, MD: US Army Research Laboratory.

Mitchell, D., Samms, C., Glumm, M., Krausman, A., Brelsford, M. and Garrett, L. 2004. *Improved Performance Research Integration Tool (IMPRINT) Model Analyses in Support of the Situational Understanding as an Enabler for Unit of Action Maneuver Team Soldiers Science and Technology Objective (STO) in Support of Future Combat Systems (FCS).* (Technical Report ARL-TR-3405). Aberdeen Proving Ground, MD: US Army Research Laboratory.

Moorhead, I., Holmes, S. and Furnell, A. 2004. *Understanding Multisensory Integration for Pilot Spatial Orientation.* (Technical Report SPC 03-3048). Farnborough, UK: Qinetiq.*

Murray, A., Katzky, R. and Khosla, P. 2003. Psychophysical characterization and tested validation of a wearable vibrotactile glove for telemanipulation. *Presence*, 12(2), 156–82.*

Pettitt, R., Redden, E. and Carstens, C. 2006. *Comparison of Army Hand and Arm Signals to a Covert Tactile Communication System in a Dynamic Environment.* (Technical Report ARL-TR-3838). Aberdeen Proving Ground, MD: US Army Research Laboratory.*

Pomranky, R. and Wojciechowski, J. 2007. *Determination of Mental Workload during Operation of Multiple Unmanned Systems.* (Technical Report ARL-TR-4309). Aberdeen Proving Ground, MD: US Army Research Laboratory.

Prewett, M., Burke, J., Yang, L., Stilson, R., Gray, A., Redden, E., Elliott, L. and Coovert, M. 2006. *The Benefits of Multimodal Information: A Meta-Analysis Comparing Visual and Visual–Tactile Feedback. Proceedings of ICMI '2006*, Banff, Canada, November 2006, 333–38.

Raj, A.K., Kass, S.J. and Perry, J.F. 2000. *Vibrotactile Displays for Improving Spatial Awareness. Proceedings of the International Ergonomics Association XIV Triennial Congress/44th Annual Meeting of the HFES*, San Diego, CA, July–August 2000, 1, 181–84.*

Rasmussen, J. 1982. Human errors: A taxonomy for describing human malfunction in industrial installations. *Journal of Occupational Accidents*, 4, 311–33.

——. 1983. Skills, rules, and knowledge: Signals, signs, and symbols, and other distinctions in human performance models. *IEEE Transactions on Systems, Man, and Cybernetics*, SMC-13(3), 257–66.

Redden, E.S. and Elliott, L.R. 2007. *Research-Based Display Design Guidelines for Vehicle Crewman and Ground Warrior Interfaces Which Enhance Situational Understanding and Decision Cycle Performance.* (Technical Report ARL-TR-4231). Aberdeen Proving Ground, MD: US Army Research Laboratory.

Redden, E.S., Carstens, C.B., Turner, D.D. and Elliott, L.R. 2006. *Localization of Tactile Signals as a Function of Tactor Operating Characteristics.* (Technical Report ARL-TR-3971). Aberdeen Proving Ground, MD: US Army Research Laboratory.

Redden, E.S., Elliott, L.R., Pettitt, R.A. and Carstens, C.B. 2009. A tactile option to reduce robot controller size. *Journal of Multimodal User Interfaces*, 2(3–4), 205–16.

Reeves, D., Winter, K., Kane, R., Elsmore, T. and Bleiberg, J. 2001. *ANAM 2001 User's Manual: Clinical and Research Modules*. (Special Report NCRF-SR-2001-1, 2001). San Diego, CA: National Cognitive Recovery Foundation.

Rupert, A.H. 2000a. An instrumentation solution for reducing spatial disorientation mishaps. *IEEE Engineering in Medicine and Biology*, 19, 71–80.

——. 2000b. Tactile situation awareness system: Proprioceptive prostheses for sensory deficiencies. *Aviation, Space, and Environmental Medicine*, 71(9), A92–A99.

Savick, D., Elliott, L., Zubal, O. and Stachowiak, C. 2008. *The Effect of Audio and Tactile Cues on Soldier Decision Making and Navigation in Complex Simulation Scenarios*. (Technical Report ARL-TR-4413). Aberdeen Proving Ground, MD: US Army Research Laboratory.*

Scalsky, D., Meshesha, D. and Struken, S. 2009. *Army Expeditionary Warrior Experiment (AEWE) Spiral E Final Report*. Alexandria, VA: US Army Test and Evaluation Command.

Scott, J.J. and Gray, R. 2008. A comparison of tactile, visual, and auditory warnings for rear-end collision prevention in simulated driving. *Human Factors*, 50(2), 264–75.*

Self, B., van Erp, J., Eriksson, L. and Elliott, L.R. 2007. Human factors issues of tactile displays for military environments, in *Tactile Displays for Military Environments*, edited by J. van Erp. [Online]. Available at: http://ftp.rta.nato.int/public/PubFullText/RTO/TR/RTO-TR-HFM-122/TR-HFM-122-03.pdf (accessed: July 19, 2012).

Sheridan, T.B. 1992. *Telerobotics, Automation, and Human Supervisory Control*. Boston, MA: MIT Press.

Shiffrin, R.M. and Schneider, W. 1984. Automatic and controlled processing revisited. *Psychological Review*, 91(2), 269–76.

Sklar, A.E. and Sarter, N.B. 1999. Good vibrations: Tactile feedback in support of attention allocation and human–automation coordination in event-driven domains. *Human Factors*, 41(4), 543–52.*

Small, R., Keller, J., Wickens, C., Socash, C., Ronan, A. and Fisher, A. 2006. *Multisensory Integration for Pilot Spatial Orientation*. (Technical Report AFRL-HE-WP-TR-2006-0166).* Wright-Patterson Air Force Base, OH: Air Force Research Laboratory Human Effectiveness Directorate.

Smith, E.A. 2006. *Complexity, Networking, and Effects-Based Approaches to Operations*. Vienna, VA: CCRP Press. Available at: http://www.dodccrp.org/files/Smith_ Complexity.pdf (accessed: July 19, 2012).

Spink, A., Cole, C. and Waller, M. 2009. Multitasking behavior. *Annual Review of Information Science and Technology*, 93–118.

Tan, H.Z., Gray, R., Young, J.J. and Traylor, R. 2003. A haptic back display for attentional and directional cueing. *Haptics-e: The Electronic Journal of Haptic Research*, 3(1). [Online]. http://www.haptics-e.org/ (accessed: July 2, 2012).*

Tang, H., Beebe, D.J. and Kramer, A.F. 1997. *Comparison of Tactile and Visual Feedback for a Multi-State Input Mechanism. IEMBS '97: Proceedings of the 19th Annual International Conference of the IEEE Engineering in Medicine and Biology Society*, Chicago, IL, October 1997, 4, 1697–1700.

Tripp, L., Warm, J., Matthews, G., Chiu, P. and Bracken, B. 2007. *+Gz Acceleration Loss of Consciousness: Use of G-suit Pressurization and Sensory Stimulation to Enhance Recovery. Proceedings of the HFES 51st Annual Meeting*, Baltimore, MD, October 2007, 1296–1300.*

van Erp, J. 2007. *Tactile Displays for Navigation and Orientation: Perception and Behavior*. Leiden, The Netherlands: Mostert and van Onderen.

van Erp, J.B.F. and van Veen, H.A.H.C. 2004. Vibrotactile in-vehicle navigation system. *Transportation Research Part F*, 7, 247–56.

van Erp, J.B.F. and Werkhoven, P. 2006. *Validation of Principles for Tactile Navigation Displays: Proceedings of the HFES 50th Annual Meeting*, San Francisco, CA, October 2006, 1687–1691.*

van Erp, J., Eriksson, L., Levin, B., Carlander, O., Veltman, J. and Vos, W. 2007. Tactile cueing effect on performance in simulated aerial combat with high acceleration. *Aviation, Space, and Environmental Medicine*, 78(12), 1128–1134.*

van Erp, J.B.F., Groen, E., Bos, J. and van Veen, H. 2006. A tactile cockpit instrument supports the control of self-motion during spatial disorientation. *Human Factors*, 48(2), 219–228.*

van Erp, J.B.F., Jansen, C. Dobbins, T. and van Veen, H.A.H.C. 2004. *Vibrotactile Waypoint Navigation at Sea and in the Air: Two Case Studies. Proceedings of Eurohaptics 2004*, Munich, Germany, June 2004, 166–73.*

van Erp, J.B.F., Meppelink, R. and van Veen, H.A.H.C. 2002. *A touchy car.* (Report No. TD-2001-0358). Soesterberg, The Netherlands: TNO Human Factors Research Institute.*

van Erp, J., Veltman, J. and van Veen, H. 2003. *A Tactile Cockpit Instrument to Support Altitude Control. Proceedings of the HFES 47th Annual Meeting*, Santa Monica, CA, October 2003, 47(1), 114–18.*

van Erp, J.B.F., Veltman, J.A., van Veen, H.A.H.C. and Oving, A.B. 2003. Tactile torso display as countermeasure to reduce night vision goggles induced drift. *Spatial Disorientation in Military Vehicles: Causes, Consequences, and Cures, RTO HFM Symposium*, La Coruña, Spain, 15–17 April 2002.

Veltman, H.J.A. and Jansen, C. 2004. The adaptive operator, in *Human Performance, Situation Awareness, and Automation: Current Research and Trends*, Vol. II, edited by D.A. Vincenzi, M. Mouloua and P.A. Hancock. Mahwah, NJ: Lawrence Erlbaum, 7–10.

Wickens, C.D. 1984. Processing resources in attention, in *Varieties of Attention*, edited by R. Parasuraman and D.R. Davies. New York: Academic Press, 63–102.

———. 1991. Processing resources and attention, in *Multiple Task Performance*, edited by D.L. Damos. London: Taylor & Francis, 3–34.

———. 1992. *Engineering Psychology and Human Performance*. New York: HarperCollins.

——. 2002. Multiple resources and performance prediction. *Theoretical Issues in Ergonomics Science*, 3(2), 159–77.

——. 2008. Multiple resources and mental workload. *Human Factors*, 50(3), 449–54.

Wickens, C.D., Goh, J., Helleberg, J., Horrey, W. and Talleur, D.A. (2003). Attentional models of multi-task pilot performance using advanced display technology. *Human Factors*, 45(3), 360–80.

Chapter 5

Tactile Displays in Army Operational Environments

Timothy L. White, Andrea S. Krausman, and Ellen C. Haas

In combat environments, critical information must be communicated to soldiers to aid them in maintaining their situational awareness (SA). It is important not only to collect intelligence that can reduce the enemy's advantage but also to communicate that information effectively in order to improve survivability as well as achieve the overall mission (Gilson, Redden, and Elliott 2007). Currently, much of this information is presented through the visual and auditory sensory channels, thereby inducing overload and performance degradation. This type of information must be communicated in such a way that visual, auditory, and cognitive overload are minimized (van Erp and Self 2008). Wickens' Multiple-Resource Theory (MRT) suggests that offloading information from overtaxed sensory modalities to other modalities can reduce workload (Wickens 2002). A viable alternative communication medium may be tactile displays, which have been shown to reduce cognitive overload by presenting information that is easy to interpret, is intuitive, and does not divert the user's attention away from the operational task at hand (Redden et al. 2006).

In general, a tactile display comprises vibrating actuators or tactors that are mounted on the body (for example, the torso, arm, or wrist) and stimulate the skin's surface. Some examples of tactile displays are shown in Figure 5.1. Advancements in tactile display technologies have yielded tactors that are relatively small, inexpensive, and robust, such as the vibrating alert feature in mobile phones and pagers (van Erp and van Veen 2004). Although these displays use a single vibrating tactor, combining several tactors into a larger display expands the capability of tactors to convey more complex information. An advantage to tactile display technology is that even basic tactile cues are intuitive and informative (Castle and Dobbins 2004). For example, a tap on the shoulder immediately captures your attention and tells you that someone is behind you and where they are located (Castle and Dobbins 2004). Also, since the information provided by tactile displays is easily discerned, it requires few cognitive resources to interpret, thereby freeing up these necessary cognitive resources for other tasks (van Erp and Self 2008).

A

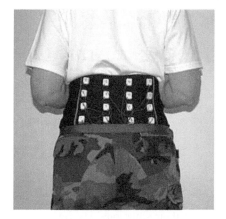

B

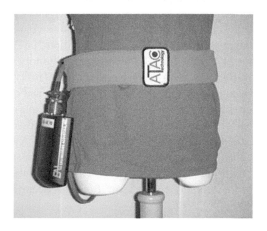

C

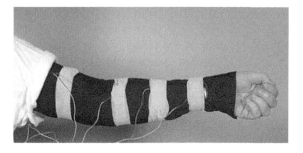

Figure 5.1   Displays for: A: the back; B: a belt; C: the arm

The skin is the largest organ of the human body and for the average adult it has a surface area of about 1.8 m^2 (Sherrick and Cholewiak 1986). There are three types of skin: glabrous, hairy, and mucocutaneous (Greenspan and Bolanowski 1996). Glabrous skin is the hairless skin that is found on the palms and soles, and mucocutaneous skin (for example, lips) is the skin that borders entrances to the body's interior (Greenspan and Bolanowski 1996). Of the three types of skin, hairy skin covers most of the human body. The skin has many sensory receptors or mechanoreceptors for receiving sensations like vibration, pressure, texture, temperature, and pain. The mechanoreceptors of the skin are sensitive to mechanical pressure or deformation of the skin (Sekuler and Blake 1990). The mechanoreceptor types are defined by their rate of adaptation to pressure and vibration and their receptive field size (Greenspan and Bolanowski 1996). Adaptation is the mechanism by which mechanoreceptors respond to sustained skin indentation (Greenspan and Bolanowski 1996). The receptive field is the area of skin that generates a response in a sensory neuron when stimulated, and this area is dependent on the intensity of the stimulus (Greenspan and Bolanowski 1996). Because of the varying characteristics of mechanoreceptors and their distribution throughout the body, the perceptual resolution and sensitivity of the skin to tactile stimuli vary at different body locations, which have implications for the placement of tactile displays on the body.

Tactile stimuli or cues can be provided as a single discrete signal or as patterns. The type of tactile stimuli provided depends on the information that is being conveyed to the user. Signals are normally used to provide an alert or direction, whereas a tactile pattern is used to communicate a more complex message, or orientation information. Patterns typically consist of multiple tactors firing one at a time or sequentially. The majority of tactile display systems that are in use employ vibrotactile stimuli (Cholewiak and Collins 2003). The responses of mechanoreceptors in the skin depend on the frequency, amplitude, and duration of the vibrotactile stimuli as well as the location at which a stimulus is applied (Jones and Sarter 2008). Though vibrations with frequencies between 20 Hz and 500 Hz can be perceived, Jones and Sarter (2008) identify optimal sensitivity with frequencies between 150 Hz and 300 Hz. At a given frequency, the amplitude level required for detection varies across body sites (Von Békésy 1959). Amplitude and frequency parameters must be chosen carefully, especially when encoding information, because of their interactions, and it is not recommended that they be manipulated simultaneously during vibrotactile communications (Jones and Sarter 2008). With regard to duration, results from one investigation showed that participants prefer that the duration of vibrotactile alerts be between 50 ms and 200 ms (Kaaresoja and Linjama 2005).

In recent years, there has been an increased interest in the use of tactile displays as an alternative means of effectively communicating information to military personnel. Although current military applications rarely use tactile displays in combat operations, research has demonstrated the benefit of using tactile displays in a variety of operational environments including dismounted applications, mounted applications, and human–robotic interface applications (for example, teleoperation, operation of single or multiple robots, and robotic swarm display applications). Use of the tactile modality has become commonplace in everyday technology (for example, cell phones) and its use will continue to expand with the advancement of tactile technology.

Dismounted Soldier Use

Dismounted soldiers are a vital component of the US Army, and they must operate in extremely demanding environments. During combat, the mission of dismounted soldiers may require them to perform a number of maneuvers such as running, jumping, climbing, balancing, kneeling, and crawling. In order for tactile display systems to be successfully implemented, tactile cues must be easily perceived under these operating conditions (Elliott et al. 2009). Futhermore, the implementation of tactile displays must allow dismounted soldiers to be ready to move their weapon to engage an enemy target while being attentive to the terrain and enemy threats (van Erp and Self 2008). In the next part of this chapter, two useful dismounted applications of tactile cueing, navigation and communication, are discussed.

Navigation

Cues for navigation can be provided visually or auditorily. However, these cueing methods are not always the best means of providing navigation information to dismounted soldiers who are exposed to noisy environments as well as situations in which it is more important for them to focus their visual attention on their surroundings rather than on a visual display (for example, a global positioning system (GPS)) (van Erp et al. 2005). In these instances, providing navigation information with tactile cues has proven to be a feasible alternative. In one study, pedestrians were able to navigate waypoints using a tactile belt display at almost normal walking speeds (van Erp et al. 2005). The Netherlands Organisation for Applied Scientific Research—Human Factors (TNO-HF) Personal Tactile Navigator System, which uses geo-location and compass data to provide information about waypoint locations, has been used in a number of investigations. Dismounted soldiers were able to easily navigate waypoints without dedicating their eyes and hands to the restrictions of relying on a visual display (Duistermaat 2005). In another series of studies that employed the Personal Tactile Navigator System, dismounted soldiers were able to successfully navigate during both day and night operations (Elliott et al. 2007).

Communication

Although vibrotactile displays are becoming increasingly common in everyday devices such as mobile phones, pagers, and game controllers, the simple vibrations used in these devices do not fully capture the potential of vibration as a means of communicating (Brewster and Brown 2004). Rather, encoding information into tactile patterns may be a method of furthering the benefits of vibrotactile technology as a communication medium. Another area of particular interest to the military is using tactile patterns to relay information to the soldier. Traditionally, military hand signals are used to visually communicate information (for example, "attention," and "move out") to dismounted soldiers during noise discipline when voice communication is prohibited. Translating hand signals into tactile patterns would allow soldiers to receive hand signals without relying on visual input or the visual channel. Pettitt, Redden, and Carstens (2006) evaluated soldiers' abilities to interpret and respond to tactile and visual hand signals. Results demonstrated that soldiers were able to receive, interpret, and accurately respond to tactile commands 38 percent faster than to the conventional hand and arm signals provided from the front and 61 percent faster than those provided from the rear. Soldiers also commented they were able to focus more attention on negotiating obstacles and on area SA when receiving tactile signals than when receiving hand signals visually.

In another study, Jones, Lockyer, and Piateski (2006) investigated the ability of participants to recognize and respond to tactile navigation patterns presented to the lower back using a 4 × 4 matrix of tactors. In their study, participants were able to recognize and correctly respond to the patterns with almost perfect accuracy. Krausman and White (2006) extended this research by performing an investigation that compared providing vibrotactile patterns with the 4 × 4 matrix on the back and on an 8-tactor belt around the torso while negotiating obstacles (Figure 5.1). Although the sample size was small and no significant differences were found between the two displays, additional research should be conducted to identify potential differences between the two torso display types when conveying tactile patterns (Krausman and White 2006).

Parameters such as frequency, amplitude, and duration of tactile signals have been used to encode tactile patterns (Brewster and Brown 2004). In military environments, some signals or messages may need to be encoded with some level of urgency. Research has shown that varying the intensity and inter-stimulus interval of tactile patterns can provide soldiers with a sense of urgency. In stationary environments, soldiers were able to detect and identify tactile patterns and their varying urgency levels with nearly 100 percent accuracy (White 2011). In a dismounted field study, findings revealed that it is best to employ an intensity level that is strong enough to be detected while performing dismounted maneuvers and to vary inter-stimulus interval to provide soldiers with a sense of urgency (White 2012).

Vehicle-mounted Use

Soldiers serving as part of an aviation crew or ground combat vehicle crew may also benefit from tactile display technology. Ground vehicle crews perform the majority of their mission from within their vehicle and, as a result, crewmembers depend heavily on visual and auditory information displays to obtain situational information. Although visual displays are familiar to most users, the tendency is to present vast quantities of visual information that can distract and overload soldiers, compromising their ability to perform safely and efficiently. While auditory displays are suitable for specific applications (see Sanders and McCormick 1993), special care must be taken when using auditory displays in noisy environments such as combat vehicles. As a result, tactile displays may be a unique and effective information channel to augment the more traditional visual and auditory displays. In addition, tactile displays are particularly relevant when the use of visual and auditory displays is not feasible, such as during covert operations when light and noise discipline are essential (Furnell, Holmes, and King 2003).

As progress is made in the development and integration of tactile technologies, using tactile displays in military combat vehicles will become more of a reality. Employing tactile displays for vehicle crewmembers should reduce both demands on, and interference with, the overloaded visual and auditory channels, thereby enhancing performance and soldier safety (Gilson, Redden, and Elliott 2007). In the following paragraphs, we describe some of the potential benefits of tactile displays for crewmembers, namely, spatial orientation, navigation, and communication (Castle and Dobbins 2004, van Erp and Self 2008).

Spatial Orientation

The military aviation environment is complex, dynamic, and stressful, both physically and mentally (van Erp and Self 2008). While pilots perform the physical task of flying an aircraft, they also must maintain their SA in rapidly changing environmental and tactical conditions, acquire targets and make decisions regarding weapon usage, perform demanding aircraft maneuvers, and maintain communication with members of their unit and air controllers (van Erp and Self 2008). Often, pilots must perform these tasks in degraded visual conditions (for example, night operations, whiteouts and brownouts). Of particular concern to the military is when the complexities of flight and degraded visual cues cause pilots to become disoriented and lose their sense of where they are in "space" relative to the ground (Castle and Dobbins 2004, van Erp and Self 2008). In the literature, this is called "spatial disorientation" (SD).

SD is a major threat to pilot safety and very costly to the military in terms of the number of lives and aircraft lost, as well as mission impact. Incidences of SD account for nearly 25 percent of US Army helicopter losses (Castle and Dobbins 2004). In fact, the prevalence of SD served as a catalyst for the development of tactile displays to aid spatial orientation (Castle and Dobbins 2004). Empirical findings support the use of tactile torso displays as an effective countermeasure for SD, since their use has resulted in a reduction in the number of aviation-related SD mishaps. Compared to other SD countermeasures, such as training and the development of advanced cockpit displays, tactile displays are considered "the most important advance in recent years with the potential to combat spatial orientation" (van Erp and Self 2008).

Several types of tactile displays (for example, belts and vests) have been developed to enhance spatial orientation under adverse operational conditions. Potentially the most familiar tactile vest is the Tactile Situation Awareness System (TSAS), developed at the US Naval Aerospace Medical Research Laboratory to aid pilot orientation (Myles and Binseel 2009). An overarching goal of the TSAS project was to minimize the occurrence of SD and ease the visual overload naturally placed on pilots by the numerous cockpit instruments that compete for their attention (Myles and Binseel 2009). The TSAS tactile display consists of 32 tactors sewn into a vest worn on the pilot's torso. Tactors are positioned inside the vest in a matrix of 8 columns and 4 rows (van Erp and Self 2008). Tactile cues generated by the TSAS assist the pilot in determining the aircraft's orientation with respect to the ground and whether or not corrective action is required (Myles and Binseel 2009). For example, a tactile signal applied to the front of the torso indicates that the pilot should raise the nose of the aircraft. Flight demonstrations with the TSAS vest confirmed that tactile displays can intuitively provide orientation information through the sense of touch, enabling the pilot to maintain control of an aircraft, even when blindfolded (Myles and Binseel 2009).

Aircraft hover and landing maneuvers may also benefit from tactile displays (Raj, Kass, and Perry 2000). In visually degraded conditions such as brownouts, whiteouts, and night operations, aircraft maneuvers become increasingly difficult since all external visual references are lost. In addition, critical information on aircraft drift is currently not provided by Army aircraft cockpit displays, so the pilot must guess as to the direction and degree of drift (Curry et al. 2008). Tactile cues from the TSAS vest have proven beneficial to pilots during precision hover and landing by providing "ground truth" with regard to drift. For instance, if an aircraft drifts too far off a targeted position, tactile cues can advise the pilot of the corrective action necessary to remedy the situation (van Erp and Self 2008).

Navigation

Another application for which tactile displays show promise is navigation. Traditionally, navigation is a visual task, although cues may be provided through spoken instructions. Recently, the effective presentation of navigation cues via the sense of touch has been developed. Tactile cues have the potential to overcome the limitations of visual and audio cues, which are compromised by poor visibility and noisy environments (Dobbins and Samways 2002). For example, tactile displays can present information such as waypoint direction cues, no-go areas, obstacle-avoidance cues, and collision-avoidance information, and can be especially relevant as vehicle crews move over unfamiliar terrain (Jones and Sarter 2008). Similarly, a tactile display can be used for navigational purposes to inform pilots when they have strayed too far off course (van Erp and Self 2008).

Waypoint navigation has been successfully demonstrated with dismounted soldiers (Elliott et al. 2006, Gilson, Redden, and Elliott 2007). It follows that tactile displays may be useful as a navigation aid in combat vehicles as well. Results of studies examining commercial in-vehicle navigation systems have concluded that although these systems were designed to assist drivers, oftentimes the visual and/or audio cues distract the driver, creating conditions in which cognitive overload may occur (van Erp and van Veen 2004, van Erp and Self 2008). For example, in a study by van Erp and van Veen (2004), three types of in-vehicle navigation displays were tested as participants drove through a simulated urban environment: visual, tactile, and multimodal. In this study, the tactile display consisted of eight tactors mounted in the driver's seat. Results demonstrated a reduction in driver workload while navigating with the tactile display as compared to the visual display. Studies have also been performed with wheeled and tracked vehicle operators using tactile navigation cues (Krausman and White 2008) as well as the use of a tactile display as a navigation aid for a blind boat driver attempting to set the blind world water speed record (Castle and Dobbins 2004). A distinct advantage of using tactile displays for navigation is that they provide "eyes free" and "hands free" navigation information that enables vehicle crew members to devote more time to other instruments and tasks when operating in demanding, complex environments (Gilson, Redden, and Elliott 2007, van Erp and Self 2008).

Communication

In addition to tactile patterns discussed earlier in this chapter (under "Dismounted Soldier Use"/"Communication"), tactile displays can effectively communicate alerts to vehicle commanders or other crewmembers during combat missions. An alert can be used to convey the appearance and/or location of a new threat in the environment (Carlander and Ericksson 2006, Glumm, Kehring, and White 2006), notify crew members (for example, platoon leader) when they receive new messages so they are not required to constantly monitor a visual display (Krausman, Elliott, and Pettitt 2005), and focus visual searches or allocate attention to specific locations on a visual display (Haas et al. 2005, Tan et al. 2003, Hopp et al. 2005). In addition, tactile cues can communicate proximity to an obstacle, another vehicle, or an imminent vehicular collision (Ho, Tan, and Spence 2005). If warranted, complex information could be conveyed to vehicle crews through tactile patterns or a tactile language as described previously.

In summary, using the sense of touch as an information channel for vehicle crews is promising, especially in situations when their visual and auditory channels are degraded (for example, in night operations, heavy fog, loud noise and so on), when cognitive capacity is lessened (for example, fatigue, stress, sustained ops), and when noise and light discipline are in effect (Gilson, Redden, and Elliott 2007). In the following part of this chapter, we address the potential for using tactile displays to support human–robotic interaction.

Human–Robotic Interface Use

Human–robot operations assume some level of human control or supervision. In robotics systems, human–robot interaction (HRI) can range from full control, supervision from far (teleoperation) or near, as well as semi-autonomous (partial control or supervision) or fully autonomous operation (supervision only when absolutely needed). In military environments, human–robotic interfaces can be found in mounted as well as dismounted systems.

To permit the soldier to remain cognizant of his or her interaction with the robotic system, the user interface requires some degree of user immersion as well as effective, intuitive human–computer interaction. Early robotic systems, including teleoperation systems in which the user controlled robots from a distance, mainly used visual feedback from a camera mounted on the robot. However, it became clear that this limited feedback resulted in high visual and mental workload and seriously hampered operator performance. Researchers found that enhancing user feedback decreases task difficulty (Lathan and Tracey 2002) and creates a greater sense of operator immersion in a robotic control environment (Burdea, Richard, and Coiffet 1996). As a consequence, researchers and developers have begun to augment the visual feedback by adding information to camera images and by including more abstract information (for example, robot status or health) in tactical displays. Some of this synthetic information is in the form of tactile displays. This part of the chapter describes research in which tactile displays are used alone or as a supplement to visual displays in a wide variety of HRI tasks.

Tactile displays can improve the quality and safety of operator control of robotic vehicles by presenting warnings and tactical information. In addition, tactile displays used in teleoperation of military robotic vehicles (mobile robots and unmanned aerial vehicles (UAVs)) can mimic relevant cues arising from the vehicle environment and make them available for off-vehicle operators in remote locations. These cues include the presence of environmental obstacles as well as position and acceleration. This part of the chapter describes applications for warnings and tactical displays, and providing environmental information.

Warning and Tactical Displays

Tactile displays are efficient and effective in providing warning information and have also been useful in tactical displays (that is, displays providing battlefield information). Within the last several years, researchers have found that tactile cues are very useful in robotic displays applications for providing warning and tactical information in demanding environments. Sklar and Sarter (1999) demonstrated that tactile cues are more effective than visual cues when informing users of unexpected changes or new system information; results were later confirmed by Calhoun et al. (2003). For example, during distributed unmanned vehicle operations, tactile cues can be used to alert team members that a new drawing has been posted on a shared electronic map (McDermott, Fisher, and Allender 2008).

In a series of studies, Chen (2008, 2009) and Chen and Terrence (2008, 2009) examined how aided target recognition (AiTR), used in a combined gunner/robotics operator task, might benefit the performance of concurrent gunner, robotic, and communication tasks in a simulated military environment. AiTR gunnery cues were delivered using tactile or tactile plus visual cueing. These researchers found that participants' robotics and communication tasks both improved significantly with AiTR. In addition, although participants' spatial ability was found to be a reliable predictor of robotics task performance, this performance gap was narrower when tactile-aided AiTR was available to assist with the gunnery task.

Chen, Haas, and Barnes (2007) performed a literature review to summarize user interface design issues (that is, video image bandwidth, time lags, lack of proprioception, frame of reference, attention switches) that might influence user performance in robotic teleoperation. They found that tactile displays, especially when combined with displays of other modalities (visual and audio), are promising technologies for providing feedback during teleoperation of HRI.

Haas and Stachowiak (2007) conducted a field study to determine the extent to which the integration of tactile and spatial auditory displays could be combined with visual information to effectively convey HRI target search information while the operator traveled in a tactical moving vehicle that experienced relatively high levels of vibration and jolt on different types of terrain (gravel and cross-country). Workload data (Haas and Stachowiak 2007) indicated that tactile and multimodal (tactile plus audio) displays had lower mean workload ratings than did audio displays alone. Human performance data (target-tracking response time), which was indicated that supplementary audio, tactile, or combined audio and tactile display modalities provide shorter response time than visual cueing alone, on gravel and cross-country terrains (Haas et al. in review). These researchers concluded that incorporating cues from both the audio and tactile modalities was advantageous to human performance and workload in an environment with strong auditory and tactile distracters; and that alone or together, advanced technologies such as tactile and spatial (3D) audio displays can supplement visual displays and enhance user performance in HRI target search tasks.

Tactile displays were explored in a study investigating the design of multimodal interfaces for swarm robotic operations (Haas et al. 2009). Robotic swarms are arrays of small autonomous robots capable of travel and operation as a unit on land and sea, and in air. In theory, the effective employment of robotic swarms can accomplish critical Army tasks, such as mapping battlefields, accompanying convoys, and locating improvised explosive devices (IEDs). The soldier–swarm interface is a critical aspect of swarm control, because it enables the soldier to remain cognizant of swarm operations and institute corrective actions during swarm supervision or over-watch. Haas et al. (2009) examined the use of different combination of visual, tactile, and spatial auditory displays when presenting geospatial and warning information from a 40-member simulated robotic swarm. In their study, 16 Marine Corps participants supervised swarm activity in a representative convoy mission, using displays that showed swarm health, swarm communication, and convoy status. The results of this study indicated that both workload and task response time decreased when tactile and audio displays were used to supplement visual displays. The authors found that tactile and auditory displays could efficiently and effectively communicate swarm-provided information to the Marines by defining specific information demanded for a supervisor of swarm operations. These results are consistent with previous studies. Haas et al. (2009) concluded that when visual displays are supplemented by tactile and audio information, users can efficiently interact with a robotic swarm participating in a representative convoy mission. The authors suggested that future research should explore tactile displays used in different swarm applications, such as mapping battlefields.

Environmental Cues

Riley et al. (2010) showed that operators using robotic systems need to be able to acquire and maintain a high level of SA with regard not only to the robotic vehicle but also to related systems, changes in the environment, and changes in the ongoing robotic missions and tasks. This is especially true in robotic teleoperation. Lathan and Tracey (2002) showed that when vibrotactile feedback was used in addition to video feedback in a teleoperation task, there were fewer collisions with walls and obstacles. Barnes and Counsell (2003) used haptic feedback to provide the robotic operator with obstacle information, thus allowing the operator to avoid collisions as well as to communicate the mobile robot's collision avoidance algorithm to the operator, which allowed the operator to override robotic collision avoidance if and when required. These researchers found that haptic feedback improved operator performance in teleoperation tasks, and recommended haptic feedback for increased fault tolerance in teleoperation systems. UAV landings alone account for up to 22 percent of all UAV mishaps (Aretz et al. 2006), leading researchers to explore the use of tactile displays for providing positional cues to UAV pilots when performing landing operations. Using a UAV synthetic task environment simulator, they found that pilots using visual cues (a map and a UAV nose camera view) as well as tactile cues incorporated into a vibrotactile vest, performed significantly better (less root mean square error from an optimal flight path), than pilots using visual displays alone.

Haptic feedback has been shown to be useful for camera control in moving UAVs. Korteling and van Emmerik (1998) found that using haptic feedback to control a camera with a joystick has the same effect as stabilizing the platform. Other researchers (Korteling and van der Borg 1997) found that a positional feedback system in which a moving joystick gives oppositional feedback about relative UAV movement provided increased operator target-tracking performance.

HRI coupled with tactile displays permits soldiers to remain cognizant of their interactions with robotic systems and maintain SA by providing information including warnings, tactical battlefield information, and environmental events.

Human Factors Issues

Suitable Body Locations

An important consideration for tactile cueing is body location. Although tactile stimuli may be perceived at many body locations, suitable body locations must be chosen based on the soldier's operational environment so that the tactile stimuli do not interfere with mission requirements. For example, the hands are not a feasible location because tasks that depend on the use and mobility of the hands may interfere with the tactile stimuli. Providing tactile stimuli to the legs is not feasible because dismounted soldiers are sometimes required to crawl. Crawling could result in physical damage to the tactile actuators or masking of tactile stimuli while this maneuver is being performed. Based on previous research, the torso has been shown to be the most feasible location to provide tactile stimuli for dismounted soldiers (White 2010). It is stable, body-centered, and three-dimensional. It has also shown promise in receiving simple discrete signals as well as more complex patterns (Gilson, Redden, and Elliott 2007, van Erp 2007).

Because crewmember tasks are typically performed while seated or standing in a vehicle (unless they dismount), identifying a suitable location for tactors is not as challenging as for dismounted soldiers. As mentioned previously, the advantages of using the torso include a large surface area and a 360° field of regard (van Erp and Self 2008). Depending on the type of information being conveyed, the arms and legs are also an option, although these areas are less sensitive than the torso (Weinstein 1968). Some of the research investigating tactile displays in vehicles has suggested that tactors be embedded into the seat (either the back or the pan), rather than mounted on the body. When considering combat vehicles, the protective body armor vests worn by crewmembers preclude the integration of tactors with the seat back. However, embedding tactors in the seat seems promising especially for navigation tasks in automobiles (van Erp and van Veen 2004, Hogema et al. 2009).

Masking

A primary concern for the integration of tactile technology for use in vehicle platforms is the inherent vibration that occurs when vehicles move over the battlefield. The vibration may interfere with perception of tactile signals, often called "masking" in the literature (van Erp and Self 2008, Furnell, Holmes, and King 2003). According to van Erp and Self (2008), masking is a change in the perception of a stimulus when a second stimulus is close in time or space. An example of masking of tactile cues is seen in Castle and Dobbins (2004), which documents the development of a tactile navigation system used by a blind boat driver in setting the world water speed record. Tactile signals were given to the blind boat driver; however, the vibrations from the tactile device were not felt because of the vibrations generated by the boat as it traveled on the water's surface. Vibrations from the large boat engines also interfered with tactile signal perception. Moving the tactors from their original location on the driver's arm to the torso enhanced perception of the tactile cues. These findings suggest that body location of the tactile display is an important consideration when evaluating tactile signal perception in moving vehicles. A thorough discussion of the impact of body location on perception of tactile stimuli is provided in Weinstein (1968), Stevens and Choo (1999), and White (2010).

Furnell, King, and Holmes (2004) investigated the effects of whole body vibration on vibrotactile thresholds and found that whole body vibration (WBV) increases the perceptual threshold to a 250 Hz tactile stimulus, but that the strength of this effect is governed by the interaction between WBV frequency and amplitude. However, the authors imply that WBV need not be problematic for tactile signal perception in moving vehicles if the frequency is low (from 0.5 Hz to 2 Hz). To counteract the impact of WBV on vibrotactile perception, Furnell, King and Holmes (2004) suggest that vehicle vibration profiles be considered during the design of tactile systems, so that tactile parameters (that is, frequency and amplitude) can be set to appropriate levels, thereby increasing the likelihood of perception and reducing the potential for masking. The authors also caution that amplitude settings for tactile cues should be high enough to avoid masking effects, yet not to the point of producing pain or discomfort for operators, as they would be likely to turn the tactile display down or off, or remove it from their body.

System Weight/Bulk

As dismounted soldiers perform various maneuvers to accomplish their mission, they must wear a fighting load. As shown in Table 5.1, the weight of a typical fighting load for a dismounted soldier is nearly 70 lb. This weight varies depending on the size of the soldier as well as any additional equipment that he or she may be required to carry. The weight of the fighting load has a direct effect on how efficiently a soldier is able to perform maneuvers. Therefore, designers must develop tactile display systems that do not increase the fighting load significantly or interfere with the perception of tactile signals. In one study, the interceptor body armor worn by participants did not interfere with the detection of tactile signals (Redden et al. 2006). A tactile display system consists of a power source (that is, a battery), a main processing unit (that is, a control box), and tactor-driven electronics (McGrath et al. 2008). However, the heaviest component of a tactile display system is the battery. The power required to drive the tactile actuators determines the type of battery necessary to operate the system. Consequently, there must be balance between the power requirements for individual tactile actuators, the number of tactile actuators necessary to successfully communicate critical information, and battery capacity in order for designers to select appropriate battery sizes.

Table 5.1 Typical fighting load

Item Description	Weight (lb)
Individual first aid kit	0.17
Belt with buckle	0.44
Underclothing and socks	0.48
Grenades, fragmentation, inert (2 each)	2.00
Medium modular integrated communications helmet	3.30
Battle dress uniform	3.80
Boots (direct molded sole)	4.10
M40 mask with hood and outserts in carrier	4.34
Rifle magazines (30 rounds, 6 each)	6.30
Canteen with cover and 1 quart of water (2 each)	6.60
M4 carbine with loaded 30 round magazine	7.50
Interceptor body armor with improved outer tactical vest (size medium)	30.00

System weight and bulk can also be an issue for air and ground vehicle crews. In current Army aircraft, cockpits are loaded with instruments and equipment and provide limited space for the pilot (van Erp et al. 2008). Future concepts for Army ground vehicles include smaller, lighter, and faster vehicles. In both situations, there is limited area for the pilot or vehicle crew. Therefore, it is imperative that tactile technologies remain as small and unobtrusive as possible and not encumber the user. A tactile display that has a minimum number of tactors yet still provides the essential information will be lighter, more robust, and easier to maintain than a larger display with more tactors (van Erp and Self 2008).

User Acceptance

User acceptance is critical to the development and fielding of any product or system, including tactile displays (van Erp and Self 2008). To enhance user acceptance of tactile displays, there are several issues to consider. For example, system designers should ensure that tactile displays are accurate, reliable, and intuitive. This will ensure that the amount of time required for training is minimized. Furthermore this will reduce the cognitive load of users. If the user must spend extra time cognitively processing tactile information, then many of the advantages of using this modality will be lost (van Erp and Self 2008). Another important consideration is comfort. Particularly since some users may be more sensitive to tactile stimulation than others, tactors must be designed so that they are detectable but not irritable. Also, whenever possible, tactile displays should be integrated into existing equipment worn on the body (that is, a t-shirt or vest) thereby reducing the requirement for soldiers to don additional pieces of equipment (van Erp and Self 2008). In addition to the vibration generated by tactors, they also produce some degree of acoustic energy, which may interfere with the soldiers' task performance. This acoustic energy can be detrimental to the user soldiers trying to maintain stealth in a combat situation. System designers must pay special attention to limiting the emission of acoustic energy (Sorkin 1987).

Information Processing

To optimize tactile displays for dismounted soldiers and vehicle crew members, there is an ongoing need to understand the implications of tactile communications on human information processing (Castle and Dobbins 2004). As tactile technology progresses, so will the desire to more fully use the tactile sense as an information channel. However, the tactile channel, like the visual and auditory channels, has a limited capacity to process information, and when capacity is reached or exceeded, performance declines. Researchers and system designers must consider how to most effectively use the tactile sense without overburdening the soldier. For example, how many tactile patterns can be perceived and reliably discriminated without significantly increasing training time or cognitive processing (Gilson, Redden, and Elliott 2007). However, increasing the number of patterns may require too much training and cognitive effort for the patterns to be effective.

Future Work

Although tactile displays are not the only solution to sensory and cognitive overload, current research has proven their potential for a number of military applications. However, further research should be done to maximize the benefits of tactile displays. Future research should establish design guidelines for effective use of tactile displays by studying how messages of varying complexity impact cognitive processing so as not to overburden the user. Currently, ARL researchers are investigating the feasibility of adding urgency to tactile patterns by manipulating the operating parameters of tactile actuators, such as frequency, amplitude, and duration. Another area that future work should focus on is the number of tactile patterns a soldier can learn. In a pilot investigation, participants were able to learn a tactile grammar consisting of 56 different tactile symbols in less than 3.5 hours (Fuchs, Johnston, Hale, and Axelsson 2008). These findings suggest that the development of a standard tactile language is feasible. However, soldiers must be able to receive, interpret, and respond to the language without increasing cognitive overload or performance degradation. In regard to the human–robotic interface, different robotic display applications, such as using tactile displays in a swarm HRI for tasks such as mapping battlefields, should be explored. Also, it is important to consider the role of tactile displays in multimodal or cross-modal interfaces, which may provide additional benefits beyond tactile cues alone (see Chapter 4).

References

Aretz, D.T., Andre, T.S., Self, B.P. and Brenaman, C.A. 2006. Effect of tactile feedback on unmanned aerial vehicle landings. *Proceedings of the Interservice/ Industry Training, Simulation & Education Conference*, Orlando, FL, December 2006.

Barnes, D. and Counsell, M. 2003. Haptic communication for mobile robot operations. *Industrial Robot: An International Journal*, 30(6), 552–63.

Brewster, S.A. and Brown, L.M. 2004. Tactons: Structured tactile messages for non-visual information display. *Proceedings of the 5th Australasian User Interface Conference*, Dunedin, New Zealand, January 2004, 15–23.

Burdea, G.C., Richard, P. and Coiffet, P. 1996. Multimodality virtual reality: Input–output devices, system integration, and human factors. *International Journal of Human–Computer Interaction*, 8(1), 5–24.

Calhoun, G., Draper, M., Ruff, H., Fontejon, J. and Guilfoos, B. 2003. Evaluation of tactile alerts for control station operation. *Proceedings of the HFES 47th Annual Meeting*, Santa Monica, CA, September 2003, 2118–2122.

Carlander, O. and Eriksson, L. 2006. Uni and bimodal threat cueing with vibrotactile and 3D audio technologies in a combat vehicle. *Proceedings of the HFES 50th Annual Meeting*, San Francisco, CA, October 2006: 1552–1556.

Castle, H. and Dobbins, T. 2004. Tactile displays for enhanced performance and safety. *Proceedings of the 11th SAFE (Europe) Symposium*, Lyon, France, April 2004. Bracknell, UK: SAFE Europe.

Chen, J.Y. 2008. Effectiveness of concurrent performance of military and robotics tasks and effects of cueing in a simulated multi-tasking environment. *Proceedings of the HFES 52nd Annual Meeting*, New York, September 2008, 237–41.

———. 2009. Concurrent performance of military and robotics tasks and effects of cueing in a simulated multi-tasking environment. *Presence—Teleoperators and Virtual Environments*, 18(1), 1–150.

Chen, J.Y. and Terrence, P.I. 2008. Effects of tactile cueing on concurrent performance of military and robotics tasks in a simulated multitasking environment. *Ergonomics*, 51(8), 1137–1152.

———. 2009. Effects of imperfect automation and individual differences on concurrent performance of military and robotics tasks in a simulated multitasking environment. *Ergonomics*, 52(8), 907–920.

Chen, J.Y., Haas, E.C. and Barnes, M.J. 2007. Human performance issues and user interface design for teleoperated robots. *IEEE Transactions on Systems, Man and Cybernetics. Part C—Applications and Reviews*, 37(6), 1231–1245.

Cholewiak, R.W. and Collins, A.A. 2003. Vibrotactile localization on the arm: Effects of place, space, and age. *Perception & Psychophysics*, 65(7), 1058–1077.

Curry, I., Estrada, A., Webb, C. and Erickson, B. 2008. *Efficacy of Tactile Cues From a Limited Belt-area System in Orienting Well-Rested and Fatigued*

Pilots in a Complex Flight Environment. (Technical Report ADA485178). Fort Rucker, AL: US Army Aeromedical Research Laboratory.

Dobbins, T. and Samways, S. 2002. The use of tactile navigation displays for the reduction of disorientation in maritime environments. *Proceedings of the Human Factors and Medicine Panel Symposium on Spatial Disorientation in Military Vehicles: Causes, Consequences, and Cures*. La Coruña, Spain, April 2002: NATO RTO HFM.

Duistermaat, M. 2005. *Tactile Land Navigation in Night Operations* (TNO-DV3 2005 M065). Soesterberg, The Netherlands: TNO Defence, Security and Safety.

Elliott, L.R., Coovert, M.D., Prewett, M., Walvord, A.G., Saboe, K. and Johnson, R. 2009. *A Review and Meta Analysis of Vibrotactile and Visual Information Displays*. (Technical Report ARL-TR-4955). Aberdeen Proving Ground, MD: US Army Research Laboratory.

Elliott, L.R., Duistermaat, M., Redden, E.S. and van Erp, J. 2007. *Multimodal Guidance for Land Navigation*. (Technical Report ARL-TR-4295). Aberdeen Proving Ground, MD: US Army Research Laboratory.

Elliott, L.R., Redden, E., Pettitt, R., Carstens, C., van Erp, J. and Duistermaat, M.E.S. 2006. *Tactile Guidance for Land Navigation*. (Technical Report ARL-TR-3814). Aberdeen Proving Ground, MD: US Army Research Laboratory.

Fuchs, S., Johnston, M., Hale, K.S. and Axelsson, P. 2008. Results from pilot testing a system for tactile reception of advanced patterns (STRAP). *Proceedings of the HFES 52nd Annual Meeting*, New York, September 2008, 1302–1306.

Furnell, A., Holmes, S.R. and King, S. 2003. Exposure to whole-body vibration increases vibrotactile perception thresholds for a 250Hz stimulus. *Proceedings of the 38th Conference on Human Response to Vibration*, Gosport, UK. Alverstoke: Institute of Naval Medicine.

Furnell, A., King, S. and Holmes, S.R. 2004. Manipulation of whole-body vibration frequency and amplitude alters vibrotactile perception thresholds for a 250Hz stimulus. *Proceedings of the 39th UK Conference on Human Response to Vibration*. Ludlow, UK, September 2004: RMS Vibration Test Laboratory.

Gilson, R.D., Redden, E.S. and Elliott, L.R. (eds) 2007. *Remote Tactile Displays for Future Soldiers*. (Technical Report ARL-SR-0152). Aberdeen Proving Ground, MD: US Army Research Laboratory.

Glumm, M., Kehring, K. and White, T.L. 2006. *Effects of Tactile, Visual, and Auditory Cues about Threat Location on Target Acquisition and Attention to Visual and Auditory Communications*. (Technical Report ARL-TR-3863). Aberdeen Proving Ground, MD: US Army Research Laboratory.

Greenspan, J.D. and Bolanowski, S.J. 1996. The Psychophysics of Tactile Perception and Its Peripheral Physiological Basis, in *Pain and Touch*, edited by L. Kruger. NY: Academic Press, 25–103.

Haas, E.C. and Stachowiak, C. 2007. Multimodal displays to enhance human–robot interaction on-the-move. *Proceedings of the 2007 Workshop on Performance Metrics for Intelligent Systems*, Gaithersburg, August 2007. New York: ACM.

Haas, E.C., Fields, M., Hill, S. and Stachowiak, C. 2009. *Extreme Scalability: Designing Interfaces and Algorithms for Soldier–Robotic Swarm Interaction.* (Technical Report ARL-TR-4800). US Army Research Laboratory, Aberdeen Proving Ground, MD.

Haas, E.C., Pillalamarri, R., Stachowiak, C. and Lattin, M.A. (2005). *Audio Cues to Assist Visual Search in Robotic System Operator Control Unit Displays.* (Technical Report ARL-TR-3632). Aberdeen Proving Ground, MD: US Army Research Laboratory.

Haas, E.C., Stachowiak, C., White, T., Feng, T. and Pillalamarri, K. In review. *Multimodal Displays for Robotic Operations On the Move.* (Technical Report). US Army Research Laboratory, Aberdeen Proving Ground, MD.

Ho, C., Tan, Z.T. and Spence, C. 2005. Using spatial vibrotactile cues to direct visual attention in driving scenes. *Transportation Research Part F: Traffic Psychology and Behavior*, 8(6), 397–412.

Hogema, J.H., De Vries, S.C., van Erp, J.B.F. and Kiefer, R. 2009. A tactile seat for directional coding in car driving: Field evaluation. *IEEE Transactions on Haptics*, 2(4), 181–88.

Hopp, P.J., Smith, C.A.P., Clegg, B.A. and Heggestad, E.D. 2005. Interruption management: The use of attention-directing tactile cues. *Human Factors*, 47(1), 1–11.

Jones, L.A. and Sarter, N.B. 2008. Tactile displays: Guidance for their design and application. *Human Factors*, 50(1), 90–111.

Jones, L.A., Lockyer, B. and Piateski, E. 2006. Tactile display and vibrotactile pattern recognition on the torso. *Advanced Robotics*, 20(12), 1359–1374.

Kaaresoja, T. and Linjama, J. 2005. Perception of short tactile pulses generated by a vibration motor in a mobile phone. *Proceedings of the First Joint Eurohaptics Conference and Symposium on Haptic Interfaces for Virtual Environment and Teleoperator Systems*, Pisa, Italy, March, 471–72.

Korteling, J.E. and van der Borg, W. 1997. Partial camera automation in an unmanned air vehicle. *IEEE Transactions on Systems, Man, and Cybernetics–Part A: Systems and Humans*, 27(2), 256–62.

Korteling, J.E. and van Emmerik, M.L. 1998. Continuous haptic information in target tracking from a moving platform. *Human Factors*, 40(2), 198–208.

Krausman, A.S. and White, T.L. 2006. *Tactile Displays and Detectability of Vibrotactile Patterns as Combat Assault Maneuvers are being Performed.* (Technical Report ARL-TR-3998). Aberdeen Proving Ground, MD: US Army Research Laboratory.

———. 2008. *Detection and Localization of Vibrotactile Signals in Moving Vehicles.* (Technical Report ARL-TR-4463). Aberdeen Proving Ground, MD: US Army Research Laboratory.

Krausman, A.S., Elliott, L.R. and Pettitt, R.A. 2005. *Effects of Visual, Auditory, and Tactile Alerts on Platoon Leader Performance and Decision Making.* (Technical Report ARL-TR-3633). Aberdeen Proving Ground, MD: US Army Research Laboratory.

Lathan, C. and Tracey, M. 2002. The effects of operator spatial perception and sensory feedback on human–robot teleoperation performance. *Presence: Teleoperation and Virtual Environments*, 11(4), 368–77.

McDermott, P.L., Fisher, A. and Allender, L. 2008. The transmission of spatial route information in distributed unmanned vehicle teams. *Proceedings of the HFES 52nd Annual Meeting*, New York, 2008, 52(4), 257–61.

McGrath, B., McKinley, A., Duistermaat, M., Carlander, O., Brill, C., Zets, G. and van Erp, J.B.F. 2008. Tactile actuator technology, in *Tactile Displays for Orientation, Navigation and Communication in Air, Sea, and Land Environments*, edited by J.B.F. van Erp and B.P. Self. (RTO Technical Report TR-HFM-122). Neuilly-sur-Seine Cedex, France: Research and Technology Organization, North Atlantic Treaty Organisation.

Myles, K. and Binseel, M. 2009. Exploring the tactile modality for HMDs, in *Helmet-Mounted Displays: Sensation, Perception, and Cognition Issues*, edited by C.E Rash, M. Russo, T. Letowski and E. Schmeisser. Fort Rucker, AL: US Army Aeromedical Research Laboratory.

Pettitt, R.A., Redden, E.S. and Carstens, C.B. 2006. *Comparison of Army Hand and Arm Signals to a Covert Tactile Communication System in a Dynamic Environment*. (Technical Report ARL-TR-3838). Aberdeen Proving Ground, MD: US Army Research Laboratory.

Raj, A.K., Kass, S.J. and Perry, J.F. 2000. Vibrotactile displays for improving spatial awareness. *Proceedings of the IEA XIV Triennial Congress/44th Annual Meeting of the HFES*, San Diego, CA, July–August 2000, 1, 181–84.

Redden, E.S., Carstens, C.B., Turner, D.D. and Elliott, L.R. 2006. *Localization of Tactile Signals as a Function of Tactor Operating Characteristics*. (Technical Report ARL-TR-3971). Aberdeen Proving Ground, MD: US Army Research Laboratory.

Riley, J.M., Strater, L.D., Chappell, S.L., Connors, E.S. and Endsley, M.R. 2010. Situation awareness in human–robot interaction: Challenges and user interface requirements, in *Human Robot Interactions in Future Military Operations*, edited by M. Barnes and F. Jentsch. London: Ashgate, 171–91.

Sanders, M.S. and McCormick, E.J. 1993. *Human Factors in Engineering and Design*. New York: McGraw-Hill.

Sekuler, R. and Blake R. 1990. *Perception*. 3rd Edition. New York: McGraw-Hill.

Sherrick, C.E. and Cholewiak, R.W. 1986. Cutaneous sensitivity, in *Handbook of Perception and Human Performance*, edited by K. Boff, L. Kaufman and J.L. Thomas. NY: Wiley, 12.1–12.58.

Sklar, A.E. and Sarter, N.B. 1999. Good vibrations: Tactile feedback in support of attention allocation and human–automation coordination in event-driven domains. *Human Factors*, 41(4), 543–52.

Sorkin, R.D. 1987. Design of auditory and tactile displays, in *Handbook of Human Factors*, edited by G. Salvendy. New York: Wiley, 549–76.

Stevens, J.C. and Choo, K.K. 1999. Spatial acuity of the body surface over the life span. *Somatosensory and Motor Research*, 13, 153–66.

Tan, H.Z., Gray, R., Young, J.J. and Traylor, R. 2003. A haptic back display for attentional and directional cueing. *Haptics-e: The Electronic Journal of Haptic Research*, 3(1). [Online]. Available at: http://www.haptics-e.org/ (accessed: July 7, 2012).

van Erp, J.B.F. 2007. *Tactile Displays for Navigation and Orientation: Perception and Behavior*. Leiden, The Netherlands: Mostert and van Onderen.

van Erp, J.B.F. and Self, B.P. 2008. *Tactile Displays for Orientation, Navigation and Communication in Air, Sea and Land Environments*. (Technical Report TR-HFM-122). Neuilly-sur-Seine Cedex, France: NATO Research and Technology Organisation.

van Erp, J.B.F. and van Veen, H.A.H.C. 2004. Vibrotactile in-vehicle navigation system. *Transportation Research Part F*, 7, 247–56.

van Erp, J.B.F, van Veen, H.A.H.C., Jansen, C. and Dobbins, T. 2005. Waypoint navigation with a vibrotactile waist belt. *ACM Transactions on Applied Perception*, 2(2), 106–117.

Von Békésy, G. 1959. Synchronism of neural discharges and their demultiplication in pitch perception on the skin and in hearing. *Journal of the Acoustical Society of America*, 27, 830–41.

Weinstein, S. 1968. Intensive and extensive aspects of tactile sensitivity as a function of body part, sex, and laterality, in *The Skin Senses*, edited by D.R. Kenshalo. Springfield, IL: Charles C. Thomas, 195–222.

White, T.L. 2010. *Suitable Body Locations and Vibrotactile Cueing Types for Dismounted Soldiers*. (Technical Report ARL-TR-5186). Aberdeen Proving Ground, MD: US Army Research Laboratory.

White, T.L. 2011. *The Perceived Urgency of Tactile Patterns*. (Technical Report ARL-TR-5557). Aberdeen Proving Ground, MD: US Army Research Laboratory.

White, T.L. and Krausman, A.S. 2012. *The Perceived Urgency of Tactile Patterns During Dismounted Soldier Movements*. (Technical Report ARL-TR-6013). Aberdeen Proving Ground, MD: US Army Research Laboratory.

Wickens, C.D. 2002. Multiple resources and performance prediction. *Theoretical Issues in Ergonomics Science*, 3(2), 159–77.

PART II
Overcoming Operational and Environmental Conditions

Chapter 6

Operations on the Move: Vehicle Movement and Soldier Performance

Richard A. Tauson

Recently, technology has allowed the use of automated communications systems in civilian and military vehicle environments. In civilian applications, this has taken the form of cell phones, navigational systems, and police and fire department automated dispatch displays. In military applications, the applications are much broader. They include command and control, indirect driving, and teleoperation of robotic systems.

The need for command and control (C2) in rapidly moving forces was demonstrated in Operation Desert Storm. Tactical Operations Centers (TOCs) (the planning and operations command center for a brigade or division) could only operate in a stationary configuration, which took two hours to set up and an hour to tear down. In Iraq, during rapid transit maneuver through the desert (sometimes called the "Hail Mary"), the slower TOC vehicles had difficulty keeping up with the M-1 tanks and Bradley Fighting Vehicles (McKiernan 1992), limiting the commander's ability to monitor and influence the battle. More recently, operations in Afghanistan have been characterized by small, dispersed, mobile units, which rely on dynamic command and control to coordinate information and mutual support (Ackerman 2002).

The Army has made several attempts to develop systems to support operations on the move. The Mounted Battle Command on the Move is an attempt to provide a division or brigade commander with the ability to move around the battlefield without losing the ability to monitor and influence operations. This program recognized the soldier's ability to do some tasks, such as monitoring voice communications and some visual information on computer displays in moving vehicles However, other tasks, such as entering data onto a C2 computer, require at least briefly stopping the vehicle. In contrast, the Future Combat System Operational Requirements Document (ORD) had numerous requirements for operations in moving vehicles, with little recognition of the limitations that vehicle movement might put on the human operators.

The introduction of information systems into moving environments has come at a cost to performance, which has been demonstrated in laboratory and in applied civilian and military environments. The impact can be on the vehicle driver using automated systems or on non-driving passengers (or crewmembers).

In the applied civilian world, the use of cell phone and texting devices has shown to cause a substantial degradation in driver performance. Green (2000) summarized a number of studies showing increased motor vehicle accident rates associated with the increased workload of interaction with dialing and talking on cell phones and with interacting with navigation devices. As one example of cell phone-induced degradation, drivers have been shown to have increased reaction time and increased rear-end collisions when talking on cell phones, regardless of age (Strayer and Drews 2004). The performance degradation created by cell phone use during driving is different in nature, but comparable in magnitude to the effect of alcohol intoxication (Strayer, Drews, and Crouch 2006). Similar driver performance degradation has been associated with texting (Hosking, Young, and Regan 2007). Adverse effects also extend to passengers, who experience increased motion sickness while watching movies in a moving vehicle (Morimoto et al. 2008, Schoettle and Sivak 2009).

This chapter describes some of the basic concepts and approaches to understanding the effect of vehicle motion on soldier performance. It describes several areas of research executed at the US Army Research Laboratory (ARL), supplemented by related studies, which provide some suggestions for designing systems intended for use in military vehicles.

General Considerations

In order to discuss the effects of vehicle motion on soldier performance, it is helpful to have a conceptual framework in which to work. Vehicle motion effects can be characterized in terms of kinetic movement (shock and vibration), confinement and isolation, and environmental effects. These effects on a person in the vehicle can each be immediate, persistent, or chronic. Immediate effects would include task degradation in reaching to touch a target (McDowell et al. 2005) or in a short-term memory task (Sherwood and Griffin 1990). Persistent effects might extend for minutes or up to 24 hours, for example, motion sickness (Ungs 1989). Chronic effects would include joint damage (Lewis 1962) or neurological injury (Donati 2001). In addition, each of these effects must be considered in terms of an interaction between the vehicle environment (including workstation design, air quality, vehicle suspension, and road conditions), the operator (demographics, training, stress, and fatigue), and the tasks to be performed in the vehicle.

Shock and Vibration

One of the primary differences between stationary and moving environments is the addition of vibration and shock to the vehicle and hence, potentially, to the soldier. When designing a vehicle workstation, one must consider the effects on the human body in terms of whole body vibration and resonant frequencies of specific body parts.

The definitive vibration standard for whole body vibration exposure is ISO/DIS 2631-5 (ISO 2004). The ISO standard allows one to calculate human exposure limits based on time and vibration characteristics, which must relate back to the task and environment in which the operator must function. Exceeding these limits may result in lumbar spine damage, central nervous system injury, and gastrointestinal irritation (Lewis 1962).

A vehicle's design should also provide limits on vibrations that match the resonant frequencies of specific body parts (Table 6.1). Resonant frequencies describe vibrations that cause a body part to oscillate at amplitudes greater than the originating vibration. Exposure to vibrations of 0–20 Hz can be amplified by the human body, with chronic exposure resulting in a variety of disorders including impaired vision, high blood pressure, bone joint and tendon disorders, and Renaud's Syndrome (Alliance Training and Consulting, Inc. 2006).

Table 6.1   Selected resonant frequencies of the human body (adapted from Duarte and Pereira 2006)

Body segment/Organ/Function	Resonant Frequency
Upper jaw	100–200 Hz
Hand (seated posture)	30–50 Hz
Eyeball—intraocular structures	20–90 Hz
Head	20–40 Hz
Lower arm (standing posture)	16–30 Hz
Spinal column (axial; standing posture)	10–12 Hz
Speech	5–20 Hz
Arm and chest wall (seated posture)	5–10 Hz
Near field vision	2–20 Hz
Leg (variable from 2 Hz with knees flexed to over 20 Hz with rigid posture)	2–20 Hz
Lungs	4–8 Hz
Abdomen	4–8 Hz
Shoulders (standing posture)	4–5 Hz

Specific Research and Applications

Command and Control Vehicle and Motion Sickness Studies

In the early 1990s, the US Army developed the XM-5 Command and Control Vehicle (C2V) in an attempt to produce a vehicle expressly designed to support C2 tasks during vehicle movement. The C2V was based on an extended Bradley Fighting Vehicle chassis, with an enclosed housing behind the cab. The housing contained four computer workstations with supporting computers, an environmental control unit, and communications systems. The frontmost workstation faced forward, with the remaining three workstations facing toward the driver's side wall, about 14° from the perpendicular to the axis of travel. The suspension had been modified to minimize the high-frequency vibration associated with tracked vehicles so as to minimize damage to the electronics.

ARL's first opportunity to evaluate this vehicle involved 14 participants who were asked to complete a cognitive task battery and a series of symptom and stress questionnaires during exposures of 30 minutes each on stationary, dirt road, and cross-country (sandy creek bed) rides. Participants were generally able to complete the cognitive batteries in all conditions. Performance and stress were slightly affected by vehicle motion, but a number of participants indicated increased discomfort in the motion sickness scales—two subjects vomited and one had to be removed from the test due to severe symptoms (nausea, cold sweats, and severe pallor) (Tauson et al. 1995).

In a subsequent, unpublished evaluation by ARL and the Army Test and Evaluation Command, National Guard artillerymen attempted similar tasks in stationary, dirt road, and moderately challenging cross-country conditions in the C2V. Complete results are not available, but one soldier became completely unconscious for well over a minute as a result of vehicle movement and had to be medically evacuated. A review of videotape of the soldier showed that the incident was probably caused by a combination of low blood pressure associated with nausea and a vaso-vagal maneuver while repressing the impulse to vomit, which further lowered his blood pressure. While not medically important in the long term, this incident can certainly be considered a significant performance decrement. Since other literature shows that cognitive performance is adversely affected by motion sickness (Dahlman et al. 2009, Wiker, Kennedy and Pepper 1983, Sherwood and Griffin 1990), the Army became concerned about the possible effect of motion on the soldiers' ability to function in moving vehicles. This concern was supported when ARL observed an exercise in which US marines were asked to operate computers in a moving tracked vehicle and 11 of 15 participants had moderate to severe motion sickness (Rickert 2000) in addition to reduced performance in typing and cursor control tasks.

Motion sickness appears to be a phenomenon involving the visual and vestibular systems. It is believed that the phenomenon is instigated by mismatches between the two sensory systems, such as vehicle movement in an enclosed space, when the inner ear detects movement while the eyes perceive a stationary environment (Reason 1978). Very low frequency vibrations (0.3–0.5 Hz) seem most likely to provoke motion sickness symptoms, which may include nausea, headache, pallor, sweating, dizziness, lassitude, drowsiness, and lack of concentration (Bittner and Guignard 1985), as well as drowsiness. Horizontal movement appears to be the most likely axis to cause symptoms (Golding, Markley, and Stott 1995) and uncontrolled movement tends to increase symptoms (Fukuda 1975), especially head movement (Johnson and Mayne 1953).

To quantify the effect of vehicle motion on motion sickness and soldier performance in the C2V, ARL, Army Test and Evaluation Command, and the National Aeronautics and Space Administration (NASA) were tasked with putting together a comprehensive study. In a within-subjects experiment, 24 soldiers rode in each seat of three different configurations of the C2V (baseline, perpendicular seating, and forward-facing). An exposure consisted of 4 hours of mixed moving (tank trails and unimproved roads) and stationary operations, during which the soldiers completed cognitive test batteries and symptom questionnaires.

The results of this study are reported in Cowings et al. 1999: 12 of the 24 soldiers reported moderate to severe symptoms. The most common symptoms were drowsiness, headache, warmth, nausea, and stomach discomfort. Actual vomiting was reported by 15 percent of the soldiers, and there tended to be recurrence among those soldiers. There was some decrease in symptoms through the first t2–3trials, followed by leveling or a slight increase in symptoms through the duration of the study. One soldier exhibited symptoms severe and consistent enough that the experimenters were obligated to ask her to withdraw from the experiment. The task performance results are at least as important in a military setting as symptoms, but translating a cognitive battery score to a practical environment is difficult. One metric, a 5 percent decrease in 5 of 7 tests (considered a performance decrement standard), was seen in 11 of 24 soldiers, with decrements of >5 percent in at least 2 tests in 22 of the 24 soldiers. Fortunately, the test battery used had been calibrated against blood alcohol levels (BAL) and an alternative way to look at the data is to say that 8 of 24 soldiers exhibited performance decrements equivalent to legal intoxication (BAL > 0.08) and 19 of 24 showed significant impairment (BAL > 0.025).

Before discussing what about this vehicle design led to a decline in soldier comfort and performance, let us look at some of the things they did right in designing this vehicle. The vehicle was equipped with an excellent Environmental Control Unit, which provided filtered, cool air to each workstation (some soldiers requested the ability to reduce airflow to the rear workstation). The workstations were equipped with commercial truck seats, with headrests and four-point restraints to reduce uncontrolled movement, which both decreases motion sickness and, as we will see, improves display and control interaction. However, the automotive test reports on the C2V show that the soft suspension successfully damped out much of the high frequency vibration, leaving a huge peak in vehicle movement below 1 Hz, maximizing the low-frequency, high-amplitude vibrations known to invoke motion sickness. This may have been magnified by the soft-suspension seats, since human performance is most affected by the frequency and amplitude of energy delivered through the seat and, to an extent, through the floor to the legs (ISO 2004). While there were other considerations involved, the difficulty that soldiers had in tolerating and performing in the C2V certainly contributed to the decision to abandon its development.

Indirect Driving

Another task that soldiers face during vehicle movement is indirect driving. In contrast to C2 tasks, which are primarily conducted by non-driving crew members, indirect driving is conducted by the driver of the vehicle. In some proposed or developmental military vehicles, the driver is expected to control the vehicle and maintain local situational awareness (SA) through camera views rather than viewing the outside environment directly through the windshield or other direct view. Indirect driving is proposed to allow the crew to be located under better protection, away from weak points such as periscopes and hatches. In the currently fielded fleet of Stryker vehicles, indirect driving is an option available to the driver, in addition to normal open-hatch driving, to support closed-hatch driving under night conditions. The driver can use a front-mounted, infrared camera display to supplement or replace normal "direct vision" driving.

Indirect driving also allows the driver to develop local SA in ways that conventional direct driving cannot, through displaying expanded visual spectra (such as infrared (IR) to see at night or through dust) or by constructive camera placement (such as rear placement for backing up or mast-mounted cameras to provide an overview). It has also been shown that, as soldiers are required to combine driving with targeting, planning, or SA tasks, they may have to rely more on automated assistance, such as Autonomous Navigation Systems (ANS) (Mitchell 2003).

The problems with indirect driving are generally associated with the visual display (that is, the camera view displayed on a monitor). Visual display and image quality can affect image interpretation, orientation, and motion sickness. Characteristics of image quality that can affect an operator's ability to execute indirect driving successfully include image acuity, field of view, image magnification, depth of field, and color perception (O'Kane 1996, van Erp, and Padmos 2003).

ARL, independently and in cooperation with the US Army's Tank Automotive Research Development and Engineering Command (TARDEC), has developed a significant amount of experience in indirect driving. In the 1990s ARL developed a test bed on a high-mobility multipurpose wheeled vehicle (HMMWV) with an enclosed workspace and three displays fed by hood-mounted cameras. This provided the operator with the ability to drive the vehicle from a "normal" driving location with traditional steering wheel and foot pedals, but with visual feedback determined by various display and camera settings instead of by direct vision through a windshield.

Display space is generally limited in an enclosed driver's station, forcing the designer into a choice between field of view (FoV) and display magnification. Given a fixed display size, limited FoV reduces lateral control of the vehicle (van Erp and Padmos 2003). However, compressing the image to increase FoV also reduces driving performance. In one study (Smyth, Gombash, and Burcham 2001), a unity display (little or no magnification) of 110° was compared to camera fields of view of 150°, 205°, and 257° (or compression ratios of 1.364, 1.864, and 2.336, respectively). While driving over a lane-following course with stationary obstacles, driver performance was best when FoV was closest to unity. Wider fields of view led to slower speeds and more obstacle strikes. However, drivers indicated that the wider FoV was helpful in navigation since it assisted with path selection. It is worth noting that all indirect driving conditions showed worse performance than direct driving. It is also interesting that, in the wider FoVs, operators perceived they were going faster, possibly because of the visual compression of the displays or the limited FoV (van Erp and Padmos 2003, Salvatore 1968). In addition, 2 of the 10 operators had to withdraw due to motion sickness in the indirect driving conditions. An additional condition in this study provided the camera input through a helmet-mounted display with unity (110°) vision. In this condition, performance was slower and less accurate (that is, more obstacle strikes) than other conditions, and 1 of the 10 participants withdrew due to motion sickness.

In a similar study, a ride motion simulator was used to create the experience of driving a Stryker wheeled infantry carrier over terrain similar to a tank trail using wide flat-panel (123°), narrow flat panel (40°), and narrow helmet-mounted (40°) displays. All displays were unity, and the wide FoV allowed better speeds and obstacle avoidance, though the narrow flat-panel display and the helmet-mounted display resulted in similar performance (Meldrum, Paul, McDowell, and Smyth 2004).

One possible way of improving the performance in the helmet-mounted display is to emulate Oving and van Erp (2001), who found that speed and maneuverability were improved by using a head-slaved camera, compared to a fixed camera FoV system. However, while Meldrum et al. (2004) discuss adding a head tracker to the system, they express concerns about latency and signal noise that would be introduced to the display.

While somewhat different from indirect driving, the same technologies have been applied to target detection using indirect vision in a moving vehicle (Smyth 2002). Again, vehicle motion and indirect vision both reduced the ability of the operator to detect targets. However, in the stationary condition, the use of sound localization cues improved performance, and the author speculates that the same effect was not found during vehicle movement because of limitations in the auditory cueing system. Another use of multimodal displays is the vibrotactile display. This consists of small, vibrating "buttons" (or tactors) placed on the operator's body to provide information by providing a tactile (high-frequency vibration) stimulus at an identifiable location. Use of tactors can include status alarms (incoming message indicator) or to supplement a display search by activating the tactor corresponding to a section of a display. One study showed improvements in vehicle navigation using vibrotactile cues to supplement visual displays (van Erp and van Veen 2004). However, vehicle movement can mask tactile cues, so both tactor and location on the body must be selected with caution (Krausman and White 2008). There is potential that multimodal displays, including directional audio signals and tactile cues, may be used in the future to reduce reliance on visual displays in moving vehicles, potentially reducing motion sickness and visual fatigue.

Teleoperations

Increasingly, the military is employing unmanned vehicles, such as the aerial Predator and the small unmanned ground robots used to scout for improvised explosive devices in Iraq and Afghanistan. Teleoperated systems provide the advantage of allowing soldiers to explore and affect areas of interest without exposing themselves to risk and, as robotic navigation systems improve, may allow one soldier to control multiple vehicles during convoys or routine movement (Muench et al. 2000).

Teleoperation is the operation of a machine, often an unmanned system, from a distance. Although sometimes the word "teleoperation" is used when the operator has direct sight of the vehicle under control, more often a vehicle being controlled from a distance in direct vision is called "remote control." Usually teleoperation implies seeing the environment through a camera view. As such, teleoperation can be considered a special case of indirect driving, with all of the interface challenges such as limited FoV, visual distortion, and motion sickness, to which may be added the challenges presented by latency and uncoupled movement.

Latency is a measure of time between when a signal is sent and when the system responds. In normal driving, this could be illustrated as the time between a driver seeing an obstacle in the street and the driver engaging the brakes causing the vehicle to finally stop, which can more generically be described as the sensory, perception, and response components of latency. In teleoperation, sensory latency is increased by the delay during which the camera signal is processed, the image is transmitted to the operator, and the signal is reprocessed and seen on the display. Response latency is increased after the operator activates a control by the information processing and transmission time required to send instructions to the teleoperated vehicle. Latencies below 10 ms are generally below human perception and are acceptable in most applications (Mansfield 1973). However, in a study supported by ARL as part of the Future Combat System program, it was shown that latencies of 250 ms during teleoperated driving could be overcome with effort, but latencies longer than 250 ms made driving much slower with more collisions (Bolling and Reudin 2009). This study was done with the operator at a stationary workstation. Currently, the combined effects of latency and uncoupled movement have not been well defined.

Uncoupled movement is a phenomenon experienced by an operator in a moving vehicle while fixating on a display showing the point of view of another vehicle

(Muth, Walker, and Fiorello 2006). In this environment, the operator's vestibular and somatosensory systems are experiencing motion from the vehicle the operator is occupying, but the operator's visual system is experiencing a combination of stationary visual cues about the immediate environment and dynamic cues about the motion of the remote vehicle the operator is controlling.

In a series of tests using teleoperated military vehicles, ARL showed that teleoperation typically takes about twice as long as conventional driving and doubles the number of obstacles hit (Scribner and Gombash 1998). However, in a subsequent study the teleoperated vehicle was programmed with a preset course and global positioning system (GPS) tracker, which allowed the operator to simply monitor the vehicle and enter course corrections as needed. This improved performance and user acceptance over teleoperation whether or not the operator needed to control the vehicle's speed (Scribner and Dahn 2008).

ARL conducted an evaluation of uncoupled movement using a driving simulator mounted in an enclosed rear seat of a HMMWV (Hill, Stachowiak, and Tauson 2004). The operators experienced substantial high incidence of motion sickness during the teleoperation task, whether the HMMWV they were in was moving or stationary, suggesting that the symptoms were caused by the simulator.

However, in subsequent studies at Clemson University, motion sickness was worse in teleoperating from a moving vehicle compared to a stationary vehicle (Muth, Walker, and Fiorello 2006). In addition, uncoupled movement in flight simulators showed degradation in cognitive tasks, even in participants who did not have motion sickness symptoms (Muth 2009). The impact of teleoperation within a moving vehicle and the associated phenomenon of uncoupled motion are not yet fully understood, but its importance may increase as more unmanned systems and virtual "viewpoints" are used by crewmembers with moving vehicles.

Principles for the Design for Human Performance in Moving Vehicles

Excluding motion sickness effects, command and control represents the simplest task set to discuss. In most cases, the operator is sitting at a workstation looking at a relatively static display, which may include text, map displays, or photographs from remote sensors. The operator has to manipulate these screens to bring up specific information and compose and send messages. By introducing vehicle motion into the soldier's C2 tasks, one imposes challenges in the areas of display readability and visual fatigue, control manipulation, and data input.

One study compared time and error detection performance while reading paragraphs set in different fonts (12, 14, and 16 point) and with varied line spacing (Burcham, Hill, and Tauson 2006). Each font and line-spacing condition was evaluated in a desktop environment and in a vehicle that was stationary (idling) or moving over road or cross-country courses. Performance was generally degraded by vehicle movement, but appeared to be better in the moving conditions in the 14-point condition (visual angle of 23.4 arc minutes). A similar study by the Naval Postgraduate School in support of the Future Combat System program found performance improved by going from a font with 20 to 32 arc minutes of visual angle (McCauley 2006). However, more research is needed to understand the requirements for computer display of information, especially maps and other graphic displays designed for use in moving vehicles.

If vehicle movement makes visually fixating on a display more demanding, it increases the difficulty of fine control and data input even more. Surprisingly, keyboard input in moving vehicles has not been a great challenge (Tauson et al. 1995) as long as the keyboard is anchored to a surface in the vehicle and the operator can brace his or her wrist or forearms on the same surface. However, other studies have shown that fine hand coordination (Cowings et al. 1999) and reaching tasks involving touch screen activation (McDowell et al. 2005) can be significantly degraded by vehicle motion.

TARDEC has developed a series of enclosed workstations, first in a Bradley Fighting Vehicle and later in a Stryker wheeled vehicle. In coordination with ARL, TARDEC has evaluated a number of tasks and interfaces, the results of which are described in Metcalfe, Davis, Tauson, and McDowell (2008). Based on past experience and existing studies, the workstation began with some basic design principles (Metcalfe et al. 2008: 18):

- Maximize hand contact with the primary steering input device by integration of mission-critical functions.
- Minimize movement distance to highly used functions that cannot be integrated with the primary steering input device in order to maximize accuracy based on Fitt's Law.
- Accommodate effects of vehicle motion with appropriate touch-screen button sizes.
- Indicate touch events based on button release ("last contact" strategy).
- Maintain multiple interface options sufficient to achieve different task goals across a range of environments.

Using the workstation designed with these principles in mind, TARDEC and ARL embarked on several evaluations of different control devices, described in Metcalfe et al. 2008. In a comparison of trackballs and touch screens to input a coordinate on an electronic map, touch screens were faster, but had more inaccuracy or errors. In part, this might be due to the size of fingers, but vehicle movement did make accurate finger placement difficult. Generally, users suggested that precise input should be made using the joystick or the keyboard. A hybrid interface was suggested, where large icons might be selected through the touch screen and precisely placed using the joystick. However, in any case where touch screens are used in moving vehicles, activation should be on last contact to allow the operator to move his or her hand on the screen until it is on the desired location. When using soft buttons on a touch-screen display, the following design guidelines are recommended:

- Provide an anchor or grasp point to support the hand.
- Enlarge button sizes to 1.5 inches wide, 1 inch high, with 0.13-inch spacing between buttons.
- Active buttons on the "last contact."[1]

One recurring theme is that performing many tasks in a moving vehicle will be associated with some degree of performance loss. In some cases, the wisest course may be to follow the example of the Mounted Battle Command on the Move, in which the ORD specifically recognizes that some tasks can be done during vehicle movement and some cannot. However, if mission requirements force soldiers to operate inside of moving vehicles, there are some design strategies that should help them.

The vehicle suspension and seating system should be designed to minimize the more resonant frequencies of the human body, especially the head, chest, and arms (Duarte and Pereira 2006), and it should minimize the frequencies near 0.5 Hz that are associated with motion sickness (Bittner and Guignard 1985).

If motion sickness is a concern, workstations should be designed to allow postural stability (possibly using a four- or five-point restraint system) and, especially, head stability. Additional strategies to minimize motion sickness might include task design (for example, not requiring excessive head movement to look at separated displays, avoiding multiple moving windows on a display), insuring adequate cool airflow, and an artificial horizon to reduce sensory conflict (Rolnick and Bles 1989).

For indirect driving or teleoperation tasks, displays should approach unity (Smyth, Gombash, and Burcham 2001), with the widest FoV possible given the workstations' constraints (Meldrum et al. 2004). If display size constrains FoV, a unity view should be supplemented by an on-demand wider FoV for navigation or cornering. Head-slaved cameras with helmet-mounted displays are promising but immature technologies, which may be useful when latency and reliability problems are overcome.

Information displays should be customized to compensate for vehicle movement. Text font sizes should be at least 14 point (Burcham, Hill, and Tauson 2006, McCauley 2006. The requirements for maps, icons, and complex dynamic displays in moving vehicles is an open question, but at a minimum such displays should be demonstrated to be useful in dynamic environments prior to fielding.

Controls should be located to minimize hand movement to reach them. When hand movement is required, an anchor point or brace should be available. Examples would be a wrist rest in front of a keyboard or a grasp bar on the edge of a touch screen. Buttons, either physical or soft buttons that must be activated during vehicle motion, should be oversized, with 1.5 inches horizontal and

---

1 Last contact activation is a system where the selected item on the screen is highlighted when the operator touches the screen, but not activated. The operator can move his finger or stylus until the desired item is highlighted and the item will only be activated when he removes his finger or stylus, breaking the "last contact" with the screen.

1 inch vertical as a good guideline. The type of control used should reflect the task requirements (Metcalfe et al. 2008).

Touch screens are good for rapid, gross selections, but are not very accurate during vehicle movement. Trackballs allow more precision, but are slower. In many cases, a combination of controls may provide an optimal interface. When touch screens must be used in a moving vehicle, in addition to anchor points for the operator's hands and larger icons, the designer should provide an interface with "last contact" activation (Metcalfe et al. 2008).

Summary

In this chapter, we examined the issue of operation "on the move" and the potential impacts of operating in moving vehicles on human performance. Specifically, we discussed three tasks that may of particular interest in military operations: command and control, indirect driving, and teleoperation. In addition, we present principles for the design of displays, controls, and other vehicle-related factors, which may help mitigate the potential performance losses resulting from operations in moving vehicles.

Clearly, there is still a great deal that is not understood and great opportunities for future work to improve workstation design to support military and civilian tasks. Interactions between latency and vehicle movement, the use of multimodal displays in moving vehicles, and the effects of vehicle-induced isolation on collaborative tasks are all areas that, hopefully, will be researched and better understood in the near future.

References

Ackerman, R. 2002. Technology empowers information operations in Afghanistan. *SIGNAL Magazine*. Available at: http://www.afcea.org/signal/articles/anmviewer.asp?a=429 &print=yes (accessed: July 17, 2012).

Alliance Training and Consulting, Inc. 2006. *Techniques to Control and/or Prevent Vibration in the Workplace*. Available at: http://www.alliancetac.com/index.html?PAGE_ID=154 (accessed: July 17, 2012).

Bittner, A. and Guignard, J. 1985. Human factors engineering principles for minimizing adverse ship motion effects: Theory and practice. *Naval Engineers Journal*, 97(4), 205–213.

Bolling, R. and Reudin, J. 2009. *Impact of Latency on Teleoperating a Large UGV*. SDD-A-595. Fort Monmouth, NJ: Network Analysis and Integration Laboratory.

Burcham, P., Hill, S. and Tauson, R. 2006. *Readability in a Motion Environment*. (Unpublished research). Aberdeen Proving Ground, MD: US Army Research Laboratory.

Cowings P., Toscano, W., DeRoshia, C. and Tauson, R. 1999. *The Effects of the Command and Control Vehicle (C2V) Operational Environment on Soldier Health and Performance.* (Technical Report ARL-MR-468). Aberdeen Proving Ground, MD: US Army Research Laboratory.

Dahlman, J., Sjors, A., Lindstrom, J., Ledin, T. and Falkmer, T. 2009. Performance and autonomic responses during motion sickness. *Human Factors*, 51(1), 56–66.

Donati, P. 2001. *Evaluation of Occupational Exposures to Hand-Transmitted Vibration: Frequency Weighting and Exposure Duration (A Preliminary Survey).* (Ec Biomed II concerted action BMH4-CT98-3291). Vancouvre, Cedex, France.

Duarte, M.L.M. and Pereira, M.B. 2006. Vision influence on whole-body human vibration comfort levels. *Shock and Vibration*, 13, 367–77.

Fukuda, T. 1975. Postural behavior and motion sickness. *Acta Otolaryngol Supp*, 330, 9–14.

Golding, J., Markley, H. and Stott, J. 1995. The effect of motion direction, body axis, and posture on motion sickness induced by low frequency linear oscillation. *Aviation, Space, and Environmental Medicine*, 66(11), 1046–1051.

Green, P. 2000. Crashes induced by driver information systems and what can be done to reduce them. *Proceedings of the 2000 International Congress on Transportation Electronics*, Dearborn, MI, October 2000. Warrendale, PA: Society of Automotive Engineers.

Hill, S., Stachowiak, C. and Tauson, R. 2004. Soldier performance in the enclosed compartment of a moving vehicle. *Proceedings of NATO Applied Vehicle Technology Panel Symposium*, RTO-MP-AVT-110, Prague, October 2004.

Hosking, S., Young, K. and Regan, M. 2007. The effects of text messaging on young novice driver performance, in *Distracted Driving*, edited by I.J. Faulks, M. Regan, M. Stevenson, J. Brown, A. Porter and J.D. Irwin. Sydney, NSW: Australasian College of Road Safety, 155–87.

ISO 2004. *Mechanical Vibration and Shock—Evaluation of Human Exposure to Whole-Body Vibration—Part 5: Method for Evaluation of Vibration Containing Multiple Shocks* (ISO 2631-5). Geneva: International Organization for Standardization.

Johnson, W. and Mayne, J. 1953. Stimulus required to produce motion sickness: Restriction of head movement as a preventative of airsickness on airborne troops. *Aviation Medicine*, 24, 400–411.

Krausman, A. and White, T. 2008. *Detection and Localization of Vibrotactile Signals in Moving Vehicles.* (Technical Report ARL-TR-4463). Aberdeen Proving Ground, MD: US Army Research Laboratory.

Lewis, J. 1962. *A Partial Review of the Literature on Physiological Disorders Resulting from the Operation of Motor Vehicles.* (Technical Memorandum 17-62). Aberdeen Proving Ground, MD: Human Engineering Laboratory.

Mansfield, R. 1973. Latency functions in human vision. *Vision Research*, 13, 2219–2234.

McCauley, M. 2006. Risk 146 Vehicle motion effects (VME): Legibility on the move and air ventilation evaluation. (Technical Report F21417-002, Contract Number CF001-0F21). Seattle, WA: Boeing Corporation.

McDowell, K., Rider, K.A., Truong, N. and Paul, V. 2005. The effects of ride motion on reaction times for reaching tasks. *Proceedings of the 2005 SAE World Congress,* Warrendale, PA, April 2005, 108–115. Detroit, MI: SAE International.

McKiernan, D.D. 1992. *Command, Control, and Communications at the VII Corps Tactical Command Post: Operation Desert Shield/Desert Storm.* Carlisle Barracks, PA: US Army War College.

Meldrum, A., Paul, V., McDowell, K. and Smyth, C. 2004. *CAT D/RMS Driving FOV Experiment: Preliminary Results and Lessons Learned.* (IVSS-2004-MAS-01). Warren, MI: US Army Tank Automotive Command.

Metcalfe, J., Davis, J., Tauson, R. and McDowell, K. 2008. *Assessing Constraints on Soldier Cognitive Performance during Vehicle Motion.* (Technical Report ARL-TR-4461). Aberdeen Proving Ground, MD: US Army Research Laboratory.

Mitchell, D. 2003. *Advanced Improved Performance Research Integration Tool performance (IMPRINT) Vetronics Technology Test Bed Model Development.* (Technical Report ARL-TN-0208). Aberdeen Proving Ground, MD: US Army Research Laboratory.

Morimoto, A., Naoki, I., Daisuke, I., Hitoshi, A., Atsuo, K. and Masui, F. 2008. Effects of reading books and watching movies on inducement of car sickness. *Proceedings of the Fédération Internationale des Sociétés d'Ingénieurs de Techniques de l'Automobile (FISITA) World Automotive Congress,* Munich, Germany, September 2008. Available at: http://www.atzonline.com/index.php %3Bdo=show/site=a4e/sid=4948189704bf58548693d5997505826/alloc=3/id=7631 (accessed: July 17, 2012).

Muench, P., Laughery, S., Eversen, J., Houle, K. and Ikramulla, F. 2000. Teleoperation convoy. *Proceedings of SPIE.* Warren, MI: Tank Automotive Research Development and Engineering Command.

Muth, E. 2009. The challenge of uncoupled motion: Duration of cognitive and physiological aftereffects. *Human Factors,* 51(5), 752–61.

Muth, E., Walker, A. and Fiorello, M. 2006. Effects of uncoupled motion on performance. *Human Factors,* 48(3), 600–607.

O'Kane, B.L. 1996. Driving with indirect viewing sensors: Understanding the visual perception issues. *Proceedings of SPIE,* 2736, 248–58.

Oving, A. and van Erp, J. 2001. Armored vehicle driving with a head-slaved individual viewing system: Field experiments. *Reports of TNO Human Factors Department of Skilled Behavior,* 1–19.

Reason, J. 1978. Motion sickness: Some theoretical and practical considerations. *Applied Ergonomics,* 9(3), 163–67.

Rickert, D. 2000. *C4I Mobile Operational Prototype (CMOP) User Jury 8 Summary Report.* Woodbridge, VA: General Dynamics Amphibious Systems.

Rolnick, A. and Bles, W. 1989. Performance and well-being under tilting conditions: The effect of visual reference and artificial horizon. *Aviation, Space, and Environmental Medicine*, 60, 779–85.

Salvatore, S. 1968. The estimation of vehicular velocity as a function of visual stimulation. *Human Factors*, 10, 27–32.

Schoettle, B. and Sivak, M. 2009. In-vehicle video and motion sickness. Ann Arbor, MI: University of Michigan, Transportation Research Institute. Available at: http://deepblue.lib.umich.edu/handle/2027.42/61931 (accessed: July 17, 2012).

Scribner, D. and Dahn, D. 2008. *A Comparison of Soldier Performance in a Moving Command Vehicle under Manned, Teleoperated, and Semi-Autonomous Robotic Mine Detector System Control Modes*. (Technical Report ARL-TR-4609). Aberdeen Proving Ground, MD: US Army Research Laboratory.

Scribner, D. and Gombash, J. 1998. *The Effect of Stereoscopic and Wide Field of View Conditions on Teleoperator Performance*. (Technical Report ARL-TR-1598). Aberdeen Proving Ground, MD: US Army Research Laboratory.

Sherwood, N. and Griffin, M. 1990. Effects of whole-body vibration on short-term memory. *Aviation, Space, and Environmental Medicine*, 61(12), 1092–1097.

Smyth, C.C. 2002. *Modeling Indirect Vision Driving with Fixed Flat Panel Displays: Task Performance and Mental Workload*. (Technical Report ARL-TR-2701). Aberdeen Proving Ground, MD: US Army Research Laboratory.

Smyth, C.C., Gombash, J.W. and Burcham, P.M. 2001. *Indirect Vision Driving with Fixed Flat Panel Displays for Near-Unity, Wide, and Extended Fields of Camera View*. (Technical Report ARL-TR-2511). Aberdeen Proving Ground: US Army Research Laboratory.

Strayer, D. and Drews, F. 2004. Profiles in driver distraction: Effects of cell phone conversations on younger and older drivers. *Human Factors*, 46(4), 640–49.

Strayer, D., Drews, F. and Crouch, D. 2006. A comparison of the cell phone driver and the drunk driver. *Human Factors*, 48(2), 381–91.

Tauson, R., Doss, N., Rice, D., Tyrol, D. and Davidson, D. 1995. *The Effect of Vehicle Noise and Vibration (Caused by Moving Operations) on Cognitive Performance in the Command and Control Vehicle*. (Technical Report ARL-MR-279). Aberdeen Proving Ground, MD: US Army Research Laboratory.

Ungs, T. 1989. Simulator induced syndrome: Evidence for long-term aftereffect. *Aviation, Space, and Environmental Medicine*, 60(3), 252–55.

van Erp, J.B.F. and Padmos, P. 2003. Image parameters for driving with indirect viewing systems. *Ergonomics*, 46(15), 1471–1499.

van Erp, J.B.F. and van Veen, H.A.H.C. 2004. Vibrotactile in-vehicle navigation system. *Transportation Research Part F*, 7, 247–56.

Wiker, S., Kennedy, R. and Pepper, R. 1983. Development of performance evaluation tests for environmental research (PETER): Navigation plotting. *Aviation, Space, and Environmental Medicine*, 54, 144–149.

Chapter 7

Night Vision Goggle Design: Overcoming the Obstacle of Darkness on the Ground

Elizabeth S. Redden and Linda R. Elliott

A cloak of invisibility is the best means of surprise and better than any armor as a means of protection. Moreover, the cloak that nature provides nightly has the advantage of being more consistent and predictable than any artificial one.

(Hart 1985)

Introduction

The need for combat superiority at night has existed since the advent of combat itself. In the 1970s, the Army began fielding night vision goggles (NVGs) for use in ground operations. These early goggles came with a host of human interface issues with the result that many warfighters wore them only intermittently (Love, Kennedy, and Strickland 1978, Redden 1996, Redden and Mills 1996, Thornton and Redden 1994, Thornton, Redden and McDonald 1995, Thornton, Redden et al. 1996, Turner et al. 1980).

Operation Desert Storm reinforced the importance of continuous operations and unit effectiveness at night because it offered the ability to apply continuous pressure, deny easy target acquisition, and gain the advantage of surprise. In the 1990s, a goal of those working in the area of night vision was to "own the night" and efforts toward optimizing the night-fighting capability of ground troops were increased. Since then, technological and human factors design improvements to NVGs have contributed to ever-increasing performance at night. Current design improvements have addressed many of the inefficiencies of early NVGs and have resulted in performance increases that have, in some cases, equaled or exceeded daytime performance. This chapter focuses on human factors design guidelines and user interface issues generated from decades of experiments and experience. They provide a foundation on which to build future studies aimed at further increasing soldier performance at night.

Night-vision Technology

The cloak of darkness has long shown the potential to be an effective cover for military maneuver. However, until relatively recently, this potential advantage has been offset by the performance degradations caused by limited visibility.

Before the 1990s, warfighters had to rely entirely on their dark-adapted eyes and ambient light and, thus, night operations carried great risks. Later, warfighters employed artificial battlefield illumination, such as torches, flashlights, searchlights, and illumination rounds fired from mortars, when they had to fight at night. However, artificial illumination often aided the enemy as well, by giving away positions and providing information about intentions. In the late twentieth century, night-vision technology developed in the United States changed this situation and made the dark battlefield the ally of US forces.

Image Intensification Systems

Combat night-vision capability began with a technology called image intensification (I^2) that was developed in the 1950s. This technology operates in the near infrared (IR) light spectrum and gathers ambient light from the moon, stars, and manmade sources of light, and intensifies it (Fulton and Mason 1982). In the mid-1960s, the United States fielded a small starlight scope, which was put to use in Vietnam. Vietnam was unprecedented in that night missions became routine. At this point in history, night-vision technology began to provide the means to operate efficiently at night.

I^2 devices were small in size and lightweight, had low power requirements and could be manufactured at a relatively low cost. These attributes enabled proliferation of I^2 goggles for head-worn, individual soldier applications. I^2 tubes provide vision near that of normal daytime vision under optimal conditions (20/25 to 20/40). Their primary weaknesses, then and now, include problems with depth perception and, in low ambient light, loss of visual night adaptation after intermittent wear. Also, active IR light and laser pointers used with the technology are visible to others wearing I^2 devices. In addition, I^2 capability is impaired during conditions of high fog or smoke.

Thermal Systems

During the 1970s, a technology that operates in the far IR light spectrum emerged. These thermal forward-looking IR detector systems pick up heat signatures and provide the advantage of seeing warm entities not only in the dark but also through many types of smoke, fog, and other obscurants. Very few systems at that time had been made into packages that could be headborne because weight and resolution sufficient for movement had not made it past the prototype stage. Thus, the majority of the thermal night-vision systems developed then were sights for weapons, driving systems, and handheld or tripod-mounted viewing systems. The 1980s and 1990s saw the fielding of many very successful thermal systems. The United States clearly "owned" the night at this point in history and possessed a clear advantage on the nighttime battlefield over adversaries. However, eventually the rudimentary technology for night-vision systems developed by foreign governments began to catch up with that of the United States.

Fusion Systems

Both I^2 and thermal technologies provide advantages depending on conditions. While I^2 provides more granular, detailed information about the battlefield, thermal systems often provide more contrast between the "hot" targets on the battlefield and cooler background scenes, greater target-detection ranges, and the ability to see through many obscurants. These mixed advantages demonstrated the need for a system that would provide the best features of both technologies by combining or fusing them together. The sensor fusion process collects data with a variety of sensor types, combines and processes them, and then presents them as an integrated product that is richer and more intelligible than its parts (Hunn 2008). This technology allows for enhanced imaging and laser aiming in smoke-obscured, dark environments. The method used to combine these images is critical to the effectiveness of the system since misregistration of the images, resolution losses, latency, and gain problems cause significant problems for a user. Also, the method used to combine images can preclude certain sensor technologies from being used (for example, I^2 tubes cannot be used in digitally fused goggles). Image fusion methods generally fall into two categories: optical overlay fusion and digital fusion. Each method has its own advantages and disadvantages for use in NVGs.

Optical overlay fusion   The turn of the twenty-first century brought the advent of optical fusion. The first optical fusion systems developed in the United States used an overlay to present two different spectral images to the observer (Bonnett, Redden, and Carstens 2005, Carstens, Bonnett, and Redden 2006, Redden, Carstens, and Bonnett 2006). One of the images (the I^2 image) was viewed directly by the eye and the other image (the thermal image) was overlaid on top of it.

Beginning in 2003, a series of three field tests were conducted on optically fused night-vision devices (NVDs) at Fort Benning, GA (Bonnett, Redden, and Carstens 2005, Carstens, Bonnett, and Redden 2006, Redden, Carstens, and Bonnett 2006). The experimental exercises included realistic tactical evaluations of the technology. Results indicated that optically fused goggles enhanced target-detection capability, especially through obscurants, without interfering with dismounted movement through rough terrain. Specifically, results indicated that optical fusion significantly increased the ability of soldiers to detect targets in an open field under smoke and no smoke conditions over the I^2 baseline goggles. In woodland environments, optical fusion did not interfere with cross-country movement times and increased the range at which thermal and human targets were detected. Deep shadows, camouflage, and smoke did not mask targets when the thermal overlay was present. In the Military Operations in Urban Terrain (MOUT) setting, optical fusion provided significantly greater target detection capability under smoke conditions than was provided by the I^2 baseline. Optical fusion was also found to be compatible with the AN/PAQ-4C and AN/PEQ-2 laser aiming devices, which previously were only compatible with I^2 devices.

The success of the optically fused devices in these experiments and other Army testing resulted in the fielding of the enhanced night vision goggle (ENVG), which is now designated as the AN/PSQ-20. It gained wide acceptance by US troops in Iraq and Afghanistan. It offered maximum resolution because a direct view system is not limited by the resolution of a display. It also provided the best of both types of spectral bands—the detail provided by an I^2 technology and the long range and vision through obscurants provided by the thermal technology. Soldiers stated that hot human targets seemed to pop right out of the scene when the optical fusion goggles were worn (Bonnett, Redden, and Carstens 2005, Carstens, Bonnett, and Redden 2006, Redden, Carstens and Bonnett 2006).

Digital fusion   Digital fusion is the Army's long-term goal, because it provides the potential to send images from the sensor to other soldiers on the battlefield via the battlefield network. It also provides more flexibility in configuring the package shape since the imaging lenses and detectors can be remotely located. Digital fusion combines digital imagery from two or more different spectral waveband images from separate imaging systems in a combined eyepiece at the pixel level.

While digitally fused NVGs show great potential, there are currently technical problems that need to be overcome. The low-light cameras that are used in digital goggles instead of I^2 tubes do not currently provide resolution that equals that achieved through a tube. Many digital systems produce parallax problems as well as latency that is observable and, in some cases, causes cybersickness in the wearer (Redden, Bonnet, and Carstens 2006, Redden, Turner, and Carstens 2006, Redden, Bonnett, and Carstens 2007, Swiecicki et al. 2009). Fusion algorithm efficiency also varies among different types of digitally fused goggles and can greatly affect performance. Finally, soldiers wearing digitally fused goggles with inherent display resolution limits tend to detect fewer targets and detect them more slowly than they do with optically fused goggles (Redden, Bonnett, and Carstens 2006).

Short-wavelength IR

A new technology that operates in the short-wavelength IR (SWIR) light spectrum is currently being investigated. The US Army is interested in SWIR technology, because SWIR works in darker conditions than I^2, which senses and amplifies reflected visible starlight or other ambient light. Night sky radiance (nightglow) emits five to seven times more illumination than starlight and nearly all of it is in the SWIR wavelengths (Leong 2007). Thus, a SWIR camera can see objects with great clarity on moonless nights. SWIR sensors also convert light to electrical signals and are inherently suitable for standard storage or transmission across networks. When used with a SWIR illuminator, SWIR provides additional capability over current I^2 systems when no ambient light is present. First, this capability is undetectable by I^2 systems as a different waveband is used. Also, unlike thermal technology, a viewer equipped with SWIR technology can see through glass.

The Urban ENVG is a prototype system being investigated by the US Army, which is being used to determine the impact of adding a SWIR capability into the existing ENVG system. The system digitally fuses the SWIR and thermal sensors and this is optically fused with the I² sensor. The technology prototype was specifically designed to serve as an evaluation platform and is not ready for use in the field.

Bonnett, Carstens, and Redden (in review) demonstrated the utility of SWIR for NVGs during a Capability Assessment Test at Fort Benning, GA. Soldiers participating in a force-on-force exercise using the SWIR were able to easily detect members of the opposing force who were using the IR illuminator on their goggles in a true dark room. The SWIR illuminator produced no visible signature. However, the SWIR goggle exhibited some of the same problems found in previously tested digital goggles such as latency, which can produce cybersickness, image offset problems, and mounting problems, which need to be addressed.

Advances in NVG technology are not sufficient to ensure that the United States will continue to "own the night." Technology and the human must form a symbiotic relationship for effective performance. Warfighter performance when NVGs are worn is a function of the goggle technology, warfighter capability and training, and their interaction (the warfighter/technology interface). Thus, human factors design issues are of utmost importance to the US Army's ability to "own the night." The following human factors design guidelines are based on "Redden's Rules of NVG Design" and the user interface issues with NVGs that have been identified and addressed over the last four decades. Many of these issues are applicable to NVGs and other NVDs regardless of the technology. Those that are specific to a technology are identified as such.

Human Factors Design Guidelines and Redden's Rules

Redden's Rule 1: Headborne Weight is a Critical Consideration in NVG Design

1a. There is no such thing as "light" infantry.
1b. The effectiveness of a soldier in combat is inversely proportional to the weight of his equipment.
1c. On any platform, the number of items tends to accumulate to fill the space available for mounting.

Because headborne weight is so critical to ground troops, it is imperative that it be given priority in the design of NVGs. After a review of ergonomic and biodynamic factors related to the introduction of head-mounted loads to land troops, Ivancevic and Beagley (2004) recommended that a head-mounted load should be chosen that has the smallest mass, is the most symmetrically balanced and aligned to the head's center of mass, and is closest to the head.

The design of night-vision systems for pilots has often included the use of a counterbalance to symmetrically balance the load of a helmet configured with NVGs. However, Ivancevic and Beagley (2004) dissuade against the use of a counterbalance for ground troops wearing NVGs. They stated that despite reducing the inertial moment tending toward forward rotation, it imposes an additional mass, thus increasing the overall inertial movement in each of the five other directions of rotation as well as compressing the neck and thus raising the risk of injury. However, the design of the AN/PSQ-20 uses the battery pack as a counterbalance. This design does not introduce increased weight by adding a counterbalance, which was cautioned against by Ivancevic and Beagley (2004), rather, it redistributes the weight of the goggles by moving it to the back of the helmet, which moves the center of gravity of the goggle closer to the center of the head. Carstens, Bonnett, and Redden (2006) reported that the improved center of gravity of the ENVG (AN/PSQ-20) somewhat compensated for its increased weight over the AN/PVS-14, thus soldiers rated it nearly as comfortable to wear as the AN/PVS-14.

There have been several studies of the effect of headborne weight on pilots but it is more difficult to find studies of headborne weight on dismounted warfighters. McLean et al. (1998) determined a maximum allowable helmet system mass of 5.5 lb for pilots, based on Newton's second law.[1] The potential to survive a crash was factored into the limit. Ground troops are not exposed to acceleration environments like those of pilots, so this limit may be too restrictive for them. However, McEntire and Shanahan (1997) predicted that chronic and not easily recognized neck injuries could occur from ground troops wearing helmets with attachments that weighed as little as 7.7 lb. Thus, the total headborne weight limit of 5.5 lb has also been used as a rule of thumb in the human factors design of headborne night-vision systems (helmet plus NVG weight). Sovelius and colleagues (2008) found that the weight of the NVGs themselves is very important because it shifts the center of gravity of the headborne weight, increasing the loads on the neck structures, and the frontal weight from the NVGs causes a further increase in the activity of cervical muscles that are already subjected to high strain.

Not only does headborne weight create the potential for injury but it has also been shown to impact individual movement techniques (IMT) of dismounted troops. Project Manager (PM) Sensors and Lasers funded the development of an IMT course by the Fort Benning Human Research and Engineering Directorate Field Office to study the effects of increased NVG weight on IMT. Figure 7.1 (derived from Redden and Bonnett 2002) shows the impact of weight on the IMT course completion times.

––––––––––––

1 Newton's second law. A body of mass m subject to a force F undergoes an acceleration a that has the same direction as the force and a magnitude that is directly proportional to the force and inversely proportional to the mass, that is, $F = ma$. Alternatively, the total force applied on a body is equal to the time derivative of linear momentum of the body.

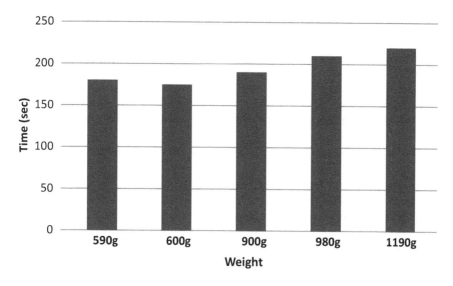

Figure 7.1   Mean time (sec) to complete the IMT course by NVG weight (g)

Another important consideration when designing NVGs is that they are not the only items planned to be placed on the helmet. Items such as identification friend or foe devices, ear protectors, head-mounted displays, visors, and so on, are all vying for a weight budget within the acceptable headborne weight limits.

Redden's Rule 2: Helmets and Mount Stability are Critical to Performance with NVGs

2a. A device is only as stable as the platform on which it is mounted.
2b. The more moving parts a mount has, the less stable the mounted object.

A stable helmet is the first building block for NVG stability. Unstable helmets result in NVGs that move around in front of the eye(s), especially as warfighters traverse rough terrain or negotiate obstacles. The Personnel Armor System for Ground Troops (PASGT) helmet was very unstable because of its webbing suspension system and its single chin strap. The Advanced Combat Helmet (ACH) and newer helmet systems have improved four-point retention and pad suspension systems that offer increased stability. However, this stability is dependent upon a good fit and proper positioning of the pads.

Once a stable helmet platform is achieved, it is important to have a stable attachment to the helmet. The mounting brackets for NVGs have been notoriously unstable on the helmet. Many of the ACH helmets include pre-drilled NVG bracket holes that alleviate much of the problem. Other potential points for instability are the goggle-to-mount mating point and the stability of the moving parts in the NVG mount.

Although increased degrees of freedom on the NVG mount create increased potential for instability, some movement of the mount is important. Vertical, tilt, fore, and aft adjustments on the mount allow the wearer to position the NVG more precisely in front of the eye. Some NVGs come with a small exit pupil, which necessitates even more precise and stable alignment in front of the eye(s).

Redden's Rule 3: Eyecup Designs Usually Create More Problems than They Solve

> 3a. The perceived usefulness of an article is inversely proportional to its actual usefulness once installed.
> 3b. The sweat from the eyecup runneth over.

The eyepiece designs of NVGs often incorporate eyecups. The purposes of these eyecups are to provide eye protection and light security, and to aid in centering the eye on the display. However, current eyecup designs create many problems for wearers of NVGs. They frequently cause fogging, sweat, and heat buildup, and are difficult to wear with glasses. If the eyecup is removable from the eyepiece, it is often taken off the NVG and lost. If the eyecup of an NVG is not removable, soldiers usually fold the eyecups back so that air can reach the lens and so that they do not touch the soldiers' face. Removal of the eyecup and folding it back result in loss of light security. Some eyecups incorporate a spiral design and if not fully engaged, they can adversely affect the full field of view (FoV) of the NVG.

Redden's Rule 4: Avoid Complexity

> 4a. If you drop a soldier in the middle of a deserted island with something as uncomplicated as an ammunition pouch, tell him not to touch it, and come back two hours later, it will be broken "because soldiers gotta mess with stuff."
> 4b. Any system or program, however simple, if looked at during combat will become very complicated.

As NVGs provide more and more capability, they will naturally tend to become more complex. It is critical that complexity be avoided. Dismounted warfighters wearing NVGs are often in life or death situations. When bullets are flying, it is very hard to remember which knob does what and how to fine-tune a lot of adjustments on the NVGs.

An adjustment that is currently present on the AN/PVS-7 and AN/PVS-14 NVGs is the diopter adjustment. The purpose of the diopter adjustment is to make the image appear sharper to the wearer, since soldiers' eyes tend to be a little different. The large diopter adjustment of +2 to −6 diopters present in early NVGs was based upon the AN/PVS-5 NVGs, which were not compatible with prescription lenses (McLean et al. 1998). Since diopters do not compensate for astigmatism and warfighters need to see when they take their NVGs off (current NVGs *are* compatible with prescription lenses), they should wear their glasses under their NVGs. This effectively negates the requirement for diopter adjustments. Also, because "soldiers gotta mess with stuff," removal of the diopter adjustment precludes soldiers from focusing their NVGs incorrectly and causing acuity problems. The US Army Aeromedical Research Laboratory recommends a small diopter value of −0.50 as providing the most comfortable vision with good acuity for most viewers, assuming they are emmetropic or are wearing corrective lenses (McLean and van de Pol 2002).

Like the diopter focus, the objective lens focus creates the potential for warfighters to focus their NVGs incorrectly. It also creates a potential requirement for them to perform adjustments during the thick of combat, when their hands need to be on their weapons and their thoughts need to be on staying alive. Because NVGs must currently be focused for specific ranges, a warfighter must change his objective lens focus between close-distance viewing (about 2 ft) out to infinity if he wants clear vision while navigating across country on foot. The depth of field of an NVG (the distance through which satisfactory resolution can be obtained when the device is in focus for a particular distance) is limited and must be changed when one's attention is moved from far to near or vice versa (for example, when using the device to look at the ground during cross-country movement and the need to view objects at a distance arises). The depth of field for an NVG is usually defined by the user and the conditions and it changes with the focal distance setting and the viewing distances (McLean 1996). Thus, if a warfighter has his NVG focused on a target at 100 m and has acceptable resolution on targets at both 75 m and 125 m, then the depth of field extends from 75 m to 125 m. If he wants to look at a depression in the ground in front of him (5 m away), then he must refocus his NVGs to clearly see the depression. There is a need to change this procedure because the warfighter needs to keep his hands on the weapon. Refocusing the objective lens is a two-hand task (one hand to hold the NVG steady in front of the eye and the other to turn the focus ring) and even if it were a one-hand task, it should not require frequent refocusing. Preset diopter focus ranges that take into account the depth of field of the goggles would help with this problem. The warfighter would only need to touch a button to change the viewing distance. These preset ranges should also be based upon the ranges at which warfighters perform the most critical tasks in combat. The ultimate objective would be to have an automatic focus built into the NVGs once technology can ensure that the item upon which the warfighter wishes to focus is the item that is in focus, and the speed of the automatic focus is such that it does not create a visual problem.

Some adjustments on NVGs are necessary for precise alignment with the eye (that is, interpupillary distance, fore/aft, and tilt) because placement of the exit pupil optic or eyepiece must accommodate variations in head size. Also, misalignment with the eye reduces the FoV provided to the eye. However, these adjustments should be set once when the NVGs are issued or first put on, and need to be accessed very infrequently during combat.

Complex menu systems for adjustment (for example, fusion mix) increase the time for adjustment and decrease performance and soldier acceptance (Redden, Carstens, and Bonnett 2006). If a menu system must be used, critical adjustments should never be placed in anything other than the top level of the menu. In general, simple and intuitive controls that can be adjusted during cross-country movement with one hand are important. Past experiments have demonstrated that in field settings, soldiers prefer knobs over buttons (Bonnett, Redden, and Carstens 2005, Redden, Carstens, and Bonnett 2006) because they are easy to find and adjust and it is easy to tactilely determine the setting rather than having to look for feedback in the NVG.

Redden's Rule 5: Monocular Designs Work Well for Dismounted Warfighters

   5a. Two eyes are better than one, especially when they're not looking at the
       same thing.
   5b. Two different perspectives are better than one.

Charles Dickens wrote, "He had but one eye and the pocket of prejudice runs in favour of two." That is also true when it comes to the question concerning whether monocular or binocular NVGs are better. Rule 5 is a controversial one, primarily because prejudice and common sense tell us that depth-perception cues such as convergence and binocular disparity are usually obtained with two eyes, and depth perception is important to wearers of NVGs. However, numerous researchers have found that people underestimate the distance to an object and overestimate the distance between objects when binocular NVGs are worn (Baburaj and Gomez 2009, Crawley 1991). Morawiec, Niall, and Scullion (2007) hypothesize that the distortions, halo effects (bright lights appear closer than dim lights even if they are at equal distances), and luminance differences (the Pulfrich effect) present in NVGs are some of the causes of reduced depth perception when they are worn. McLean and Rash (2001) also found that stereopsis from binocular NVGs does not provide any significant additional depth perception information over the strong monocular cues provided (for example, motion parallax).

It is particularly important for dismounted warfighters to accurately judge the depth of depressions in the terrain over which they must traverse and the distance to objects which they are trying to reach. Infantry soldiers generally prefer the monocular configuration because of the criticality of depth perception to dismounted movement (Redden and Bonnett 2002, Thornton, Redden, and McDonald 1996). Furthermore, McLean and Estrada (2000) found that monocular users can fuse images from their aided and unaided eye and this provides some color and depth perception under certain light conditions. When a monocular NVG is used in sufficient ambient light, it also provides a wider FoV than a binocular NVG because the unaided dark-adapted eye has a full temporal FoV and the aided eye has a slightly depressed temporal FoV on the side of the display (McLean 1996).

It is also important to note that two things must be present for the monocular design to be better than the binocular design. First, the gain on the NVGs must be turned down so that binocular rivalry does not occur. (Binocular rivalry is the conflict between aided and unaided viewing and usually results in the aided eye overpowering the scene from the unaided eye). Second, the area to be viewed with the unaided eye must be close enough to the eye to be seen with ambient light. For example, ambient light is usually insufficient to see a target or to see the distance needed to drive a vehicle with unaided vision (Redden and McDonald 1995). On the other hand, ambient light can often be sufficient to see the ground while walking or to see an object within arm's length that the soldier needs to pick up. Thus, monocular goggles are recommended for dismounted warfighters but not for those mounted in vehicles.

Before monocular NVGs were fielded, soldiers used the AN/PVS-7 binocular intermittently, like daytime binoculars (Thornton and Redden 1994, Turner et al. 1980). They typically wore them around their necks and used them when they needed to see something at a distance (beyond the range of their unaided eyes). This interfered with the dark adaptation of their eyes. A monocular NVG leaves one dark-adapted eye uncovered and the ambient light in most situations provides the viewer with the ability to see the ground and close objects with the unaided eyes (Redden and McDonald 1996, Thornton, Redden, and McDonald 1996).

The benefits provided by having an unaided dark-adapted eye were demonstrated by field trials run by the Fort Benning HRED Field Element (Redden, Turner, and Carstens 2006). Soldiers negotiated the IMT maneuver course with two different monocular NVGs, with their unaided eyes covered with a patch and without a patch so they could use their unaided eyes along with their aided eyes. They were significantly slower with both types of NVGs (A was an optically fused NVG and B was a digitally fused NVG) when they did not have the use of their unaided eyes (they wore a patch) than when they did have the use of their dark-adapted unaided eyes (no patch was worn) (Figure 7.2). (Note: These field trials took place over the period of a month so all phases of the moon were available.) The benefits of a dark-adapted eye are apparent because if for some reason the NVG becomes inoperable or the soldier is unable to use the NVG, the soldier has a dark-adapted eye if the NVG is a monocular.

Not all research points to the conclusion of Rule 5. First, CuQlock-Knopp et al. (1995, 1996) found that binocular NVGs performed better and were preferred to monocular NVGs. They stated that the binocular performed particularly better under low light-level conditions. However, the different results could be attributed to the fact that the research was performed with monocular NVGs that did not possess a gain control so that the wearers could turn the gain down sufficiently to preclude binocular rivalry. Second, the benefits of a monocular design do not appear to extend to mounted performance. In a safety certification test for driving with the AN/PVS-14 goggles, Redden (2002) found no differences between the

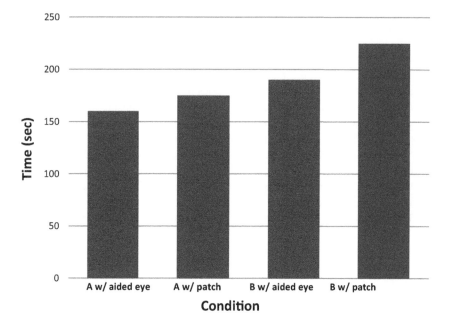

Figure 7.2   Benefits of a dark-adapted unaided eye during nighttime movement

binocular ANVIS and the monocular AN/PVS-14 for times and driving errors during completion of a hardtop and cross-country course when trained military drivers used high mobility multipurpose wheeled vehicles (HMMWVs) and 5-ton trucks. The subjective ratings for the ANVIS were slightly better than for the AN/PVS-14, but not significantly so. However, the monocular did allow them to see the instrument panels in the vehicles. Although the AH-64 pilots in their study rated the monocular slightly better for depth perception than the binocular, McLean and Estrada (2000) found that the binocular OMNI IV ANVIS was slightly preferred over the monocular AN/PVS-14 overall for pilotage. It appears that for situations other than dismounted movement in which the unaided eye does not provide benefit, there is no clear benefit for a monocular NVG.

Redden's Rule 6: Increased FoV Comes with a Price

    6a. Clarke's Law of Revolutionary Ideas: Every revolutionary idea evokes three stages of reaction:
       – It is completely impossible—don't waste my time.
       – It is possible, but it is not worth doing.
       – I said it was a good idea all along.
    6b. Nothing in life is free.

The argument is not that increases in FoV should never be made. Our normal FoV is almost 120° and most current NVGs have a FoV of just 40°, which reduces peripheral vision and causes the wearer to have to scan back and forth. The argument is that we are currently in the second stage of Clarke's Law. It is possible to design NVGs with an increased FoV; however, it is not worth doing because of the costs (weight, resolution, and money), and warfighters have demonstrated that their scanning ability can somewhat compensate for a reduced FoV. McLean and his colleagues (1998) stated that FoV and resolution are trade-offs against one another; increasing FoV results in a loss of NVG resolution above eye-limiting values. In a 1975 study in which a 40° NVG was compared to a 60° NVG, aviators preferred higher resolution over increased FoV (Sanders et al. 1975). Redden and McDonald (1996) found that NVGs with increased FoVs (60°) performed more poorly than those with 32° and 40° FoVs. (Although there were significant differences on six of eight tasks, the 40° FoV did better or tied on all of them.) The probability that would happen if all the NVGs were really the same is very low ($\alpha = 0.016$)). Tasks included time to sweep and clear rooms in a MOUT force-on-force exercise, time to navigate a dismounted cross-country course, navigation errors, number of targets detected on the cross-country course, number of errors on a hardtop driving course using the Ranger Special Operations Vehicle and motorcycle, and number of targets detected during a wide FoV firing scenario. The performance decrement was attributed to reduced resolution and increased headborne weight (see Redden's Rule 1). Likewise, Redden (2003) found that a 70° and a 95° NVG performed more poorly than a 40° NVG on a variety of infantry tasks (see the following paragraph) because variables such as exit pupil size, weight, configuration, resolution, ease of focusing, and eyecups were degraded when the FoV was increased. Problems in those areas adversely impacted the performance of the larger FoV NVGs and had more effect on performance than the increased FoV.

Redden (2003) performed an extensive field experiment using field limiting daytime goggles to document performance increases (or decreases) attributed to increases in FoV. Resolution, weight, and other variables were all held constant by the use of the field limiting goggles. Results demonstrated that as the FoV of the goggles increased (40°, 60°, 70°, 80°, 90°, and 95°), tactical patrolling times remained constant, as did the number of targets detected on a woodland patrol, close task performance, target detections, and IMT course times. Land navigation and MOUT IMT course times decreased gradually as the FoV increased. Only cross-country driving times significantly improved with all FoV increases, demonstrating that mounted and dismounted tasks respond differently to FoV increases just as they do to NVG configuration (that is, monocular vs. binocular).

Redden's Rule 7: Designers of NVG Eyepieces Must Understand the
Environment and Mission of the Dismounted Warfighter

    7a. The beauty of the eyepiece is in the eye of the holder (wearer).
    7b. "You can't depend on your eyes when your imagination is out of focus."
        (Mark Twain)
    7c. The eyepiece is the window to the night.

One size does not fit all when designing eyepieces for NVGs. Small problems with eyepieces experienced by mounted warfighters are often magnified when dismounted warfighters use the same eyepieces. Small exit pupils can cause severe problems when NVGs are worn by dismounted warfighters. Movement across rough terrain, negotiating obstacles, and helmet instability can combine to cause warfighters to lose their sight pictures as the eyepiece moves away from the center of their eyes. Non-pupil-forming eyepieces should be used for dismounted warfighter NVGs if possible because of the large eye-location volume provided behind the optics (Rash et al. 2009). Also, it is difficult to make accurate interpupillary adjustments if the eyepiece has a small exit pupil. If a pupil-forming eyepiece must be used, it should have a minimum of a 14-mm exit pupil (Rash et al. 2009). It is also important that NVGs for warfighters have sufficient eye relief to accommodate glasses (see above: Redden's Rule 4).

Dismounted warfighters wearing monocular NVGs should turn down the gain of the goggle as much as possible so that the aided eye does not overpower the vision from the unaided eye and create binocular rivalry. Redden, Bonnett, and Carstens (2007) found digital algorithms that automatically adjusted the gain to compensate for changing light levels sometimes resulted in complete loss of detail and a black foreground with light in the background, which overpowered all detail needed for negotiating terrain. Like automatic focus, automatic gain adjustment should not be included before all the problems are worked out.

Redden's Rule 8: Offsetting Sensors Comes with a Price

    8a. A man with one scale knows how much he weighs; a man with two scales
        is never sure.
    8b. Some may think that perception is reality, but reality is reality, so don't
        follow what an offset sensor says when you're near a cliff.
    8c. Who are you going to believe, me or your own eyes? (Groucho Marx)

Designers and proponents of digital NVGs often state that a digital design allows the sensors to be placed on the side (or top) of the helmet, closer to the center of gravity because digitally fused NVGs provide scene information via a display rather than a direct view sensor. These offsets appear to be desirable for several reasons. First, placing the sensors closer to the head results in a better center of gravity and less strain on the warfighters' necks. Second, offsetting the sensors reduces the forward projection of the NVGs and the potential to hit or snag the NVG on objects during movement. However, the benefits from offsetting the sensors come with a price that must be understood.

The effects of these offsets were addressed in four US Army Research Laboratory (ARL) Human Research and Engineering Directorate (HRED) experiments. Redden, Turner, and Carstens (2006) found that digitally fused goggles with viewpoint offsets performed more poorly than optically fused goggles without offsets in a wide range of infantry activities (individual movement course trials, cross-country woodland patrols, laser target trials, grid location tasks, and trials in which aiming lights were mounted on the M4 carbine). Soldiers complained that the offset was one of the problems that created these results

Redden, Turner, and Carstens (2006) and Redden, Bonnett, and Carstens (2006) found the use of digitally fused goggles with image offsets produced hand–eye dexterity problems. During this experiment, soldiers wearing NVGs with I^2 sensors mounted in three different locations were asked to touch the gray square in the middle of a 5 × 5-in grid. Figure 7.3 demonstrates the mean locations of the soldier touches on the grid. NVG A had the sensor mounted on the right side of the helmet. NVG B had the sensor mounted directly over the right eye. The sensor for NVG C was mounted on top of the right eye on the helmet.

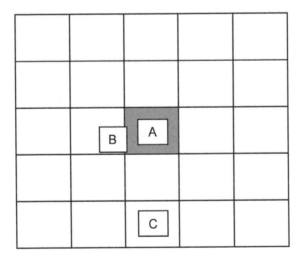

Figure 7.3   Mean location of grid touches

Both CuQlock-Knopp et al. (2001) and Redden, Bonnett, and Carstens (2006) found that wearing NVGs with hyperstereo viewpoint offsets resulted in an increase in the magnitude and direction of errors in throwing compared to no-hyperstereo viewpoint offsets.

Several researchers have found that mobility is especially sensitive to performance degradation caused by sensor offsets (Bonnett, Redden, and Carstens 2005, Redden, Bonnett, and Carstens 2006, Thornton, Redden, and McDonald 1996). A vertical viewpoint offset (on top of the helmet) resulted in soldiers complaining of a "floating feeling," some difficulty walking, and difficulty negotiating stairs (Mullins 2009, Thornton, Redden, and McDonald 1996, Bonnett, Redden, and Carstens 2005).

Because benefits derived from an improved forward projection can be obtained by offsetting the sensors over an inline system, methods for compensating for the offset are being considered by the Army. Compensation methods will also assist in aligning the thermal and I^2 images in optically fused NVGs.

Redden's Rule 9: Physical Problems are often Associated with Digital NVGs

9a. The more deadly the situation, the shorter the timeframe to react, the faster you need to move, the more likely you'll get sick.
9b. Anything that can go wrong will go wrong at the worst possible time.
9c. Cybersickness is a virtual indignity.
9d. Every time I see things your way, I get a headache.

In an experiment with two different digital NVGs and an I^2 goggle (AN/PVS-14), soldiers were asked to report any physical problems they experienced during the course of the various infantry activities (Redden, Bonnett, and Carstens 2006). Table 7.1 depicts the number of physical problems experienced with the three different NVG types. These physical problems have also been reported in other experiments (Redden, Bonnett, and Carstens 2007, Swiecicki et al. 2009).

Table 7.1   Physical problems experienced with the NVGs

	I^2	Digital A	Digital B
Cybersickness	0	26	8
Eyestrain	9	29	32
Headache	4	8	11

The physical problems have been attributed to several different factors. First, latency (the time between when the head moves and when the presented image changes to reflect this movement) and slow update rates (the frequency at which new display image frames are presented) are present in most of the digital NVGs, which cause the picture to lag behind reality. This is especially noticeable if the soldier must turn around quickly. Cybersickness can be caused by this mismatch between the visual and the vestibular sensory inputs. Offset sensors sometimes give some soldiers the feeling that they are floating above the ground and can cause eyestrain. The increased weight often associated with digital NVGs (discussed under Redden's Rule 1) can also create physical problems such as neck strain and headaches.

Rule 10: NVG Operation Must Be Trained

10a. Throwing an NVG at a warfighter and showing him where the on/off control and the battery compartment are located is not training.
10b. You can't overspend on training.

In the mid-1990s, findings by the Dismounted Battlespace Battle Lab (DBBL) indicated that at that time, there was not enough training on night-fighting equipment and that a training package needed to be developed to assist the field when planning night-fighting training. This prompted the DBBL to focus on ways to increase night-fighting training and effectiveness. In 1996, Thornton et al. developed and assessed a comprehensive set of tactics, techniques, and procedures (TTPs) for training and employing night-vision equipment to increase the effectiveness, lethality, and survivability of soldiers operating at night. In 1997, McDonald, Kohlhase, Waldheim, and Barracks reported on DBBL's development of a night-fighting training facility at Fort Benning GA. This facility was built as part of a strategy to overcome existing training deficiencies in NVDs during ground combat operations. Because of the DBBL's work on owning the night, training for NVGs increased, as did the US Army's total night-fighting capability. However, the facility has since been closed and training time has since decreased, resulting in a need for a resurgence in night-fighting training.

Effective NVG usage is not simple to learn; there are many facets of use that must be taught and practiced. For example, DeVilbiss, Antonio, and Fiedler (1994) found that an NVG adjustment class was needed before aircrew members were able to achieve optimal performance with the NVGs. McLean (1996) suggested that soldiers should practice adjusting the manual gain, the eyepiece, and the objective lens focus; learn to adjust the helmet bracket and mount and align the goggle with the eye; and also learn to fuse the display and the unaided eye images together. Redden and Mills (1996) found no correlation between soldiers' daytime and nighttime (NVGs and a laser aiming light) qualification scores, and that in order to fire effectively at night, soldiers need the same amount of training at night as they need during the day.

Digital NVGs bring additional training requirements. Soldiers need to be trained in image interpretation with I^2, thermal, and fused devices. This training should include hands-on practice with fusion mix. They need to learn the optimal fusion mix for different tasks. For example, long-range target engagement is usually best with the majority of the mix thermal and just enough I^2 to see the laser-aiming light. Carstens, Bonnett, and Redden (2006) found a potential for the thermal scene to overshadow the richer information provided by the I^2.

The presence of a "hot spot" or thermal image appears to draw the soldiers' attention away from cold images that can be seen with only the I^2. Therefore, patrolling primarily needs the rich, granular information provided by an I^2 for movement with just enough thermal to detect a camouflaged enemy.

Conclusions

The National Rifle Association says "guns don't kill people, people do." Likewise, NVGs do not see in the night, people do, and we have to remember that the ground warriors, as users of NVGs, are central to their performance. Lack of consideration of ground warriors' capabilities, limitations, and missions results in ineffectiveness. Redden's Rule 11 is a general rule that encompasses all human factors about NVGs. It says, "Perform good human factors research; the most dangerous person to good NVG human factors design is the person who thinks she or he already knows it all." The reason we conduct research is to get answers. Sometimes the answers may seem counterintuitive (that is, monocular is better than binocular for dismounted warriors), but the purpose of research is to take the speculation out of the design of NVGs. Good human factors research, coupled with technology advances, will continue to ensure that Hart's "cloak of darkness" continues to work for us, rather than against us.

References

Baburaj V.P. and Gomez, G. 2009. Analysis of problems with distance determination using night vision goggles. *Indian Journal of Aerospace Medicine*, 53(2), 1.

Bonnett, C.C., Carstens, C.B. and Redden, E.S. In review. *Urban Enhanced Night Vision Goggles Capability Assessment.* (Technical Report ARL-TR-xxxx). Aberdeen Proving Ground, MD: US Army Research Laboratory.

Bonnett, C.C., Redden, E.S. and Carstens, C.B. 2005. *Enhanced Night Vision Goggles Limited User Evaluation.* (Technical Report ARL-TR-3536). Aberdeen Proving Ground, MD: US Army Research Laboratory.

Carstens, C.B., Bonnett, C.C. and Redden, E.S. 2006. *Enhanced Night Vision Goggle Customer Test.* (Technical Report ARL-TR-3839). Aberdeen Proving Ground, MD: US Army Research Laboratory.

Crawley, J.S. 1991. *Human Factors of Night Vision Devices: Anecdotes from the Field Concerning Visual Illusions and Other Effects.* (USAARL Report No. 91-15). Fort Rucker, AL: US Army Aeromedical Research Laboratory.

CuQlock-Knopp, V.G., Myles, K.P., Malkin, F.J. and Bender, E. 2001. *The Effects of Viewpoint Offsets of Night Vision Goggles on Human Performance in a Simulated Grenade-Throwing Task.* (Technical Report ARL-TR-2407). Aberdeen Proving Ground, MD: US Army Research Laboratory.

CuQlock-Knopp, V.G., Sipes, D.E., Bender, E. and Merritt, J.O. 1995. *A Comparison of Monocular, Binocular, and Binocular Night Vision Goggles for Traversing Off-Road Terrain on Foot.* (Technical Report ARL-TR-747). Aberdeen Proving Ground, MD: US Army Research Laboratory.

CuQlock-Knopp, V.G., Sipes, D.E., Torgerson, W., Bender, E. and Merritt, J.O. 1996. *Human Off-Road Mobility, Preference, and Target-Detection Performance with Monocular, Binocular, and Binocular Night Vision Goggles.* (Technical Report ARL-TR-1170). Aberdeen Proving Ground, MD: US Army Research Laboratory.

DeVilbiss, C.A., Antonio, J.C. and Fiedler, G.M. 1994. Night vision goggle (NVG) visual acuity under ideal conditions with various adjustment procedures. *Aviation, Space, and Environmental Medicine,* 65, 705–709.

Fulton, R.W. and Mason, G.F. 1982. Night vision electro-optics technology transfer: A decade of activity. *The Journal of Technology Transfer,* 7, 35–53.

Hart Liddel, B.H. 1985. Development of night action. *Marine Corps Gazette,* March, 13–18.

Hunn, B.P. 2008. *The Human Factors of Sensor Fusion.* (Technical Report ARL-TR-4458). Aberdeen Proving Ground, MD: US Army Research Laboratory.

Ivancevic, V. and Beagley, N. 2004. *Determining the Acceptable Limits of Head Mounted Loads.* (Technical Report DSTO-TR-16577). Australian Government Department of Defence Land Operations Division: Defence Science and Technology Organization.

Leong, H.C. 2007. *Imaging and Reflectance Spectroscopy for the Evaluation of Effective Camouflage in the SWIR.* Thesis. Monterey, CA: Naval Post Graduate School.

Love, K., Kennedy, J.D. and Strickland, E.E. 1978. *Concept Evaluation Program of Employment of Night Vision Goggles.* Fort Benning, GA: US Army Infantry Board.

McDonald, K., Kohlhase, G., Waldheim, R. and Barracks, J. 1997. *Concept Experimentation Program of Night Fighting Training Facility.* (TRADOC Project No. 95-CEP-0407). Fort Benning, GA: US Army Infantry School Dismounted Battlespace Battle Lab.

McEntire, B.J. and Shanahan, D.F. 1997. *Mass Requirements for Helicopter Aircrew Helmets.* (USAARL Report No. 96-17). Fort Rucker, AL: US Army Aeromedical Research Laboratory.

McLean, W.E. 1996. *ANVIS Objective Lens Depth of Field.* (USAARL Report No. 96-17). Fort Rucker, AL: US Army Aeromedical Research Laboratory.

McLean, W.E. and Estrada, A. 2000. *Feasibility of Using the AN/PVS-14 Monocular Night Vision Device for Pilotage.* (USAARL Report No. 2000-18). Fort Rucker, AL: US Army Aeromedical Research Laboratory.

McLean, W.E. and Rash, C.E. 2001. Visual performance, in *Helmet Mounted Displays: Design Issues for Rotary-Wing Aircraft,* edited by C.E. Rash. Bellingham, Washington: SPIE-The International Society for Optical Engineering Press Book, 153–66.

McLean, W.E. and van de Pol, C. 2002. *Diopter Focus of ANVIS Eyepieces Using Monocular and Binocular Techniques*. (USAARL Report No. 98-28). Fort Rucker, AL: US Army Aeromedical Research Laboratory.

McLean, W.E., Rash, C.E., McEntire, J., Braithwaite, M.G. and Mora, J.C. 1998. *A Performance History of AN/PVS-5 and ANVIS Image Intensification Systems in US Army Aviation*. (USAARL Report No. 2002-08). Fort Rucker, AL: US Army Aeromedical Research Laboratory.

Morawiec, G. Niall, K.K. and Scullion, K. 2007. *Distance Estimation to Flashes in a Simulated Night Vision Environment*. (TR-2007-143). Defence Research and Development Canada.

Mullins, L.L. 2009. *Effects of Viewpoint Offset for Use in the Development of Advanced Indirect View Sensor Systems*. (Technical Report ARL-TR-4775). Aberdeen Proving Ground, MD: US Army Research Laboratory.

Rash, C.E., Bayer, M.M., Harding, T.H. and McLean, W.E. 2009. Visual helmet-mounted display, *in Helmet-Mounted Displays: Sensation, Perception and Cognition Issues*, edited by C.E. Rash, M.B. Russo, T.R. Letowski and E.T. Schmeisser. Fort Rucker, AL: US Army Aeromedical Research Laboratory.

Redden, E.S. 1996. *Concept Evaluation Program Test: MP Night Fighting Tactics, Techniques, and Procedures*. (TRADOC Project No. 96-CEP-7405). Fort Benning, GA: US Army Infantry School Dismounted Battlespace Battle Lab.

——. 2002. *Safety Assessment of Wearing the AN/PVS-14 Monocular Night Vision Device (MNVD) and AN/AVS-6 Aviator's Night Vision Imaging System (ANVIS) During 5-Ton and HMMWV Night Driving*. (Technical Report ARL-TR-2580). Aberdeen Proving Ground, MD: US Army Research Laboratory.

——. 2003. *Field of View Military Utility Experiment*. (Technical Report ARL-TR-2985). Aberdeen Proving Ground, MD: US Army Research Laboratory.

Redden, E.S. and Bonnett, C.B. 2002. *Baseline Evaluation of Soldier Performance on the Individual Movement Technique and the Military Operations in Urban Terrain Courses*. Unpublished report to PM Sensors and Lasers.

Redden, E.S. and McDonald, K. 1995. *Experimentation and Evaluation of Night Driving Devices*. (TRADOC Project Number 94-CEP-251). Fort Benning, Georgia: US Army Infantry School Dismounted Battlespace Battle Lab.

Redden, B. and McDonald, K. 1996. *Concept Evaluation Program Test of Image Intensification Device Field of View*. (TRADOC Project No. 95-CEP-0446). Fort Benning, GA: US Army Infantry School Dismounted Battlespace Battle Lab.

Redden E.S. and Mills, S. 1996. *Concept Evaluation Program Test of MP Night Fighting Tactics, Techniques, and Procedures* (TRADOC Project Number 96-CEP-7405). Fort Benning, GA: US Army Infantry School Dismounted Battlespace Battle Lab.

Redden, E.S., Bonnett, C.C. and Carstens, C.B. 2006. *Digitally Enhanced Night Vision Goggles Limited User Evaluation*. (Technical Report ARL-TR-3831). Aberdeen Proving Ground, MD: US Army Research Laboratory.

Redden, E.S., Bonnett, C.C. and Carstens, C.B. 2007. *Enhanced Night Vision Goggles, Digital: Limited User Evaluation II.* (Technical Report ARL-TR-4233). Aberdeen Proving Ground, MD: US Army Research Laboratory.

Redden, E.S., Carstens, C.B. and Bonnett, C.C. 2006. *Enhanced Night Vision Goggles Customer Test: Part II.* (Technical Report ARL-TR-3995). Aberdeen Proving Ground, MD: US Army Research Laboratory.

Redden, E.S., Turner, D.D. and Carstens, C.B. 2006. *The Effect of Future Forces Warrior Planned Sensor Offset on Performance of Infantry Tasks: Limited User Evaluation.* (Technical Report ARL-TR-3764). Aberdeen Proving Ground, MD: US Army Research Laboratory.

Sanders, M.G., Kimball, K.A., Frezell, T.L. and Hoffmann, M.A. 1975. *Aviator Performance Measurement During Low Altitude Rotary-Wing Flight with the AN/PVS-5 Night Vision Goggles.* (USAARL Report No. 76-10). Fort Rucker, AL: US Army Aeromedical Research Laboratory.

Sovelius, R., Oksa, J., Rintala, H., Huhtala, H. and Siitonen, S. 2008. Neck muscle strain when wearing helmet and NVG during acceleration on a trampoline. *Aviation, Space, and Environmental Medicine,* 79(2), 112–16.

Swiecicki, C.C., Bonnett, C.C., Redden, E.S. and Carstens, C.B. 2009. *Enhanced Night Vision Goggles (ENVG) Digital (D) Limited User Evaluation.* (Technical Report ARL-TR-4923). Aberdeen Proving Ground, MD: US Army Research Laboratory.

Thornton, C.H. and Redden, E.S. 1994. *Operation Night Eagle Company and Battalion Own the Night Advance Warfighting Experiment.* (TRADOC Project Number 93-CEP-0185). Fort Benning, GA: US Army Infantry School Dismounted Battlespace Battle Lab.

Thornton, C.H., Redden, E.S. and McDonald, K. 1995. *Concept Evaluation Program Test of Night Firing, Phase I.* (TRADOC Project Number 92-CEP-0921). Fort Benning, GA: US Army Infantry School Dismounted Battlespace Battle Lab.

Thornton, C.H., Redden, E.S. and McDonald, K. 1996. *Advanced Image Intensification (AI²) Battle Lab Warfighting Experiment.* (Battle Lab Project Number 0017). Fort Benning, GA: US Army Infantry School Dismounted Battlespace Battle Lab.

Thornton, C.H., Redden, B., McDonald, K. and Williams D. 1996. *Concept Evaluation Program Test of Own the Night (OTN) Tactics, Techniques, and Procedures.* (TRADOC Project Number 95-CEP-0417). Fort Benning, GA: US Army Infantry School Dismounted Battlespace Battle Lab.

Turner, D.D., Redden, E., Wasson, J.R. and Farrell, D.K. 1980. *Operational Test I of Low Cost Night Vision Goggles.* (TRADOC Project Number OTN-680). Fort Benning, GA: US Army Infantry Board.

Chapter 8

The Effects of Encapsulation on Dismounted Warrior Performance

Lamar Garrett, Debbie Patton, and Linda Mullins

Background

Soldiers, first responders, and search-and-rescue personnel are among an elite group whose job requirements include exposure to hazardous and toxic elements (that is, laser, ballistic, chemical, biological, radiological, and nuclear (Chemical, Biological, Radiological, and Nuclear (CBRN)), and climate). These elements, not to exclude various personal, situational, and organizational factors, can affect their ability to perform cognitive and physical duties effectively. An encapsulation ensemble meant to protect against these dangers can degrade performance. The lack of integration of such protection with existing equipment contributes to unnecessary weight and bulk. These professionals are required to wear encapsulated systems for varying durations of time and under uncertain conditions. A systematic approach to decreasing casualties and enhancing performance within CBRN-related environments is overdue.

The issue addressed in this chapter is what little information or guidance exists to properly evaluate the entire "encapsulation ensemble" while performing allocated tasks that must be carried out in the anticipated operational environment. The risk of missing the very critical communications information, alerts, and cues that guide such performance increases with information overload and perceptual restriction. Correspondingly, the fatigue induced by the system coupled with the risks of delayed reaction time may lead to decreased performance accuracy (such as operating remote vehicles and devices) posing potentially fatal problems for personnel and those they protect.

Encapsulation is defined as the inclusion of "one thing within another thing so that the included thing is not apparent".[1] Soldier encapsulation is defined as enclosing the soldier's body in such a manner that all skin is protected from exposure to the elements of the battlefield. The Department of Defense performed numerous chemical protective clothing studies during the 1990s. Unfortunately, most of that research focused on performance in a taxing environment (such as sustained operations, continuous operations, heat, and so on), or using new tools or systems (such as medical scissors, microclimate control systems, and so on)

---

1   See "encapsulation," at http://www.whatis.com (2010) (accessed: July 25, 2012).

rather than the effects of the suit on basic military skills (Headley, Hudgens, and Cunningham 1997, Fatkin and Hudgens 1994, Davis, Wick, Salvi, and Kash 1990, Blewitt et al. 1994). This type of information is critical to achieving effective mission performance, as well as the survival capabilities for the next generation of dismounted soldiers. Although research has been conducted on individual items of combat equipment and various components of dismounted soldier systems, very little performance-based research has been conducted using a system approach to validate soldier–equipment compatibility. Integration of the protective mask, chemical protective clothing (including gloves), boots, and individual combat equipment (that is, a helmet) is required when soldiers are operating in a suspected contaminated environment (Davis et al. 1990).

To illustrate, in the past, individual soldier combat equipment was developed using a variety of incompatible solutions. Reasons range from differing maturity of separate technology solutions, to varying delivery timelines of the equipment modules, to varying levels of experience and training backgrounds of evaluation personnel. This lack of integration has contributed to excessive equipment bulk and weight, leading to heat stress, reduced agility, mobility, and survivability, as well as less than optimal mission performance. The principal concerns expressed by soldiers wearing various individual combat equipment ensemble in the field, involves design integration, comfort level and, ultimately, its impact on mission performance.

Using a human system integration (HSI) approach, one can hypothesize that encapsulation will impact individual performance; as information-processing increases, cognitive performance decreases; and as the encapsulation configuration's weight is increased, mobility and agility decrease and heat stress increases, leading to degraded mission performance. In this chapter, we examine the impact of three protective ensembles on the performance of common soldier tasks such as shooting, grenade-throwing, and individual movement techniques, as well as cognitive readiness and cognitive performance. Encapsulation systems by themselves raise internal body temperatures by 10° (HQ, DA, and AF, 2003). This rise in temperature while encapsulated contributes to a decreased readiness of our soldiers. Yet we still expect our soldiers to perform duties successfully while encapsulated. By employing the HSI method of measurement, we gain a better understanding of how using a systematic approach can identify compatibility issues, such as weight distribution between configurations, task completion time, fatigue, and cognitive readiness. This chapter focuses on the human factors and equipment compatibility issues identified during research efforts conducted at the US Army Research Laboratory (ARL), Human Research and Engineering Directorate's (HRED) Known Distance (KD) Range, Wirsing Course, and Shooting Performance Facility—and includes implications for the design and development of soldier protective ensembles, and recommendations for future work.

Using these state-of-the-art facilities, we can examine soldier performance on aim, auditory ability, dexterity, mobility, cognitive stress, and performance.

Examples of performance measured in these are call-sign recognition, live-fire trials, grenade throw, individual movement techniques, small-arms field strip exercise, reaction to recorded sounds and, finally, equipment compatibility. Equipment compatibility looks specifically at the interaction of sighting systems (iron sights versus close-combat-optics), protective mask fit, the combat drinking system for CBRN protection, and psychological effects of the system while performing these tasks through monitoring stress perceptions as measured by the Readiness Assessment Monitoring System (RAMS) developed by ARL and cognitive performance as measured by the Cognitive Performance Assessment for Stress and Endurance (CPASE).

Encapsulation Assessment

In order to assess the effects of encapsulation on soldier performance with a system of systems design, we must determine the compatibility of encapsulation configurations in relation to soldiers' clothing and equipment; determine the impact of encapsulation on individual performance during mobility and portability maneuvers; investigate the ability of soldiers to process information during movement; determine if encapsulation adversely affects individual shooting performance; assess methods for measuring effects of encapsulation on dismounted warrior mission performance; and assess soldiers' stress perceptions and evaluate aspects of cognitive performance during encapsulation. Figure 8.1 illustrates the three ensembles discussed in this chapter.

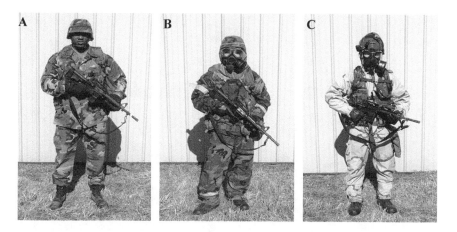

Figure 8.1   NBC Configurations for: A: Baseline, B: Current, C: Future

Test Facilities

The field ranges employed to look at these effects include an obstacle course, a cross-country course, and a small-arms shooting performance facility. The obstacle course (Figure 8.2) consists of 20 individual obstacles spread over a serpentine course of about 500 meters. The course design is one that requires soldiers to alternate between load carriage methods to negotiate the various obstacles (that is, switch from shoulder slung carry to hand carry, and so on). The course design makes use of most, if not all, of the participants' muscle groups while managing the load being carried from varying body postures. The obstacles were chosen to subject the participants to the kinds of maneuvers they should expect to perform executing an assault mission task in combat, such as running, jumping, climbing, balancing, negotiating buildings, stairs, and windows, and crawling. If the encapsulation ensemble systems impact individual execution of these typical assault obstacles it will be discernible (time and observation) during obstacle course runs. Beginning and ending times are recorded and analyzed to look at deltas in performance compared to a baseline.

Figure 8.2   500-meter KD Range

Individual performance measures enable experimenters to monitor participant consistency and collect data to discriminate differences between various systems, load configurations, or both. If the item(s) the research participants are wearing or carrying are incompatible with their clothing, their equipment, or themselves, the problems will be most noticeable when negotiating the course. Another characteristic intrinsic with the variety of obstacles is its suitability to estimate the ruggedness of individual clothing, equipment, and man-portable weapon systems.

Participants completed this course faster wearing the baseline configuration compared to Future Warrior encapsulation ($F = 30.22$, $p < .015$). Additionally, the participants completed the Individual Movement Techniques faster while wearing the baseline (46 seconds) compared to the Future Warrior System (64 seconds).

The cross-country course is a path through the woods forming a loop, starting and ending at KD Range. The path is unobstructed in places, but elsewhere crosses marshes, thick foliage, and fallen trees and is designed to simulate dismounted warrior movements executed during a "movement to contact" mission task. The overall course length is 4 kilometers. The course is designed to give research participants a chance to evaluate the comfort and utility of the test loads while marching at a moderate pace over generally flat natural terrain. Additionally, the cross-country course can disclose problems involving loads that would not appear during a march over open terrain.

Operational Tasks

A target detection task was used to assess the effects of encapsulation on situational awareness (SA) during movement along the cross-country course. The task consisted of detecting wooden silhouette targets placed along the course. The targets were placed such that there were 8 targets during the first 3 days and 16 targets during the last 3 days of the investigation. Relative to the path, each target was placed at a location 15 meters or 30 meters to the left or the right. The target positions were counterbalanced from left to right and spaced at equal meters apart over the length of the course. The target location changed for each mission to minimize learning effects. SA was defined as the number of targets detected. Participants completed the course more quickly in the baseline configuration (38 minutes) than in the Current Warrior System (45 minutes), which by comparison was completed more quickly than the Future Warrior System (49 minutes).

To assess the effects of encapsulation on speech recognition during movement along the cross-country course, radio call signs were presented auditorily to the participants throughout the duration of their movement along the 4-kilometre cross-country course. These callsigns come from the 66-item version of the Callsign Acquisition Test (CAT) (Rao and Letowski 2003). The CAT was developed by the Army Research Laboratory ARL to evaluate speech recognition using a militarily relevant vocabulary. The results indicate that the baseline configuration yielded the greatest number of correct identifications (approximately 58 percent).

Encapsulation resulted in a 43-percent decline. There were no significant differences in score between the Current and Future Warrior Systems.

Small-arms operations inherent to the infantry mission were tested using the small-arms shooting performance facility (Figure 8.3). This live-firing range is subdivided into four firing lanes (A, B, C, and D) designed to present to a single shooter, located at a fixed firing position, pop-up targets placed at 50 m, 75 m, 100 m, 150 m, 200 m, 250 m, 300 m, 400 m, and 500 m. These programmable pop-up targets feature the silhouette of a human facing the shooter. A bullet passing through the target electronically registers a "hit" and lowers the target. The precise location of bullet penetration is recorded, allowing the experimenter to determine whether the shot fired was a lethal hit or too far to one side. This type of data is not currently collected elsewhere within the Army.

Figure 8.3   Small arms shooting performance facility

Soldiers fired M4 carbines from a central location on the firing range, centered among the four firing lanes. The center firing position was used so that targets could be presented across the entire range. Using the whole range offered targets at a wider field of view (FoV) than any one of the four lanes alone could provide.

This four-lane method helped assure that if there were any differences among configurations with respect to FoV, these differences would become apparent with targets presented at increasingly wider FoVs.

Post-hoc analysis (Table 8.1) revealed that more targets were hit while wearing the baseline configuration compared to either Current or Future Warrior configurations. Current and Future Warrior configurations shot with almost equal accuracy; there were no significant differences between these two configurations.

Table 8.1   Tukey HSD post hoc analysis of equipment configuration on target hit percentage

Configuration		Mean difference	Std. error	p
Baseline	Current	.12	.021	.000
	Future Warrior	.11	.020	.000
Current	Baseline	−.12	.021	.000
	Future Warrior	−.01	.020	.805
Future Warrior	Baseline	−.11	.020	.000
	Current	.01	.020	.805

Stress Assessment and Cognitive Readiness

In order to understand the cognitive readiness of the encapsulated soldier during performance of soldier tasks, it is crucial to assess their responses to stress. Fatkin and Patton (2008) talk extensively about how the quality of soldier performance can be affected by various personal, situational, and organizational factors, particularly within dynamic and stressful environments. They define the stress experience as the interaction of individual appraisals and situational factors that require adapting to in order to achieve effective soldier performance. The ARL-HRED developed a standardized assessment paradigm: the Standardized Stress and Readiness Assessment (SARA) and the RAMS. This paradigm is used to assess stress perceptions and readiness of soldiers functioning under challenging and often extreme circumstances. Diagnosing the nature of the stress-performance relationship can guide changes that will enhance one's cognitive readiness. Cognitive readiness is the optimization and enhancement of human cognitive performance (Cosenzo, Fatkin, and Patton 2007). It is the critical component of effective mission performance, especially when performing multiple functions and adapting to dynamic threats.

To examine the psychological effects of encapsulation, psychological questionnaires were selected from the RAMS and administered according to the SARA methodology. Soldiers completed these questionnaires just prior to and immediately following task completion, while wearing the differing encapsulation configurations. Using this method we are able to compare the specific psychological effects of encapsulation. In this encapsulation research, we used the Subjective Stress Scale (SUBJ), the Specific Rating of Events (SRE), and the Multiple Affect Adjective Checklist-Revised (MAACL-R) with explicit instructions (that is, how do you feel "right now" or how did you feel "during the event you just completed"). The SUBJ and SRE measure "generalized" stress perceptions related to specific events. The SUBJ (Kerle and Bialek 1958) requires the participant to choose one item from a list of words that describe his or her current feelings; whereas the SRE requires the participant to choose a value on a scale from 0–100 that best describes how he or she feels during a specific time (Fatkin, King, and Hudgens 1990). The MAACL-R (Lubin and Zuckerman 1999) measures an individual's stress response to a particular event and measures five response components: anxiety, depression, hostility, positive affect, and sensation-seeking.

Analyses revealed that the soldiers experienced significantly higher levels of stress during the cross-county and obstacle tasks when compared to the pre data collection and live-fire event. Both encapsulation configurations created significantly higher SUBJ stress ratings during the cross-country tasks compared to the baseline, $t(11) = -3.362$, $p = .006$; $t(11) = -3.50$, $p = .005$. The same was true for the cross-country tasks compared to the obstacle course $t(11) = -3.842$, $p < .003$; $t(11) = -4.414$, $p = .001$.

Additionally, SUBJ ratings of stress for the Future Warrior configuration were significantly higher than Current ($t(11) = -1.80$, $p < .01$) and baseline ($t(11) = -1.24$, $p < .01$) configurations. Likewise, the SRE measure level of stress for the Current configuration was significantly higher than the baseline condition, $t(11) = -.53$, $p < .01$).

Similarly, the MAACL-R Multivariate Analysis of Variance (MANOVA) revealed a significant main effect for Configuration (Wilks' $\lambda = .42$, $F(2,10) = 6.92$, $p = .013$). However, main effects were *not* found for Session (pre and post) (Wilks' $\lambda = .67$, $F(3,9) = 1.50$, $p = .281$). These results indicate that the encapsulation ensemble is the cause for the stress experienced. As measured by the MAACL-R Depression subscale, we know that while wearing the Current and Future Warrior encapsulation ensembles, soldiers' perceptions of failure to perform well, or possibly the belief that the ensemble was not adequate to allow good performance while completing their tasks, is significantly higher compared to baseline ($t(11) = -2.96$, $p = .013$; ($t(11) = -3.037$, $p = .011$). Also, soldiers' frustrations, as measured by the MAACL-R Hostility subscale, while wearing these ensembles are also significantly higher compared to the baseline ($t(11) = -2.56$, $p = .025$; $t(11) = -4.15$, $p = .002$). These higher hostility levels indicate frustration caused poorer performance. These frustration levels may be due to the different weights and designs of each configuration, which in turn slowed the soldier. Another contributing factor was that the mask from one of the configurations fogged so much that the soldier could not see his own feet. These levels of frustration are comparable to the other comparative military scenarios, particularly when new equipment is being researched, such as new encapsulation suits.

In order to put these stress levels into perspective, we compared stress responses from this research effort with other military-relevant encapsulated research efforts using the same RAMS components. The groups include soldiers performing patient litter decontamination (Patient Decontamination), where the participants wore the current Mission Oriented Protective Posture (MOPP4) gear at the time the research was conducted during day operations; Chemical Defense Training Facility (CDTF) with students who had six hours of MOPP4 training to decontaminate weapons and vehicles in a live agent environment; and Special Forces Assessment and Selection, where participants were in training to be selected for a Special Forces assignment.

To understand if the stress a soldier is experiencing affects his or her performance on a given task, we correlate the stress measures with the performance during a specific task. Using the same data as discussed previously, we performed Pearson correlations with performance times on the cross-country and obstacle courses. There were significant positive correlations between stress perception levels and performance on the obstacle course for the Current configuration only. Soldiers who took longer to complete the course reported higher overall stress levels on the SRE scale ($r = 0.58$, $p = 0.046$) than those who completed the course in less time. Also, soldiers who experienced high levels of frustration took more time to complete the obstacle course than those who reported lower frustration levels ($r = 0.84$, $p = 0.001$).

At specific time points during this research effort (that is, during the cross-country and obstacle courses), participants did demonstrate significant differences in stress levels related to wearing the different configurations (Current and Future Warrior). However, in many cases, these differences were not significant when compared with other encapsulated, high-stress situations (Patient Decontamination and CDTF). The SUBJ and SRE questionnaires demonstrated sensitivity to the general levels of stress the participants were feeling. The MAACL-R showed sensitivity as a measure of the soldiers' stress perceptions between encapsulation configurations and indicated significant differences in the perception of negative affect and hostility between the experimental condition and the non-stress day. Past research has shown that soldier hostility levels as reported from the Hostility subscale of the MAACL-R have typically correlated with the workload frustration scale of the National Aeronautics and Space Administration-Task Load Index (NASA-TLX). In a study conducted at the Naval Air Warfare Center, an assessment was made regarding the efficiency of advanced man-mounted cooling systems in minimizing thermal stress (Kaufman and Fatkin 2001). The frustration levels reported by members of the US Marine Corps Chemical Biological Incident Response Force (CBIRF) wearing Current CBIRF and US Navy chemical protective outer garments were also correlated with their ratings on the NASA-TLX workload frustration scale.

When considering appropriate countermeasures to mitigate the negative effects of encapsulation, experience needs to be considered. Early investigations conducted within the Army program, the Physiological and Psychological Effects of NBC and Extended Operations on Crew Program, at the CDTF, indicated that amount of practice, mission rehearsal, and experience significantly affected reports of anxiety or uncertainty. During the CDTF research, three separate anxiety measures demonstrated that a junior enlisted group experienced a moderate level of stress, while an experienced group did not report a level of anxiety that was significantly different from an independent control group.

During the live-fire exercise portion of this research, soldiers reported significantly lower SRE and SUBJ ratings than the other military scenarios and their stress levels across the board are lower during the live-fire scenario. This is believed to be associated with the nature of live-fire scenarios because soldiers have a desire to fire live ammunition. In this session, their responses on the subjective measures indicated this was a positive experience for them. Previous research using psychological and physiological measures conducted on this live-fire range (Fatkin et al. 1991) has shown that testosterone levels are high, and vigor and vigilance reign. This general positive affect has been termed the *"HOOAH Effect"* (Fatkin and Patton 2008).

Cognitive Performance Assessment

A better understanding of the relationship between stress and performance will lead to methods that enhance soldier performance in stressful conditions. The CPASE was developed as a battery of cognitive measures that are sensitive indicators of human performance in stressful environments (Mullins 1996). This battery of tests requires 6 min to administer and examines aspects of short-term memory, logical reasoning, mathematical calculation and perception, and spatial processing functioning (Mullins 2002). These correspond to military operational tasks such as map-reading, navigation, communications, operations, and decision-making. Multiple versions of the cognitive battery have been developed to minimize learning effects over repeated testing. The following cognitive performance tests are administered:

- *Verbal Memory:* Short-term memory is tested using lists taken from a word usage text (Thorndike and Lorge 1944). Each list consists of 12, 1- or 2-syllable words with the most common usage rating (100 or more per million). Research participants have 1 minute to study the list and 1 minute for recall.
- *Logical Reasoning:* This reasoning test is based on Baddeley (1968). It tests the participants' understanding of grammatical transformations on sentences of various levels of syntactic complexity. Each item consists of a true/false statement such as "A follows B----AB" (false) or "B precedes A----BA" (true). The test is balanced for the following conditions: positive versus negative, active versus passive, precedes versus follows, order of

statement letter presentation, and order of letters in letter pair (equivalent to balancing for true/false). Letter pairs were selected to minimize acoustic and verbal confusion. Research participants have 1 minute to complete as many of the 32 items as possible.

- *Addition:* This task adapted from Williams and Lubin (1967), is used to test working memory. Each item consists of a pair of 3-digit numbers that were selected from a random number table. The task is subject paced. Research participants have 30 seconds to complete as many of the 15 problems as possible.

- *Spatial Manipulation:* Spatial skills are tested using a mental rotation task adapted from Shepherd's work (1978). A 6 x 6 grid is enclosed within a hexagon measuring 2.8 centimeters. Areas of the grid are filled in to create random patterns. To the right of each test pattern are three similar patterns. One of the three patterns is identical to the test pattern except that it has been rotated. The task is to select this pattern. Each test consists of 18 items balanced for the number of grids filled in (7, 9, or 11), pattern density (adjacent blocks filled in versus one break between pattern blocks), and rotation of the correct answer (90°, 180°, or 270°). Research participants have 2 minutes to complete as many of the items as possible.

CPASE was administered to evaluate changes in soldiers' cognitive performance while wearing the three encapsulation configurations under the various experimental conditions. There were no significant encapsulation effects for post obstacle course or post live-fire testing. Significant post cross-country encapsulation effects were found for three of the four cognitive tests: logical reasoning, addition, and spatial manipulation. Logical reasoning had a significant decline in performance from baseline ($M = 14.0$), with lower scores for Current ($M = 11.5$; 18 percent decline) and Future Warrior ($M = 12.0$; 14 percent decline) configurations. The baseline ($M = 8.5$) measure for addition was significantly different from the Future Warrior configuration ($M = 6.9$; 19 percent decline). Spatial manipulation had a significant decline in performance for the Future Warrior ($M = 13.6$) from the baseline ($M = 15.5$) and Current ($M = 14.6$) configurations (12 and 7 percent decline, respectively). However, cognitive performance differences occurred for the assessment following the cross-country course (on average it took approximately 40 min to complete). Logical reasoning had a decline in performance from baseline ($M = 13.94$) with lower scores for Current ($M = 11.5$; 18 percent decline) and Future Warrior ($M = 12.0$; 14 percent decline) configurations. The baseline ($M = 8.5$) measure for addition was different from the Future Warrior configuration ($M = 6.9$; 19 percent decline). Spatial manipulation had a decline in performance for the Future Warrior configuration ($M = 13.6$) from the baseline ($M = 15.5$) and Current ($M = 14.6$) configurations (12 and 7 percent decline, respectively). Encapsulation effects may have been found only for the cross-country condition because it was physically demanding and there were other cognitive tasks (target detection and speech intelligibility) associated with this condition.

Another explanation may be related to the additional time and exertion required to complete the cross-country condition compared to obstacle course or weapons-firing conditions. An effect of cross-country walking for a prolonged amount of time possibly resulted in increased body temperature. The thick textile required for encapsulation ensembles tends to hinder body cooling due to evaporation of body sweat. This hinders the body's ability to thermoregulate and the cumulative effects over time result in heat storage in the body, creating thermal stress. Faerevik and Reinertsen (2003) found that decrements in cognitive performance were correlated with increases in body temperature. Unfortunately, it was not feasible to monitor individuals' core temperature during this study, but further research in this area is warranted. The significant changes in cognitive performance found in this study may be due to increased temperature brought on by the higher physical load associated with the cross-country condition. If increased core body temperature is related to cognitive performance declines, this offers a powerful predictor for field performance. This physiological measure could potentially be used by commanders as an indicator or warning to watch for declines in individual mission performance.

Cognitive Performance across Sessions

The analysis across sessions for verbal memory found significant main effects between the pre and post measures for the cross-country and obstacle course. Participants performed slightly higher on the pre-test condition. The mean for the pre measure was 7.35, the post cross-country was 6.39, and the post obstacle course was 6.40. There were significant main effects for the logical reasoning task between the pre and post measures for cross-country, the obstacle course, and the weapon firing. Participants performed slightly higher on the pre-test condition. The mean for the pre measure was 13.78, the post cross-country was 12.53, the post obstacle course was 12.75, and the post weapon firing was 12.21.

For the logical reasoning task, cross-country condition had a significant interaction for session (pre and post) and equipment configuration. Performance while wearing the baseline configuration was higher than either of the encapsulation configurations, with the following means: baseline mean 13.94, Current mean 12.56, and Future Warrior mean 12.96 (baseline versus Current $p = .005$: baseline versus Future Warrior $p = .006$). For the pre-post by equipment configuration, there was a slight decline in performance from baseline for the encapsulation configurations.

For the addition task following the obstacle course there were significant main effects for session (pre and post). Performance was higher for the post measure (pre $M = 7.31$; post $M = 7.85$). This effect is most likely due to practice, which is consistent with other findings. Williams and Lubin (1967) found the addition task to be especially susceptible to practice effects and ideally recovery measures should be collected to correct for these changes.

Discussion

Given that battlefield encapsulation conditions are becoming increasingly necessary, it is important for safety, sustainability, and survivability, that the overall system performance be assessed in the aggregate. The objectives of this research were to investigate the effects of encapsulation on mission performance of dismounted soldiers and select methodologies that will be used for future research of encapsulation effects on mission performance. The results demonstrate that (1) mobility tasks take longer while encapsulated, due to increased weight and heat stress; (2) the cross-country course can distinguish differences between encapsulation systems both physically and psychologically; (3) individual shooting performance is degraded due to visual restriction associated with encapsulation systems; (4) encapsulation systems produce more psychological stress; and (5) cognitive performance is significantly degraded while encapsulated.

We conclude that the methodology used to evaluate the encapsulation systems was sensitive enough to detect differences in performance both between non-encapsulated and encapsulated systems, as well as for comparisons between encapsulated systems. The entire encapsulated system (encapsulation, weapons, and other soldier subsystems) best represents the soldier in their environment. When evaluating individual modules (for example, mask, ensemble, and drinking system) without having the complete system, it is difficult to relate the results to the system as a whole. There is no other way to determine either positive or negative interactive effects among modules, such as the joint effects of gloves (potential to restrict tactile feedback), clothing (binding at body joints can prohibit natural or comfortable posture), and protective masks (can prohibit quick and clear target-sighting, identification friend or foe, and so on)

The instrumented facilities allowed consistent and balanced evaluations across the subject pool. It is notable that France has constructed "twin" cross-country, mobility–portability, and shooter performance facilities modeled after the ARL-HRED facilities. Australia and Canada are also pursuing construction of facilities based on the ARL-HRED facilities.

Given that battlefield encapsulation is becoming increasingly necessary, it is important that safety, sustainability, and survivability of the overall system performance be assessed in the aggregate at facilities such as the ARL-HRED cross-country, mobility–portability, and shooter performance facilities.

While the evaluation process was time- and personnel-intensive, clear distinctions can be made in the effects of ensembles on the common soldier tasks. It is not necessary to evaluate various components separately and then to attempt to integrate the data into a meaningful whole, and it is not clear that statistical or operational procedures support such a modular approach. Two key variables to the success of the study were the test facilities and experimenter personnel. The instrumented facilities allowed consistent and balanced evaluations across the subject pool and the research team was highly experienced in using the facilities and measures, and was eminently prepared for a systems-level evaluation.

References

Baddeley, A. 1968. A three-minute reasoning task based on grammatical transformation. *Psychonomic Science*, 10, 341–42.

Blewett, W.K., Ramos, G.A., Redmond, D.P., Cadarette, B.S., Hudgens, G.A., Fatkin, L.T. and McKiernan, K. 1994. P2NBC2 test: The effects of microclimate cooling on tactical performance. (Technical Report, 148). Aberdeen Proving Ground, MD: US Army Chemical and Biological Defense Agency.

Cosenzo, K.A., Fatkin, L.T. and Patton, D. 2007. Ready or not: Enhancing operational effectiveness with the use of readiness and resiliency metrics. *Aviation, Space, and Environmental Medicine, Special Supplement on Operational Applications of Cognitive Performance Enhancement Technologies*, 78(5), B96–106.

Davis, E.G., Wick, C.H., Salvi, L. and Kash, H.M. 1990. *Soldier Performance of Military Operational Tasks Conducted While Wearing Chemical Individual Protective Equipment (IPE): Data Analysis in Support of the Revision of the US Army Field Manual on NBC Protection (FM 3-4)*. (BRL Report TR-3155). Aberdeen Proving Ground, MD: US Army Ballistic Research Laboratory.

Faerevik, H. and Reinertsen, R.E. 2003. Effects of wearing aircrew protective clothing on physiological and cognitive responses under various ambient conditions, *Ergonomics*, 46(8), 780–99.

Fatkin, L.T. and Hudgens, G.A. 1994. *Stress Perceptions of Soldiers Participating in Training at the Chemical Defense Training Facility: The Mediating Effects of Motivation, Experience, and Confidence Level* (Technical Report ARL-TR-365). Aberdeen Proving Ground, MD: US Army Research Laboratory.

Fatkin, L.T. and Patton, D. 2008. Ready or not: Mitigating the effects of stress through cognitive readiness, in *Stress and Performance*, edited by P.A. Hancock and J.L. Szalma. Aldershot, UK: Ashgate, 209–229.

Fatkin, L.T., Hudgens, G.A., Torre, J.P., Jr, King, J.M. and Chatterton, R.T., Jr. 1991. Psychological responses to competitive marksmanship, in *Effects of Competition and Mode of Fire on Physiological Responses, Psychological Stress Reactions, and Shooting Performance*, edited by J.P. Torre, Jr., S. Wansack, G.A. Hudgens, J.M. King, L.T. Fatkin, J. Mazurczak and J.S. Breitenbach. (Report HEL TM 11-91). Aberdeen Proving Ground, MD: US Army Research Laboratory.

Fatkin, L.T., King, J.M. and Hudgens, G.A. 1990. *Evaluation of Stress Experienced by Yellowstone Army Fire Fighters*. (Technical Memorandum No. 9-90). Aberdeen Proving Ground, MD: US Army Research Laboratory.

Headley, D.B., Hudgens, G.A. and Cunningham, D. 1997. The impact of chemical protective clothing on military operational performance. *Military Psychology*, 9(4), 359–74.

HQ, DA and AF (Headquarters, Department of the Army and Air Force). 2003. Heat Stress Control and Heat Casualty Management. (Technical Bulletin,

Medical 507, TB MED 507/AFPAM 48-152), March. Falls Church, VA: Office of The Surgeon General.

Kaufman, J.W. and Fatkin, L.T. 2001. *Assessment of Advanced Personal Cooling Systems for Use with Chemical Protective Outer Garments.* (Report No. NAWCADPAX/TR-2001/151). Patuxent River, MD: Naval Air Warfare Center Aircraft Division.

Kerle, R.H. and Bialek, H.M. 1958. *The Construction, Validation, and Application of a Subjective Stress Scale* (Staff Memorandum Fighter IV, Study 23). Presidio of Monterey, CA: US Army Leadership Human Research Unit.

Lubin, B. and Zuckerman, M. 1999. *Manual for the MAACL-R: Multiple Affect Adjective Check List—Revised.* San Diego, CA: Educational and Industrial Testing Service.

Mullins, L. 1996. *Cognitive Performance Assessment for Stress and Endurance (CPASE).* (Unpublished manuscript). Aberdeen Proving Ground, MD: US Army Research Laboratory.

——. 2002. Cognitive performance assessment for stress and endurance. *Proceedings of the 46th HFES Annual Meeting,* Baltimore, MD, September–October 2002, 925–29.

Rao, M.D. and Letowski, T.R. 2003. Speech intelligibility of the call sign acquisition test (CAT) for Army communication systems. *Proceedings of the Audio Engineering Society 114th Convention,* Amsterdam, The Netherlands, March 2003.

Shepherd, R.N. 1978. The mental image. *American Psychologist,* 33, 125–37.

Thorndike, E.L. and Lorge, I. 1944. *The Teacher's Word Book of 30,000 Words.* New York: Columbia University.

Williams, H.L. and Lubin, A. 1967. Speeded addition and sleep loss. *Journal of Experimental Psychology,* 73(2), 313–17.

Chapter 9

Soldier Auditory Situation Awareness: The Effects of Hearing Protection, Communications Headsets, and Headgear

James E. Melzer, Angelique A. Scharine, and Bruce E. Amrein

Hearing is the primary human sense that functions throughout the full sphere around an individual. Often the first warning and source of information about the identity and location of an event is through hearing. Because operationally important sounds can range from the very intense to the very quiet, soldiers need to be able to hear, locate and interpret sound sources across these intensity levels. Unfortunately, they are commonly exposed to very loud noises from weapons and vehicles that can cause temporary or permanent hearing threshold shifts, and this reduces soldiers' sensitivity to sounds, thus adversely affecting their ability to maintain auditory situation awareness (Scharine, Letowski, and Sampson 2009). In spite of these threats to hearing, soldiers are often reluctant to wear hearing protection, feeling that it reduces their sensitivity to quiet auditory cues in their environment and makes communication difficult. Hearing protection must be carefully selected to allow soldiers to hear their environment while still protecting against noise-induced hearing loss. While communications systems are essential for soldiers, a poorly designed headset can interfere with auditory situation awareness (SA)[1] and compromise the efficacy of hearing protection. Poorly designed headsets may also make communications difficult by transmitting noise from the environment. Auditory SA can further be impaired by other soldier gear, particular helmets, which can interfere with spatial localization cues. Thus, one must carefully consider the operating environment when designing or selecting hearing protection, communications headsets, and other headgear for soldiers.

---

1   SA is being aware of what is going on around you and how it might impact you in the future (Endsley 1995). The ground soldier uses all five senses to gather information, especially in an urban environment. Something that degrades a soldier's ability to sense his environment could compromise his safety (National Research Council 1995).

Auditory Situation Awareness

Auditory SA for the soldier means being able to detect, identify, and localize sound, and to communicate well either face-to-face or via radio (Scharine, Letowski, and Sampson 2009). The intensity levels in the soldier's natural environment can vary widely from the quiet rustle of leaves or the breaking of a twig to the loud impulse noises of explosions, blasts, and gunfire to the high-intensity steady-state noises of vehicles and generators. Exposure to high steady-state noise environment can adversely affect the soldier's ability to hear (in both the short and long term), as well as his[2] ability to communicate verbally.

A Soldier's Soundscape

Hearing protection is mandatory for armored vehicle crew members and dismounted soldiers when exposed to hazardous noise levels such as gunfire. However, dismounted soldiers have traditionally not supported wearing full hearing protection because they felt it was uncomfortable and caused them to lose auditory SA (Scharine, Letowski, and Sampson 2009). These concerns are legitimate given that soldiers rely heavily on their hearing for SA, especially in urban environments, where buildings can shadow, reflect, and reverberate sound.

Noise Levels in the Military Environment and the Need for Hearing Protection

Even a single sound—if loud enough—can cause permanent damage (Price 2005). Exposure to impulse noise[3] above approximately 140 dBP[4] or steady-state noise above 90 dB can induce a temporary threshold shift, a temporary hearing loss that results in an increased auditory threshold that normally disappears in 16 to 48 hours. It may also be accompanied by tinnitus, a ringing in the ears (NIDCD Information Clearinghouse 2011). Chronic exposure or exposure without sufficient recovery time can lead to permanent hearing loss. Other physiological and psychological effects of noise exposure include increased heart rate and respiration, as well as reduced cognitive processing ability (Dunlap 2002, Kardous et al. 2003).

Table 9.1 shows examples of typical threat noise levels that soldiers may be exposed to on the modern battlefield (for more detail see Amrein and Letowski, 2012; NATO, 2010; Ribera et al., 1995).

---

2 Soldiers are predominately male; therefore, the terms "he," "his," etc., will be used throughout this chapter. However, the issues and effects discussed in this chapter are not gender specific and affect men and women equally.

3 Impulse noise is defined as short pulse, < 1 sec, with a fast rise time that is more than 20 dB over the ambient noise. Total duration is typically < 200 ms. Impulse noise from a rifle shot lasts 4 ms to 15 ms.

4 P refers to "peak sound pressure".

Table 9.1   Steady-state and impulse noise levels from various military equipment types

Noise type	Vehicle or weapon	Noise level
Steady-state	HMMWV	94 dB in vehicle (moving)
	M1A2 Abrams	115 dB in vehicle (moving)
	CH-47 helicopter	107 dB in cockpit
Impulse	M16A2 rifle	157 dB at shooter's ear
	M249 machine gun	159.5 dB at gunner's ear
	Javelin missile	172.3 dB at gunner's ear
	Multi-role anti-armor antipersonnel weapon system (MAAWS)	190 dB at gunner's ear

Given the adverse physical, psychological, and performance effects of loud noise exposure (Hatfield and Gasaway 1963), the US National Institute for Occupational Safety and Health (NIOSH) has established a permissible exposure level (PEL) of 90 dB (A-weighted) for steady-state noise, above which hearing protection is required (NIOSH 1998). Both the US Army and Air Force have set their PEL at 85 dB (A-weighted) (DA Pamphlet 40-501 1998; AFOSH 48-20 2006). In each case, exposure at the PEL is limited in duration to eight hours (time-weighted average). Higher noise levels are permissible for shorter periods of time, as determined by an intensity/duration trading or exchange ratio. As sound levels increase above the PEL by the value of the trading ratio, the permissible exposure duration must be cut in half. The US Army and Air Force (DA Pamphlet 40-501 1998) establish this ratio at 3 dB. Therefore, the exposure limit is four hours at 88 dB, two hours at 91 dB, etc. Other regulating agencies use different trading ratios or exchange rates; for example, the US Occupational Safety & Health Administration (OSHA) uses an exchange rate of 5 dB and the US Navy, 4 dB (Department of the Navy 2000). Double hearing protection (wearing both an earplug and an earmuff simultaneously) is required by the US Army for steady-state noise over a PEL of 103 dB (A-weighted) and up to 108 dB (A-weighted) (DA Pamphlet 40-501 1998).

OSHA states that exposure to impulse noise should not exceed 140 dBP; however, there is less than complete agreement as to the exact method of calculating the effects of impulse noise exposure because such effects accumulate with time. The US Army requires hearing protection for impulse noises between 140 and 165 dBP (DA Pamphlet 40-501 1998). Double protection is required for impulse noises exceeding 165 dBP, but less than the Z curve,[5] specified by Military Standard 1474D (1997) and shown in Figure 9.1.

---

5   The X, Y, and Z criteria refer to lines referencing peak loudness level limits as a function of "B-duration." For levels between Y and Z, hearing protection is mandatory and

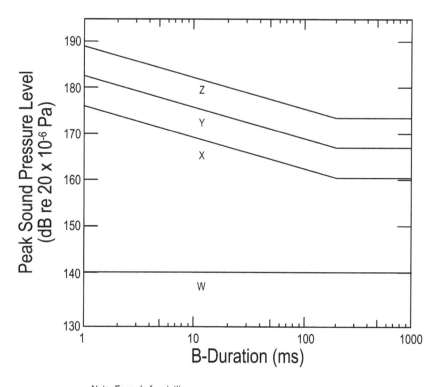

Note: Formula for plotting curves

$PPL_X = 160.5 + 6.64 \log_{10} 200/T$, for $T \leq 200$; $PPL_X = 160.5$, for $T > 200$

Where PPL_X = peak sound pressure level of curve X in dB

also, curve Z is 13 dB above curve X and curve y is 6.5 dB above X.

Figure 9.1   Peak sound pressure levels and B-duration limits for impulse noise, including the "Z-curve" (MIL-STD-1474D 1997)

Hearing loss from acoustic trauma is a significant problem for the US military. More than 11 percent of military personnel are affected with noise-induced hearing and balance disorders. This is further evidenced by the results of physical examinations given to soldiers returning from overseas deployments. Of those tested, some 28 percent have hearing loss (Schultz 2004). During fiscal year 2008, hearing-related (tinnitus or hearing loss) service-connected disabilities for US veterans receiving compensation exceeded 1,000,000 cases, with over 115,000 new cases. This represents a 5.8 percent increase over fiscal year 2007. Over 11 percent

---

daily exposure limits are five exposures for single hearing protection and 100 exposures for double hearing protection. The B-duration refers to the time from the impulse event's initiation to when it is returned to 10 percent of the peak level.

of all individual service-connected disabilities involve impairment of auditory acuity—an 8.8 percent increase over fiscal year 2007 (Department of Veterans Affairs 2008).

Threshold of Hearing

The first element of auditory SA is detection of the sound. The range of frequencies heard by humans through air conduction extends from about 20 Hz to 20 kHz (Møller and Pedersen 2004, Scharine, Henry and Binseel 2005). For low-level sounds in quiet environments, detection is dependent on a listener's hearing threshold. This threshold is defined as the minimum detectable sound level and depends on the individual's hearing sensitivity, as well as the frequency, duration, and spectral complexity of the stimulus. An individual's threshold (defined as the level for which 50 percent of instances of that stimulus are perceived) for a particular stimulus fluctuates on average 1.9 dB to 5 dB (Delany 1970, Henry et al. 2001, Robinson 1986). Among persons with "normal" hearing, intersubject standard deviations range from 3 dB to 6 dB for low and middle frequencies with the intersubject data variability increasing with stimulus frequency.

Detection is more complicated in noisy environments, because it depends on the frequency spectra of the target sound and background noise. The threshold of sound detection is affected by the extent that the energy of the signal is in the same frequency region[6] as that of the noise and is above the hearing threshold, an effect which increases linearly as the noise level increases. Using the simplest example of white noise, which contains equal amounts of energy at all frequencies, one can predict that if a 25 dB sound is just detectable in the presence of 30 dB of noise, the sound level must be increased to 45 dB to be just detectable in the presence of 50 dB of noise. Thus, rather than speak of absolute threshold levels, it is sometimes more meaningful to talk of signal-to-noise ratios (SNRs). When the sound and the noise are of equal intensity, the SNR is 1:1. Because sound levels are expressed logarithmically, a 1:1 SNR is often referred to as a SNR of 0 dB. In the previous example, an SNR of −5 dB is required to detect the sound.

Auditory Localization

We perceive the direction of sounds by processing temporal, intensity, phase, and spectral differences between the sounds reaching our left and right ears. These differences result from the interference of the sound wave with the head, pinnae,

------

6   In psychoacoustic literature, this frequency range is called the "critical band" and the physiological mechanism is called the "auditory filter." For a more detailed explanation of these concepts, please see Chapter 3 of Moore's *An Introduction to the Psychology of Hearing*, 6th Edition (2012).

and torso, a transform called the Head Related Transfer Function (HRTF), which consists of three components:

- Interaural Time Delay (ITD)—the difference in time between the sound's arrival at the closest ear and the sound's arrival at the farthest ear;
- Interaural Intensity Difference (IID)—the difference between the sound intensity in the closest ear and the sound intensity in the farthest ear, which is often shadowed by the head and the torso;
- Monaural cues from spectral filtering by the pinnae—frequency-dependent attenuation, or possibly amplification, of sound caused by sound interaction with the outer ear depending on the sound's fore/aft or up/down origin.

Binaural cues (ITDs and IIDs) are ambiguous with respect to differentiating front from rear and specifying levels of elevation. This information is thought to be conveyed through the detection of a directionally dependent spectral "notch" in complex, broadband sounds (Blauert 1996). Because the HRTF differs from person to person, it is difficult to create a generic HRTF that will accurately restore "natural hearing" for all users when wearing hearing protection or communications devices (Houtsma 2004, Chapin et al. 2003), though there are ongoing efforts to address this (McIntire et al. 2008). Localization accuracies are generally given as follows (Chapin et al. 2003, Oldfield and Parker 1984, Butler 1986):

- Just Noticeable Difference (JND) in azimuth in frontal areas is 2° to 3°;
- JND in azimuth at 90° (lateral–side) is approximately 20°;
- JND in elevation at 90° (lateral–side) is on the order of 20°.

Many hearing protection devices cover the ears or change the shape of the pinnae, which in turn changes the HRTF that a listener has grown accustomed to all his life, changing the listener's localization accuracy (Shinn-Cunningham 2000). It is possible for a user to readapt after exposure to a new HRTF, and in doing so, restore his previous localization accuracy. For example, in a study of adaptation to modified HRTFs, subjects were fitted with inserts that modified their pinnae, which they then wore continuously for 20–40 days of "training." Testing showed that the subjects were able to learn to localize with their new "ears." When the inserts were removed, all subjects still retained their original localization ability—thus demonstrating a dual adaptation state (Hofman, Riswick, and van Opstal 1998, Wightman and Kistler 1998). However, in a similar study, Russell (1977) found that listeners who wore earmuff-style devices for a three to five day period showed no adaptation to a changed HRTF. Inaccurate sound localization may be inconvenient for a civilian, but in combat it can be fatal; therefore, equipment that results in minimal changes to the soldier's HRTF is preferable.

Localization with Hearing Protection

In general, localization will be affected by anything that changes the characteristics of the individual's HTRF. As long as the sound in question is well above the threshold of hearing, the two binaural cues are relatively unaffected by hearing protection. However, monaural cues are affected by anything that changes the shape of the ear canal or the region about the ears. Thus, both over over-the-ear and in-the-ear type hearing protection can interfere with localization based on monaural cues. Earmuff-style protection still may provide inherent ILD and ITD cues, but because the pinnae are covered, spectral filtering cues may be lost, potentially causing fore/aft or up/down ambiguities. In-the-ear devices tend to perform a bit better, but can still cause significant front-to-back localization errors.

Thus, it is expected that localization would be better for earplug hearing protection, as compared to earmuffs, due to the reduced change to the profile of the outer ear. Simpson et al. (2005) tested localization ability for a visual cue paired with an auditory cue and for a visual cue alone. During the trials, measurements were made of the response time that test participants needed to estimate the location of the visual target. The auditory signal served as a "cue" for the location of the visual target. As expected, the response time was slowest for the un-cued visual-only target. During the auditory trials listeners wore nothing in or over the ears (unoccluded), earplugs, earmuffs, or a combination of both. Of the auditory trials, the response time was slowest for the double-hearing protection condition, with performance similar to that of the visual-only target trial. That is, participants were localizing on the basis of visual information alone and were made functionally deaf by the double-hearing protection. For the single-hearing protection conditions, performance was fastest for the earplugs, being only slightly slower than for the unoccluded condition. The set size of target locations was varied so that targets could come from 5, 20, or 50 possible locations, but response times showed no effect based on set size. Measurement of the extent of head movements during the visual search for the target (a measure of uncertainty in this task) showed a similar pattern; more movements were observed for the visual-only and dual-hearing protection conditions and less for the unoccluded audiovisual condition.

Similar results were found by Scharine (2005), who tested the auditory localization ability of listeners wearing each of five types (two earmuffs and three earplugs) of hearing protection/hearing restoration devices. These devices differed from those used in the Simpson et al. (2005) study in that they all included a mechanism for sound restoration in quiet or low noise conditions, so all sounds were easily detectable. There was very little difference between the earplugs—all of which achieved restoration via a microphone on the outside of the plug. One set of earplugs had independent left and right gain controls. When not set to the same gain, right–left errors resulted, suggesting the impact of disrupting the binaural IID cue. Of the two tested earmuffs, performance was worse for the device that allowed "open-ear" hearing via a "pop out" mechanism—a large odd-shaped "pinna" that created unusual spectral changes.

Both earmuffs (one with microphone restoration, the other with the "open-ear" device) impaired localization more than any of the earplugs. However, when the front-to-back errors (these are the errors primarily caused by loss of monaural cues) were removed from the data set, there was very little difference between any of the hearing protectors and all were under 10° of localization error. This suggests that the differences between earplugs and earmuffs are determined by the changes to the monaural cues.

Localization with Helmets and Helmet-Mounted Accoutrements

Much like various forms of hearing protection, helmets change the profile of the ear region and thus the nature of monaural spectral cues to localization. The previously fielded US Army helmet, the Personnel Armor System for Ground Troops (PAGST), covered most of the ear and had a hard inner surface. The currently fielded helmet is the Advanced Combat Helmet (ACH), which covers only about half of the outer ear and has a soft padded inner surface. Measurements of helmet attenuation have shown that most helmets have very little effect on the overall ability to detect sounds (Scharine, Fluitt, and Letowski 2009). However, there are measurable frequency-dependent differences, which therefore affect the monaural spectral cues used to disambiguate front from back origin. Two laboratory studies comparing the PASGT helmet and the ACH showed that these differences in monaural cues alter localization ability (Scharine 2005, Binseel et al. n.d.). There were large differences in the testing conditions used in these two studies, but the average azimuth errors were quite similar across both studies. The average azimuth error for the bare head was approximately 36°. Average azimuth errors for the PASGT helmet and the ACH were approximately 47° and 42°, respectively. In general, the pattern of errors as a function of azimuth was consistent with the interpretation that the majority of differences are due to an increase of front–back reversals for the helmets and especially so for the PASGT helmet.

Binseel et al. (n.d.) also measured elevation error and found no significant differences between the two helmets; the average error was 23°–24°, which was an increase in error of 4° over that measured with the bare head. Another study (Scharine 2009) showed that these differences are exacerbated by normal (non-anechoic) acoustic conditions and that localization ability decreases as ear coverage increases. Others have found similar results for head-mounted accoutrements (helmet, chemical/biological agent protective mask, night vision goggles, and sand/wind/dust goggles (Chapin et al. 2003). One of the prototype helmets tested by Scharine (2005) was designed with uncovered ears, but with rings around the ears. This helmet amplified sounds near 1 kHz (Scharine, Fluitt, and Letowski 2009) and increased average azimuth error by about 7° over that of the ACH. These studies suggest that it is best to leave the ears uncovered whenever possible and to avoid geometries that change the profile of the region around the ears.

Speech

The average human threshold of hearing as a function of the stimulus frequency is shown in Figure 9.2. Speech generally contains frequencies between approximately 150 Hz and 4,000 Hz, depending on the gender of the talker and the size of their vocal apparatus. Ideally, the speech signal should be at least 6 dB above the ambient noise and at least 15 dB above the listener's threshold of hearing. Recognition becomes more difficult when the spectral and temporal characteristics of the masker noise are similar to those of speech. In Figure 9.2, the solid curve shows the minimum audible sound level as a function of frequency for binaural listening as reported in the ISO standard (ISO 389-7 2005). The dashed line shows the corresponding pressure levels measured monaurally at the eardrum (Killion 1978). Differences between the two lines can be attributed to approximately 2 dB of binaural summing and frequency-dependent amplification by the pinnae between 1 kHz and 4 kHz.

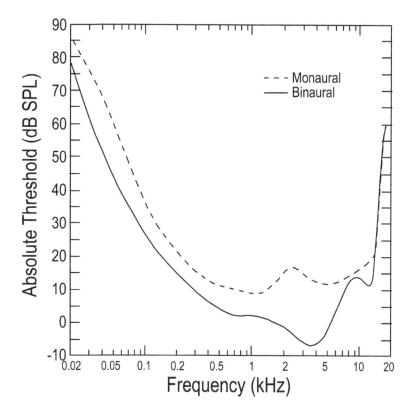

Figure 9.2   Minimum audible sound level as a function of frequency for monaural and binaural listening

It is frequently necessary to evaluate the quality of speech communication as a function of equipment transmission or environmental conditions. MIL-STD-1472G (2012) specifies speech intelligibility criteria for communications systems using three methods of scoring intelligibility—the articulation index (AI), modified rhyme test (MRT), and the phonemically balanced word test.

The MRT is the only speech measure that satisfies both military and ANSI-ASA S3.2 (2009) standards for the measurement of intelligibility over communications systems. However, it is lengthy (50, 6-word sets for a total of 300 monosyllabic words) and the words used are dissimilar to the kind of communications that occur in military operational environments. As an improvement, the US Army Research Laboratory developed a military-specific speech intelligibility measure called the Callsign Acquisition Test (CAT) (Blue and Ntuen 2003) consisting of 126 alphanumeric callsigns (constructed from combinations of the letters A–Z and the numbers 1–6 and 8; seven is not used because it has two syllables and nine is not used because it is pronounced "niner"). It has been validated through comparisons with the MRT and other phonemically balanced word lists (Gripper 2006, Gripper and Rao 2009). Since the CAT, in its longest version, contains far fewer items than the MRT, it can be administered more quickly. Perhaps more importantly, the words are standard terms used in military communications, making the test easily accepted by soldier participants. There is currently some discussion of removing the requirement for a specific speech measure from ANSI S3.2 and instead allowing the use of "operationally relevant" speech materials. If this change is made, there may be a shift toward using the CAT. Even if this shift occurs, it would be very likely that the MRT will continue to be used for some time in order to allow for comparison with previous evaluations.

At times it is desirable to evaluate the effect of the communications system on a person's ability to hear face-to-face communications. In these cases, speech measures such as the MRT and the CAT can be used. The Speech Intelligibility Index (ANSI S3.5 1997) outlines perceptual speech measures like the MRT and also describes two common predictive indices of speech intelligibility: the AI and speech transmission index (STI). Both are quick to measure and easily implemented in computer software or handheld meters. ANSI S3.5 (1997) provides information about the relationship between a particular AI or STI score and performance on common speech tests, and MIL-STD-1472G (2012) gives performance criteria in terms of both AI and MRT. Thus, a particular communications and hearing protection system can be evaluated in terms of its effect on speech communication both over the radio and in face-to-face communications.

It should be noted, however, that none of these standards can predict performance for all persons or in all contexts. Each measure is only as accurate as the degree to which the speech material used is representative of the relevant operational communications, both in size and content of vocabulary. Their use is valid primarily for comparing speech intelligibility across different communication systems. Some factors that affect speech communications are not present in the test environment; a test participant can hold their breath and remain motionless to prevent noises created by the use of chemical–biological gear from interfering with test performance. Equipment that performs well in laboratory-based intelligibility testing may perform poorly during operational use due to problems caused by incompatibility with other gear, or a system may be rejected because it is too uncomfortable for long-term wear.

Hearing Protection

There are several considerations when choosing hearing protection. For the ground soldier, hearing protection can be in either earmuff or earplug form and should have some form of hearing restoration in order to maintain auditory SA as discussed earlier. Hearing protection and restoration can be achieved through several mechanisms: passive attenuation, level dependent filters, electronic shutoffs, compression, and active noise reduction. All methods achieve different goals and have different advantages and disadvantages that will be discussed here.

Earplug hearing protection   Earplugs are a lightweight and cost-effective way of providing hearing protection and can either be passive or have active hearing restoration. If two microphones (one per ear) are used and have a common gain control, the user retains some of the binaural localization cues such as ITDs and IIDs.[7] Also, because the device is typically mounted inside the ear canal, the users retain some of the monaural spectral cues (which provide front to back disambiguation and elevation information) from pinnae filtering. Foam- or silicone-based inserts fit a wide range of users. Custom-fitted inserts can provide an even better fit, but these are more expensive. A disadvantage of earplugs is that they suffer from potential hygienic cleanliness problems so they must either be replaced frequently or cleaned and a particular set of plugs cannot be "shared" between users. Further, some adaptation is required in order to become accustomed to having one's ears occluded.

Passive earplug devices   Passive earplug devices provide uniform attenuation of external sounds with a Noise Reduction Rating averaging approximately 25 dB. These are lightweight and inexpensive. They come either in "one size fits all" or in a limited range of sizes. They can be reusable (e.g., silicone triple flange) or disposable (e.g., expanding foam), and are quite inexpensive.

Level-dependent mechanical earplug devices   Some of the more sophisticated hearing protective devices are passive, level-dependent devices that make use of special baffling or small holes in the device to allow normal conversational

---

7   In devices with electronic talk-through microphones, level differences will be maintained as long as the gain is set the same in both ears or controlled jointly. If the gain is set independently in each ear, binaural difference cues will be altered.

sounds to pass through with minimal attenuation. In the presence of high-level impulse noise, such as an explosion or gunfire, the shockwave creates turbulent airflow, which reduces the magnitude of the sound to a more acceptable level. While more expensive than simple passive linear earplugs, they are considerably less expensive than electronic devices, usually costing less than US$10 per pair.

Electronic earplugs with talk-through microphones   Analogous to commercially available hearing aids, electronic earplugs are placed independently in each ear, with a powered microphone/speaker in each unit. The microphone volume is controlled by level-dependent circuitry, which limits or compresses the impulse sound to a safer level. This way, the wearer can conduct a normal conversation with little or no attenuation, but when the circuitry senses the onset of a loud impulse, the wearer is protected. Another benefit is that it is possible to amplify external sounds, thereby extending the user's hearing distance. Since the device is located within the ear, it tends to have less impact on localization than over-the-ear-type hearing protection.

Earmuff-type hearing protection   Circumaural earmuff-style hearing protection differs from in-canal protection in that it covers the entire ear. Most standard ear cups provide good attenuation for frequencies above 500 Hz (Berger 2002). Low frequency attenuation is a function of the volume of the ear cup; consequently, large ear cups are required to achieve greater low frequency attenuation. However, earmuff attenuation is reduced whenever the seal around the ears is broken, and this can occur due to physical activity, perspiration, or wearing eyeglasses. Earmuff-style hearing protection can also be uncomfortably hot in warm weather and can interfere with other gear such as helmets and protective masks.

Level-dependent protection can be provided by equipping the earmuffs with talk-through microphones. Combined with some form of compression or shut-off feature, it can protect the user against sounds that exceed a certain threshold (typically 85 dB–90 dB). A variable gain feature can also be provided that allows the user to set the amplification at their preferred level, perhaps enhancing their perception of sounds. However, because increasing the gain of the talk-through channels increases the level of not only the target sounds but also the ambient noise, most users prefer to set the gain to a simple unity gain restoration of natural sound (Scharine, Henry, and Binseel 2005).

Talk-through microphones can provide some degree of localization, as long as the left and right channels are transmitted independently to their respective ears. However, because the earmuff, unlike the earplug, completely covers the ear, the wearer loses the fine monaural cues that provide fore/aft and up/down information.

Active Noise Reduction (ANR)

While passive hearing protection is effective, ANR can provide a measure of increased protection for high levels of ambient steady-state noise, especially at lower frequencies. ANR works by the superposition principle, where noise is actively cancelled by introducing "anti-noise" that is 180° out of phase, causing destructive interference with the original (Moy 2001).

ANR is not effective over the entire human auditory range because as the frequency content of the threat sound increases, so does the anti-noise frequency. The frequency range for which ANR is effective has an upper limit of approximately 500 Hz for earmuffs (Berger 2002) and 2 to 3 kHz for earplugs (Adaptive Technologies, Inc. 2007). These limits are due to the lag in creating the anti-noise, because the generation system required in earmuffs cannot effectively keep up with the incident sounds at frequencies above 500 Hz. Smaller systems can be used in earplugs, but these still lag the incident sound at frequencies between 2 kHz and 3 kHz. Above these limits, it is necessary to rely on the passive noise reduction capabilities of the protective device, recalling that the size of the ear cup determines the effective lower limit on frequency attenuation. However, since ANR provides protection at the lower frequencies, it allows a significant reduction in the size of the ear cup.

Communications Systems

Communication capability can be provided with a headset that receives and transmits sound via either air conduction or bone conduction. Because hearing protection is required for many military operations, communications headsets must be considered in conjunction with both hearing protection and hearing restoration, and thus are discussed here as "communications systems." Like hearing protection, traditional air-conduction headsets can be in either earmuff or earplug form. Hearing restoration is usually incorporated within the electronics of the headset, and achieved through the use of talk-through microphones that pass ambient sounds to the listener through the communications loudspeaker. Sounds at dangerous levels are prevented from damaging the soldier's hearing by electronic compression or shutoffs. On the other hand, bone-conduction headsets have vibrators placed on the head, but not over the ears, so any form of hearing protection can be used as needed. Depending on which headset is chosen, it may be necessary to augment it with earplugs for hearing protection. If the loudspeaker needs to be located in the ear canal, it is difficult to comfortably provide hearing protection while maintaining the ability to listen to face-to-face communications. This combination of communications equipment and hearing protection also may adversely impact the user's sound localization capability (Figure 9.3).

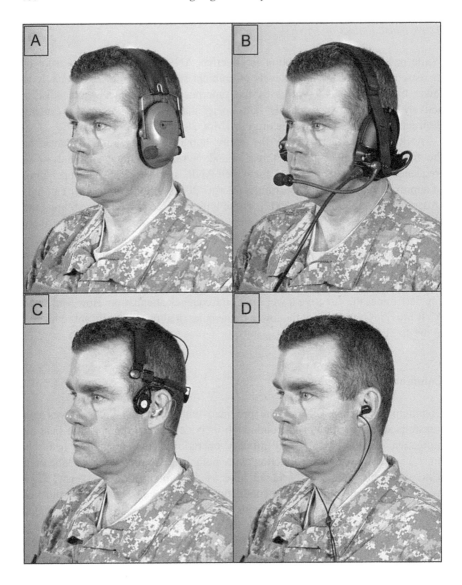

Figure 9.3   Assorted communications and hearing protection systems

A description of the various communications and hearing protection systems shown in Figure 9.3 follows:

   A. Passive earmuff system providing talk-through restoration of sounds less than 82 dB A-weighted—shown without the boom microphone and communications cable.

B. Earmuff system providing both passive and active attenuation, and talk-through restoration (<82 dB A-weighted).

C. Earplug/bone conduction combination: Non-linear earplugs provide passive attenuation of sounds above approximately 115 dB A-weighted. The bone conduction headset provides communications via vibrators located on the temporal bone and a contact microphone located on top of the head.

D. Earplug communications headset providing both passive and ANR hearing protection. Talk-through hearing restoration is provided via microphones located on the outside of the earplugs and small contact microphones inside the ears provide communications over the radio.

Earmuff Communication System

When the communications loudspeaker is located inside of circumaural earmuffs, the passive protection of the muff shields the listener from ambient noise, thereby increasing the SNR of the communications channel. For military and aviation environments where low frequency noise is abundant, passive attenuation can be augmented with ANR (Berger 2002).

Earplug Communication System

If a communications loudspeaker is placed on the inside of an earplug, it too can provide good speech intelligibility combined with good hearing protection. Hearing restoration is achieved through the use of an external microphone for talk-through. Because the earplug is located within the ear canal—which has less of an effect on monaural cues—sound localization ability is improved over that of the earmuff design. As with earplug-style hearing protection, earplug-based communications systems have potential hygienic cleanliness problems and comfort issues. Foam-based inserts fit a wide range of users and can be replaced easily. Custom-fitted inserts are usually more comfortable but are more expensive and require more extensive logistical support in terms of having trained technicians available to properly fit users.

Air Conduction Microphones

In most commercially available communications systems, whether earmuff or earplug style, an air conduction microphone is placed on a boom that can be adjusted to be in front or to the side of the mouth. Noise cancellation can be implemented to reduce the impact of ambient noise. Although these function reasonably well, they can be impractical because they move whenever bumped, so the user must frequently readjust the microphone position in order to maintain communications. Furthermore, they do not work well with chemical-biological protection, so a separate microphone located under the protective ensemble must be used.

Bone Conduction System

A bone conduction system works through the placement of small vibrators on the surface of the head. These transmit information to the user by vibrating the skull. (For a complete discussion of bone conduction, see Henry and Letowski (2007)). Because these vibrators are placed directly on the head and not over the ears, the ears remain unoccluded and, therefore, auditory SA is unaffected. If hearing protection is required, it can be chosen independently of the headset, allowing for greater individual choice on the part of the user. Hearing protection is crucial to the function of bone conduction in noisy environments because it creates a resonant chamber within the ear canal that amplifies the transmitted signal, increasing the loudness of the signal (Dirks and Swindeman 1967, Fagelson and Martin 1998). Similarly, a contact microphone can be placed on the head for transmission of speech. Because the microphone only transmits bone-conducted sound, it has the added benefit of not requiring noise cancellation to remove background noise from the transmitted signal.

Bone conduction is a relatively new technology, so—as of this writing—there are few current examples or bone-conduction system evaluations. One reason for this is that it has been difficult to demonstrate or test the full functionality of bone conduction, because it is difficult to assemble a single system with optimal head mounting, good output/input sensitivity, ruggedness, and compatibility with military radios. None of these issues are insurmountable; however, they must be resolved before bone conduction becomes a commonly accepted form of communications headset for military applications (Henry, Letowski, and Tran 2009). Additional issues associated with bone-conduction headsets include:

- Bone conduction requires good physical contact with the user's head to ensure good speech intelligibility (McBride, Letowski, and Tran 2008). Differences in head shape require easily adjusted devices in order to conform to the individual user's head while not interfering with the wearing of a helmet.
- In low levels of noise (80 dB), performance with bone conduction headsets was similar to performance with circumaural air-conduction headsets (Scharine, Henry, and Binseel 2005). There has been concern about the output of some of the commercially available bone-conduction vibrators in regard to their usefulness in high-noise environments. Henry and Mermagen (2004) found that listeners using bone conduction paired with hearing protection achieved levels of 85–90 percent correct on speech intelligibility measures when exposed to steady-state vehicle noise at 102 dB (riding in the back of a M113 armored personnel carrier). Although 90 percent accuracy is lower than the 95 percent level found for the combat vehicle crewman (CVC) helmet used in many armored vehicles, the CVC is designed for and used only in the noisiest tracked vehicles.
- Vibrator coupling to the head is also affected by its location on the head. Until recently it was unclear which locations were optimal. Work by

McBride, Letowski, and Tran (2008) investigated the signal strength and intelligibility of various vibrator locations on the head and found the best performances when the vibrator was placed on the condyle, mastoid, or temple bone.

- There are trade-offs between sensitivity and intelligibility for microphone placement; the locations where the contact microphone was most sensitive were not necessarily where speech intelligibility was greatest (Tran, Letowski, and McBride 2008). Although the condyle and mastoid were most sensitive, the ideal location for placement of a contact microphone was on the forehead, where the signal was slightly less intense, but provided the greatest speech intelligibility.

3-D Audio and Acoustically Transparent Helmets—Future Direction

When the soldier is wearing truly effective hearing protection or is fully encapsulated for protection against chemical or biological threats, hearing restoration is critical to maintaining auditory SA. Research has shown that spatial or three-dimensional (3-D) audio can dramatically improve safety and performance and decrease workload by superimposing geospatial directionality on radio communications and using the audio cues redundantly with visual cueing to direct the user's attention for alerts and warnings (see Bolia 2004 for an excellent review). The benefit of spatial audio displays is to increase SA and decrease workload by decreasing audio clutter, providing an intuitive spatial location for warnings and alerts, and redundantly coding external threats and navigational waypoints as an audio cue to direct visual attention. This is especially true when used in conjunction with a visual display, because 3-D audio cueing reduces search time and improves SA for the user (Bolia, D'Angelo, and McKinley 1999, Flanagan et al. 1998, Houtsma 2004). This technique has been called a "transparent helmet," because the helmet has been made acoustically transparent by virtue of the 3-D hearing restoration. Such an advantage would seem natural for a soldier, though much improvement is needed with regard to spatialization protocols.[8]

In addition to providing cues for the location of visual events, spatial hearing provides the listener with a major advantage for discriminating sounds in the presence of noise. Providing a spatial separation between audio sources of interest and interfering noise improves the listener's ability to detect and understand the audio content, much like the so-called Cocktail Party effect, where we can listen to specific conversations within a crowded room simply by taking advantage of binaural differences arriving at each of the ears and then using those differences to selectively attend to a specific source (Cherry 1953). Similarly, spatial hearing

---

8   There is no agreed standard for assigning radio channels or warnings to either relative or absolute geospatial locations, nor is there is a standardized set of non-speech audio warnings and alerts.

improves the understanding of speech when there are competing sources such as multiple talkers for two reasons:

- First, there is a "better ear" advantage, meaning that the target-to-masker ratio will be higher than average in one ear. Thus, the listener can attend to that ear and ignore the other. While most evident at higher audio frequencies, up to a 25 dB in target-to-masker ratio can be found with mid-frequency speech from a nearby talker, and this can result in a huge improvement in comprehension (Shinn-Cunningham 2003).
- Second, the human auditory system uses localization cues as a method of parsing the auditory scene into source objects (see Bregman 1994). This allows a listener to use the spatial cues inherent in the signal to selectively attend to the target audio source and ignore the competing masking source.

This facilitative effect of spatial separation of competing speech streams is significant. Abouchacra et al. (2009) found that simply presenting half of the competing streams of speech information to one ear and the remainder to the other greatly reduces the masking effect of competing talkers. Further dividing the information, such that the target speech occupies a unique location in space, increases intelligibility by approximately 5 to 10 percent (Abouchacra et al. 2009). Another simple demonstration of increased speech recognition due to spatial separation was described by Shilling et al. (2003) in a study of the identification of target messages in a multi-talker speech environment. They positioned the talkers at −135°, −45°, +45°, and +135° relative to forward-facing. Identification accuracy was 43 percent when competing talkers were spatially segregated, versus only 17 percent with no spatial segregation. Thus, a simple spatial segregation of multiple talkers more than doubles performance. Ericson and McKinley (1997) describe a series of experiments that investigate the effectiveness of binaural, spatial hearing to segregate a target message as compared to other factors such as quality of speech, gender, level, and attention. While the degree of facilitation varies depending on the contextual features, binaural hearing provides a relatively large advantage in low speech-to-noise conditions.

While spatial hearing can provide advantages, there are costs such as additional processing, power, and—ultimately—weight. For vehicle-mounted soldiers, command center personnel, or airmen, the technology can be implemented as a part of the existing communications and display hardware. However, for the dismounted soldier who must carry all his equipment, there are more benefits from his own binaural—real ear—hearing. The current technology for head orientation trackers and signal processing has not reached the maturity necessary to provide the instant auditory SA needed for survivability. This may seem implausible given the apparent fidelity of many 3-D audio systems on the market. However, although it is simple to create sounds that appear to come from different locations, it is difficult to map them to the correct location given the need for accurate tracking, processing, and individualized HTRFs. Research and development efforts continue in an effort to provide natural hearing to those who must be completely encapsulated by chemical–biological gear and ballistic protection.

Summary: Evaluating and Choosing Communications and Hearing Protection

From an examination of the results of numerous studies, it is very clear that providing adequate communications while simultaneously protecting the soldier's hearing from impulse and steady-state noise is not a simple task. Sound attenuation is necessary to protect the soldier's hearing in the short term as well as the long term. The soldier must be able to hear and communicate clearly in the presence of high-level impulse and steady-state noise and must be able to accurately localize sounds in the environment. Generally, it is felt that an over-the-ear (occluding) device is not appropriate because of weight, bulk, and comfort issues, a conclusion supported in several studies cited herein. It is likely that the solution to protecting soldiers' hearing while providing good SA will be an integration of the positive aspects of passive and active communications and hearing protection systems. Much work remains to reach the ultimate goal of providing effective communications and hearing protection, while preserving good auditory localization.

References

Abouchacra, K.S., Letowski, T.R., Besing, J. and Koehnke, J. 2009. Clinical application of the synchronized sentence set (S3). *Proceedings of the 16th International Congress on Sound and Vibration*, 1–8.

Adaptive Technologies, Inc. 2007. *Patent #11/122739 Electronic Earplug for Monitoring and Reducing Wideband Noise at the Tympanic Membrane.* Blacksburg, VA: US Patent Office.

AFOSH 48-20 2006. *Occupational Noise and Hearing Conservation Program.* Air Force Occupational Safety and Health Standard. Washington, DC: Department of the Air Force.

Amrein, B. E. and Letowski, T. R. 2012. *High Level Impulse Sounds and Human Hearing: Standards, Physiology, Quantification.* (Technical Report ARL-TR-6017). Aberdeen Proving Ground, MD: US Army Research Laboratory.

ANSI-ASA S3.2. 2009. *Method for Measuring the Intelligibility of Speech Over Communication Systems.* New York: American National Standards Institute.

ANSI S3.5. 1997. *Methods for the Calculation of the Speech Intelligibility Index.* (R2007). New York: American National Standards Institute.

Berger, E.H. 2002. Active noise reduction (ANR) in hearing protection: Does it make sense for industrial applications? Paper presented at the 27th Conference of the National Hearing Conservation Association, Dallas, TX, February 2002,

Spectrum Supplement 1, 20, 10. Available at: http://www.aearo.com/pdf/hearingcons/anr.pdf (accessed: 27 July, 2012).

Binseel, M.S., Scharine, A.A., Mermagen, T. and Letowski, T.R. n.d. (Unpublished laboratory data). *Sound Localization Ability of Soldiers Wearing Infantry Helmets*. Aberdeen Proving Ground, MD: US Army Research Laboratory.

Blauert, J. 1996. Spatial hearing, in *The Psychophysics of Human Sound Localization* by J. Blauert. Revised Edition. Cambridge, MA: MIT Press.

Blue, M. and Ntuen, C.A. 2003. Performance intensity (PI) function of Callsign Acquisition Test (CAT). *Proceedings of the Industrial Engineering Research Conference*, Oregon, 2003, 1–4.

Bolia, R.S. 2004. Spatial audio displays for military aviation. *International Journal of Aviation Psychology*, 14, 233–38. Special issue.

Bolia, R.S., D'Angelo, W.R. and McKinley, R.L. 1999. Aurally aided visual search in three-dimensional space. *Human Factors*, 41, 664–69.

Bregman, A.S. 1994. *Auditory Scene Analysis: The Perceptual Organization of Sound*. Cambridge, MA: MIT Press.

Butler, R.A. 1986. The bandwidth effect on monaural and binaural localization. *Hearing Research*, 21, 67–73.

Chapin, W.L., Jost., A.R., Cook, B.A., Cook, B.A., Surucu, F., Foster, S., Bolas, M., McDowall, I., Lorimer, E.R., Zurek, P.M., Desloge. J.G., Beaudoin, R.E., Shinn-Cunningham, B. and Durlach, N. 2003. *Concept and Technology Exploration for Transparent Hearing*. Project Final Report, Scorpion Audio Team.

Cherry, E.C. 1953. Some experiments on the recognition of speech, with one and with two ears. *Journal of the Acoustic Society of America*, 25, 975–79.

DA Pamphlet 40–501. 1998. *Medical Services: Hearing Conservation Program*. Washington, DC: Department of the Army.

Delany, M.E. 1970. *On the Stability of Auditory Threshold*. NPL Aero Report Ac44. London, GB: National Physical Laboratory.

Department of the Navy 2000. *Hearing Conservation Program*. Washington, DC: United States Marine Corps.

Department of Veterans Affairs 2008. *Annual Benefits Report—Fiscal Year 2008*. Available at: http://www.vba.va.gov/REPORTS/abr/2008_abr.pdf (accessed: 27 July, 2012).

Dirks, D. and Swindeman, J. 1967. The variability of occluded and unoccluded bone-conduction thresholds. *Journal of Speech and Hearing Research*, 10, 232–49.

Dunlap, D. 2002. *Noise Reduction for Improved Communications*. (Armed Forces International). Available at: http://www.armedforces-int.com/article/noise-reduction-for-improved-communications.html (accessed: 27 July, 2012).

Endsley, M. R. 1995. Toward a theory of situation awareness in dynamic systems. *Human Factors*, 37, 32–64.

Ericson, M.A. and McKinley, R.L. 1997. The intelligibility of multiple talkers separated spatially in noise, in *Binaural and Spatial Hearing in Real and*

Virtual Environments, edited by R. Gilkey and T.R. Anderson. Mahwah, NJ: Lawrence Erlbaum Associates, 701–724.

Fagelson, M. and Martin, F. 1998. The occlusion effect and ear canal sound pressure level. *American Journal of Audiology*, 7, 50–54.

Flanagan, P., McAnally, K.I., Martin, R.L., Meehan, J.W. and Oldfield, S.R. 1998. Aurally and visually guided visual search in a virtual environment. *Human Factors*, 40, 461–68.

Gripper, M. A. 2006. *Evaluations of the Callsign Acquisition Test (CAT) in acoustic environments*. Unpublished Dissertation. Greensboro, NC.

Gripper, M.A. and Rao, M.D. 2009. A speech intelligibility test for military applications: Callsign acquisition test. *Proceedings of the 15th International Conference on Industry, Engineering and Management Systems*, Cocoa Beach, FL, March 2009, 1–9.

Hatfield, J.L. and Gasaway, D.C. 1963. *Noise Problems Associated with the Operation of US Army Aircraft*. (USAARL Report 63-1). Fort Rucker, AL: US Army Aeromedical Research Unit.

Henry, J.A., Flick, C.L., Gilberty, A., Ellingson, R.M. and Fausti, S.A. 2001. Reliability of hearing thresholds: Computer-automated testing with ER-4B canal phone™ earphones. *Journal of Rehabilitation Research and Development*, 38, 1–18.

Henry, P.P. and Letowski, T.R. 2007. *Bone Conduction: Anatomy, Physiology, and Communication*. (Technical Report ARL-TR-4138). Aberdeen Proving Ground, MD: US Army Research Laboratory.

Henry, P.P. and Mermagen, T. 2004. Bone conduction communication in a military vehicle. *Proceedings of the NATO Vehicle Habitability Conference*, Prague, October 2004, 14-1–10.

Henry, P.P., Letowski, T.R. and Tran, P. 2009. *Comparison of Bone-Conduction Technologies*. (Technical Report ARL-TR-4705). Aberdeen Proving Ground, MD: US Army Research Laboratory.

Hofman, P.M., Riswick, J.G. and van Opstal, A.J. 1998. Relearning sound localization with new ears. *Nature Neuroscience*, 1, 417–21.

Houtsma, A.J. 2004. Spatial hearing, hearing impairments, and hearing protection. *US Army Medical Department Journal*, April–June, 30.

ISO 389-7. 2005. *Acoustics—Reference Zero for the Calibration of Audiometric Equipment. Part 7: Reference Threshold of Hearing Under Free-field and Diffuse-field Listening Conditions*. Geneva: International Organization for Standardization.

Kardous, C.A., Willson, R.D., Hayden, C.S., Szlapa, P., Murphy, W.J. and Reeves, E.R. 2003. Noise exposure assessment and abatement strategies at an indoor firing range. *Applied Occupational and Environmental Hygiene*, 18, 629–36.

Killion, M.C. 1978. Revised estimate of minimal audible pressure: Where is the "missing 6 dB"?, *Journal of the Acoustical Society of America*, 63, 1501–1510.

McBride, M., Letowski, T.R. and Tran, P. 2008. Bone conduction reception: Head sensitivity mapping. *Ergonomics*, 51, 702–718.

McIntire, J.P., Havig, P.R., Watamaniuk, S.N.J. and Gilkey, R.H. 2008. Aurally aided visual search performance in a dynamic environment. *Proceedings of the SPIE: Head- and Helmet-Mounted Displays XIII, Design and Applications*, 6955.

MIL-STD-1474D 1997. *Design Criteria, Noise Limits*. Washington, DC: Department of Defense.

MIL-STD-1472G 2012. *Human Engineering*. Washington, DC: Department of Defense.

Møller, H. and Pedersen, C. S. 2004. Hearing at low and infrasonic frequencies. *Noise Health*, 6, 37–57.

Moore, B.C.J., 2012, *An Introduction to the Psychology of Hearing*. Emerald: Cambridge, U.K.

Moy, C. 2001. *Active Noise Reduction Headphone Systems*. [Online: HeadWize-Technical Papers]. Available at: http://www.headwize.com/tech/anr_tech.htm (accessed: July 27, 2012).

National Research Council 1995. *Human Factors Tactical Display for the Individual Soldiers. Panel on Human Factors in the Design of Tactical Display Systems for the Individual Soldier*. [Online: The National Academies Press]. Available at: http://www.nap.edu/catalog.php?record_id=9107 (accessed: July 27, 2012).

NATO RTO 2010. *Hearing Protection– Needs, Technologies, and Performance*. (Technical Report TR-HFM-147). NATO: Neuilly-Sur-Seine, France.

NIDCD Information Clearinghouse 2011. *Noise-Induced Hearing Loss*. [Online: National Institute on Deafness and Other Communication Disorders]. Available at: http://www.nidcd.nih.gov/health/hearing/pages /noise.aspx (accessed: July 27, 2012).

NIOSH 1998. US Department of Health and Human Services, Criteria for a Recommended Standard, Occupational Noise Exposure, Revised Criteria 1998, Chapter 3: Basis for the exposure standard. [Online]. Available at: http://www.nonoise.org/hearing/criteria/criteria.htm#chap3 (accessed: 27 July, 2012).

Oldfield, S.R. and Parker, S.P.A. 1984. Acuity of sound localisation: A topography of auditory space. I. Normal hearing conditions. *Perception*, 13, 581–600.

Price, G.R. 2005. *Weapon Noise Exposure of the Human Ear Analyzed with the AHAAH Model*. [Online]. Available at: http://www.arl.army.mil/www/default.cfm?page=351 (accessed: July 27, 2012).

Ribera, J. E., Mozo, B. T., Mason, K. T. and Murphy, B. A. 1995. *Communication and Noise Hazard Survey of CH-47D Crewmembers*. (USAARL Technical Report 96-02). = Ft. Rucker, AL: US Army Aeromedical Research Laboratory.

Robinson, D.W. 1986. Sources of variability in normal hearing sensitivity. *Proceedings of the 12th International Congress on Acoustics*, Toronto, July 1986, B11–1.

Russell, G. 1977. Limits to behavioral compensation for auditory localization in earmuff listening conditions. *Journal of the Acoustical Society of America*, 61, 219–20.

Scharine, A.A. 2005. The impact of helmet design on sound detection and localization. *Journal of the Acoustical Society of America*, 117, 2561.

——. 2009. *Degradation of Auditory Localization Performance Due to Helmet Ear Coverage: The Effects of Normal Acoustic Reverberation.* (Technical Report ARL-TR-4879). Aberdeen Proving Ground, MD: US Army Research Laboratory.

Scharine, A.A., Fluitt, K. and Letowski, T.R. 2009. The effects of helmet shape on directional attenuation of sound. *Proceedings of The Sixteenth International Congress on Sound and Vibration,* Krakov, Poland, July 2009.

Scharine, A.A., Henry, P.P. and Binseel, M.S. 2005. *An Evaluation of Selected Communications Assemblies and Hearing Protection Systems: A Field Study Conducted for the Future Force Warrior Integrated Headgear Integrated Process Team.* (Technical Report ARL-TR-34). Aberdeen Proving Ground, MD: US Army Research Laboratory.

Scharine, A.A., Letowski, T.R. and Sampson, J.B. 2009. Auditory situation awareness in urban operations. *Journal of Military and Strategic Studies*, 11, 1–28.

Schultz, T.Y. 2004. Troops return with alarming rates of hearing loss. *Hearing Health*, 20(3), 18–21.

Shilling, R., Morgan, D., Mosbruger, M., Beilstein, D. and Orichel, T. 2003. Enhancing information fusion using spatial auditory displays and videogame interfaces. Abstract for Office of Naval Research (ONR) Workshop on Attention, Perception and Modeling for Complex Displays, Rensselaer Polytechnic Institute, Troy, NY, June 2003.

Shinn-Cunningham, B.G. 2000. Adapting to remapped auditory localization cues: A decision-theory model. *Perception and Psychophysics*, 62, 33–47.

——. 2003. Spatial hearing advantages in everyday environments. *Proceedings of the ONR Workshop on Attention, Perception and Modeling of Complex Displays*, Troy, NY, June 2003.

Simpson, B.D., Bolia, R.S., McKinley, R.L. and Brungart, D.S. 2005. The impact of hearing protection on sound localization and orienting behavior. *Human Factors*, 47, 188–98.

Tran, P., Letowski, T.R. and McBride, M. 2008. Bone conduction microphone: Head sensitivity mapping for speech intelligibility and sound quality. *Proceedings of the International Conference on Audio, Language and Image Processing*, Shanghai, China, July 2008, 107–111.

Wightman, F. and Kistler, D. 1998. Of vulcan ears, human ears and "earprints." *Nature Neuroscience*, 1, 337–39.

Chapter 10

Human Factors in Military Learning Environments

Valerie J. Berg Rice and Petra E. Alfred

I never teach my pupils; I only attempt to provide the conditions in which they can learn.

Albert Einstein (1879–1955)

Design is a plan for arranging elements in such a way as best to accomplish a particular purpose.

Charles Eames (1907–1978)

Human factors engineering focuses on the design of products, places, and processes to fit human capabilities and confines.[1] Human factors engineers work in diverse settings including industry, transportation, hospitality, the military, health care and, more recently, education (Alfred et al. 2009, Rice 2009). As the quotations from Einstein and Eames (above) indicate, human factors engineers can impact the education or training environment through design to best accomplish the desired purpose—learning. The human factors profession and its practitioners are essential to the engineering process, as each fresh invention, new technology, and discovery about the physical, cognitive, and emotional being of the human compels the human factors engineer to consider design issues and implications. Thus, the profession of human factors continues to grow and mature, as our knowledge of humans and our awareness of the impact of elements and objects on human functioning develops, and our creation of new products and processes used by humans matures. This knowledge of human performance and of human interaction with the environment can be used to impact individual learning.

Learning requires the integrated use of the brain and body, to include breathing, posture, movement, thinking, and even sleep, to optimize learning and retention. There are numerous opportunities in the fields of education and training to improve effectiveness, efficiency, and safety through applied human factors research and design (Table 10.1). It is likely that all areas of interest could be investigated, and perhaps improved, through the proper application of human factors principles. The purpose of this chapter is to enlighten the reader to a few areas of military

---

1 "Human factors engineering" and "ergonomics" are considered synonymous terms throughout this chapter. "Human factors engineer" and "ergonomist" are also considered interchangeable terms.

education and training, hence learning and retention, in which human factors principles and techniques might be applied and to provide examples illustrating the concepts. Chapanis (1991) defines human factors as follows:

> Human Factors/Ergonomics is a body of knowledge about human abilities, human limitations, and other human characteristics that are relevant to design. Human Factors Engineering (ergonomics implementation) is the application of human factors information to the design of tools, machines, systems, tasks, jobs, and environments for safe, comfortable, and effective human use.

Table 10.1   Potential human factors engineering issues in military education and training

Products	Places	Processes/Procedures

• Chairs
• Computers and computer interfaces
• Desks
• Software
• Student training materials
  − Textbooks
  − Instruction manuals
• Websites
	• Cafeterias
	• Classrooms
	• Hallways and meeting rooms
	• Lecture halls
	• Offices and office space
	• Outdoor campus environment
	• Physical fitness training environments: track and field, playgrounds
	• Simulated environments
		• Billing
		• Evaluation methods
		• Gaming instruction
		• Instructor selection
		• Methods to prevent attrition
		• Student remediation programs
		• Student selection
		• Student training
		• Teaching techniques
		• Test procedures
		• Training methods
		• Transfer of training
		• University level evaluations (accreditation)

The processes employed by human factors engineers are to *analyze, design,* and *test and evaluate*. During the initial analysis, the goal is to define and understand the problem. While this sounds straightforward, it is often fraught with complexity, and an incomplete analysis will result in an inadequate design solution. Extreme care must be taken to identify the problem or problems to be solved and make it very clear which problems an intervention will target, as individuals will differ in their conception of which issue should take priority. Once the analysis is complete, a potential design solution is identified. The design can be unique or an alteration of an existing design. The design can apply to a product such as a machine, a tool, or technology; an environment or setting, such as a classroom, a recreation facility or a hallway; or a procedure or process, such as student intake procedures, the processing of grades, or student programs of instruction. Design can occur at a micro or a macro level. "Microergonomics" is the term generally used to describe the development of a particular product or process, while "macroergonomics" is the evaluation and design (or redesign) of a system (or systems)—considering the impact of each system on each other, as well as the impact of the systems on individuals within the system (and vice versa). After the analysis and the identification of potential design solution, the third step in the process is to test and evaluate the design solution. The design solution is put into place and evaluated according to predetermined criteria, including effectiveness, safety, impact, and ease of use. The process is iterative, as areas of undetected difficulty are often found during the test and evaluation process. This is particularly true in a macroergonomic process evaluation.

Tools and Techniques

Numerous human factors tools and techniques exist, and several are useful for performing analysis and assessment in academic settings. They include interviews, focus groups, surveys, and usability testing; these of course are all supported by detailed literature reviews and applied research. Of these techniques, conducting interviews and focus groups are generally not included in human factors engineering curriculums (HFES 2010). However, these are crucial skills, as a richness of detail can be gained that is not available through other means. In attrition research, for example, dissimilar explanations for academic difficulty can be identified by different assessment techniques when working with faculty, students, management, and support staff. Table 10.2 shows that target groups in a military education environment have different perceptions of the issues that influence academic attrition. Each assessment technique lends itself more easily to some responses over others (although all responses are correct). Table 10.2 also shows the top factors identified as contributing to academic attrition according to a focus group with commanders, supervisors, and instructors; interviews with students who failed the course; and a written survey of students attending the course.

Usability testing applies to much more than software. In military healthcare academic settings, usability projects can range from evaluating the design of PowerPoint slide presentations to websites, from the development of written instruction manuals to game-based instruction, and from choosing between lecture-based and blended learning modes of instruction to determining whether training should be performed with living humans or simulated. For example, within the military healthcare environment, the completion of continuing education training is imperative in a number of topic areas. Usability-testing of continuing education programs with the target audience can help guide product development so it "resonates with" the audience, influencing them to think about the content and apply the information in their work. Such testing also helps to ensure the materials and mode of presentation fit the target audience in terms of their preferences, comprehension, and acceptance of the material. To demonstrate, Figure 10.1 shows the responses of drill sergeants to five sequential computerized continuing education training modules on musculoskeletal injury prevention, and their perception of the utility of this information.

Table 10.2   Identification of issues influencing attrition using different assessment techniques and target groups

Focus group with commanders, supervisors and instructors	Interviews with failed students (soldiers)	Survey of students (Soldiers)
• Behavioral issues • Cognitive ability • Inability to attend, concentrate and follow through • Life skills (time management, study skills, etc.) • Life stage/maturity • Medical issues (illnesses, injuries) • Self-motivation • Stress	• Course material • Drill sergeants • Family issues • Instructors(differences in presentation styles and requirements) • Recruiters (misrepresentation of course) • Sleep (too little) • Support systems (lack of...) • Time/fast pace of course	• Continuance commitment • Education level • Overcoming hardships prior to active duty • Parents divorced • Physical fitness • Self-efficacy specific to current training

Increased Ability to make Improvements in IP

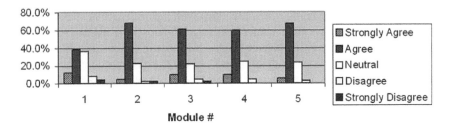

Will use Information in the future

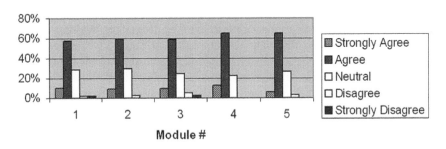

Figure 10.1   Drill sergeant usability test responses to a computerized musculoskeletal continuing education injury prevention program

While other tools with specific applications are useful, skills in these basic techniques are paramount. In addition, all ergonomic work must be participatory in nature, and partnering can occur with faculty, staff, and students, as each group will have a distinct perspective and unique concerns (DeVilbiss et al. 2010, Rice, Pekarek et al. 2002). Other human factors techniques employed in educational applications include job or workload analysis, flow charts, judgmental rating scales, training assessment, anthropometric and biomechanic analysis, videotaping, modeling, simulations, and participatory ergonomics.

Identifying the Target Audience

In an educational setting, there are a number of target audiences. They include students, instructors, management, administrative and custodial support staff, and the employers of graduates. The student population can include those who are applying for entry into the programs, resident students, off-site interns, distance-learning students, and those who attend continuing education courses on- or off-site. In most large-scale learning environments with high academic demands, such as medical military occupational specialty (MOS) training, one of the foremost concerns is attrition among resident students. This is the focus of a line-of-study for human factors engineers and faculty members at the US Army Medical Department Center and School, with particular attention to the largest training contingent, the 68W Health Care Specialist Advanced Individual Training (AIT) course. A similar diversity exists among the instructor, management, administrative staff, and employers.

Regardless of the target audience on which a project focuses, most assessments and interventions will impact each of the target groups, thus necessitating a systems approach. For example, when personal factors are identified for students who are at risk of failing a course, it is the instructors, managers, and administrative support who must implement the requisite changes. In one study, fear of failure was found to be associated with grade point average for health-care specialists (combat medics); however, those who scored high in both fear of failure and questions about motivation and confidence had the highest pass rates, demonstrating the powerful concepts of failure avoidance and success orientation (Rice et al., 2009). A number of suggestions arose from the study for assisting students whose motivation is to avoid failure, one of which involved instruction in coping skills. Instructors will execute the majority of the suggestions (after receiving instruction themselves), with the time and resources provided by management. Administrative and custodial staff will contribute by providing scheduling, paperwork, handouts, and cleaning services for the extra instruction. Students and staff will benefit from the knowledge gained, as well as the assumed higher pass rates, and future employers or supervisors will gain workers with a greater sense of self-regulation. Thus, nearly every target audience in the education and training environment will be affected by the human factors evaluation and design solutions.

The primary outcome of a successful educational program is a competent graduate. While grades are considered indicative of a person's learning and retention, and pass rates would seem to measure the ability of instructors to teach the educational material, the real test of a successful education and training program lies in the job performance of its graduates. Measuring transfer of training is difficult, as graduates disperse across the globe, yet it is the true measure of a program's effectiveness.

Designing for Places and Spaces

Design of an environment should reflect the purpose, mission, and culture behind the activities conducted within the space. According to the Ergonomics Society (Europe):

> Ergonomics is about "fit": the fit between people, the things they do, the objects they use and the environments they work, travel and play in. If good fit is achieved, the stresses on people are reduced. They are more comfortable, they can do things more quickly and easily, and they make fewer mistakes.[2]

For example, an integrated medicine office where patients learn self-regulation via biofeedback should reflect a calm, comfortable, soothing atmosphere. Similarly, a military learning environment should depict the seriousness of the learning task, be free from distractions (including uncomfortable furniture), and enable all teaching and learning tasks to be accomplished within its boundaries. On the Fort Hood Resiliency Campus, soldiers and their family members learn the art and science of being physically, mentally, and spiritually resilient, especially when confronted with difficulties. In designing the campus, the commander purposefully replaced furniture that gave the appearance of a military office with furnishings that depicted a more "homey," comfortable environment. In the building where Spiritual Fitness teaching and counseling takes place, the building includes alcoves for small gatherings, meeting rooms for education, and a library—even the lighting and background music convey a calming atmosphere. Out the back door is a garden with water features, assisting in the creation of an "oasis" on post where soldiers and family members can go to build themselves emotionally, socially, physically, and spiritually (Figure 10.2). In short, the design of the environment should make the teaching and learning process as efficient as possible for the information being conveyed. Learning environments should encourage students to learn through both formal and informal means and be flexible enough to support different purposes over time.

---

2   See http://www.humanics-es.com/def-erg.htm (accessed: July 19, 2012).

Figure 10.2 Spiritual fitness garden at the resiliency campus at Fort Hood (photo by Michael Heckman, Fort Hood Public Affairs)

Offices

Office design should enable well-organized accomplishment of tasks without the office worker incurring work-related musculoskeletal discomfort or overuse injury. Such injuries occur in academic settings (Fisher, Konkel, and Harvey 2004), and the risk factors and symptoms are comparable to those incurred among office workers in other businesses (Rice and Mays 2003). For example, at the US Army Medical Department Center and School, researchers used the Job Requirements and Physical Demands survey to identify work-related symptoms of instructors, administrators, managers, and support staff (Rice and Mays 2003; Rice, Mays, and Gable 2009). Of the 414-person sample, 54 percent identified work-related shoulder and neck symptoms and nearly 50 percent reported back/torso pain, eyestrain, and headaches. Those with more intense symptoms also reported experiencing high psychosocial stressors. Human factors engineers can assist with identifying those personnel at greatest risk for injury and the tasks associated with those risks. Using their findings, they can design programs to teach ergonomic principles, such as how to align and adjust one's chair, desk, input devices, and monitor to reduce the potential for pain or injury, as well as how to sequence tasks and design a workstation to improve efficiency.

Classrooms

Most military classrooms use chairs and tables or desks that can be rearranged for a lecture, small group interaction, or a demonstration. In order to accommodate these changes, chairs are often selected for their durability and storage (stacking), with the thought that training and education are short term and will not permanently impact a student's posture or musculoskeletal system. Students are typically permitted to stand quietly during instruction, should they tire of sitting or need to stand to remain alert. Some military training occurs over a long duration, with students spending eight hours a day in class attending programs that last up to or longer than a year, particularly during AIT. While no research is available on the comfort level of military students in the classroom and the impact of their comfort on learning, prolonged sitting can result in discomfort, distraction, and the need to move. In one study on nurses who reported a work-related musculoskeletal disorder, 20 percent reported having their first symptoms as students (Tinubu et al. 2010). If adjustable chairs cannot be purchased, students should be encouraged to move about during breaks and to stretch tight muscles. Having adjustable sit/stand tables permits service members to stand occasionally, while still being able to take notes or follow handouts readily; however, these are seldom available.

In addition to the physical design of tables and chairs, attention must be paid to visual and auditory perception in classroom environments. While it seems like common sense to maximize visual and auditory communications between instructors and students, many military academic classrooms are deep (rather than wide) to accommodate large numbers of students. This design is not conducive to interaction. Visual line of sight between instructor and student is important and is often moderated through the use of video terminals, visible to students. This is particularly important during demonstrations in larger classes; video cameras and camera angles should be adjusted to enable all participants an equally adequate view.

Understanding your target audience can assist with determining the design alternatives that may benefit individuals or groups. In a study conducted at the US Army Medical Department Center and School, auditory learners were found to have lower grades than visual learners (Rice et al., 2006). A second study showed auditory information-processing to be lower among operating room technician students who failed their training, compared with those who received a grade of B or above (Alfred and Rice 2010). Having a visual line of sight, proximity to the instructor, a seat away from ambient or other noise (air conditioners or heating vent and hallways), and access to an audio or visual tape for review purposes may be particularly important for auditory learners and those with lower auditory processing abilities (Alfred and Rice 2009, 2010).

As our youth become more agile and accustomed to learning through the Internet and other technologically advanced interactive programs, it becomes more important to investigate and include such technologies in a blended learning environment in military classrooms. Embedding technology in the classroom can range from aids to communication, such as tools enabling instructors to ask questions about the lecture and for student responses to immediately be displayed to the class (voting devices), to discrete educational software (DES) and gaming programs that "teach" a narrowly prescribed topic (Murphy et al. 2001). DES programs include integrated learning systems, computer-assisted instruction, and computer-based instruction. A number of studies demonstrate student learning gains and improvements in standardized and national tests following the introduction of technologies into elementary, middle, and high school classrooms (Murphy et al. 2001, O'Dwyer et al. 2005, Schacter 1999, Sivin-Kachala and Bialo 2000). In civilian institutions, although increased use of technology and a move toward collaborative learning has occurred, instructor education and classroom design have not appreciably changed as new technologies have been added to existing spaces (JISC 2006). This appears equally true for military learning environments, as implementing new concepts requires innovative instructors, time, and funding—a difficult task in a fast-paced, military learning environment.

Simulated Environments

Creation of an immersive simulated environment involves all of a person's senses—smell, sight, touch, taste, and hearing. One of the best sources of information for creation of a simulated environment is to create an experience based on the situation or setting being imitated. The human factors engineer can assist with the interviews, focus groups, and surveys necessary to identify the tasks and conditions to be targeted, and designed into the simulation. The human factors engineer might serve as a team member and assist with identifying each element within an environment that is important to the goal of learning and how much fidelity is necessary, conducting user tests of mock settings, identifying methods of student evaluation, evaluating transfer-of-training, and comparing the utility of a simulated environment with that of a traditional classroom learning method. One difficulty with the introduction of technology and simulation is that sometimes it is adopted for the sake of moving with the times and appearing innovative and fresh. However, if learning, retention, and on-the-job performance are equivalent to or less than learning under less expensive and intensive conditions, the expenditures may not be warranted.

Certainly, the fidelity of a simulated environment must be such that the learner experiences tasks, sights, and sensations that are similar to those he or she will eventually encounter on the job (Figure 10.3). This is critical in the military, where the first introduction to a battlefield or working with wounded warriors could evoke a shocked response, potentially slowing the reactions and interactions of the soldier, and resulting in human error as he or she attempts to perform the necessary and potentially life-saving tasks under duress. These occurrences are exactly what immersive simulations or virtual environments seek to prevent, as they target the most frequent and most difficult tasks.

Figure 10.3 **Early use of a simulator may include target accuracy information, followed by increased fidelity as a student perfects basic skills**

Product Design

Products used in military training and education settings include chairs, desks, workstations, computers, and software; the analysis and assessment of these products are often associated with human factors engineers/ergonomists. Military academic environments also include handouts, training manuals, texts, and audiovisuals. Although these products are not typically associated with human factors, they can be designed to more effectively fit student use. For example, texts and handouts should match our knowledge of the human brain and learning. That is, many can be streamlined, customized for a particular use, altered to fit the capabilities of the user (student), and organized to promote integrated, associative learning. To provide a simple example, PowerPoint slides in many military environments are often extremely busy, complicated displays that sometimes obscure the points to be made. Slides are meant to supplement lectures and do not have to stand alone. Therefore, they should be bulleted, that is, they do not need to contain full sentences. Additional detail can be provided in the notes section for use by the instructor. For most purposes, they should contain no more than 7 ± 2 points due to human limits on information-processing (Miller 1956), use a simple font without serifs to enhance readability, never contain a font size that cannot be easily read by individuals at the back of a room (typically 18 point), have sufficient contrast between the background and font, and contain only those complexities of appearance, sound, and "busyness" that help to convey the message at hand. Color choices should be complimentary and limited. Backgrounds should be similar for each slide in terms of brightness, so the eyes do not have to accommodate to a light-appearing slide, following a darker slide.

In military classrooms, most seating and office products are purchased from civilian vendors as "off the shelf" products. Human factors engineers can assist faculty and staff with methods for selecting the products most likely to meet user requirements, as well as applying ergonomic principles for safe and efficient use. Additionally, the human factors professional can assist safety personnel with evaluating injury and illness records and conducting surveys to describe the populations, tasks, and products most often associated with physical discomfort and injury. As an example, one of the tasks most associated with upper limb pain among the staff at the US Army Medical Department Center and School is the use of the mouse during computing (Rice, Mays and Gable 2009). The solution, in most cases, was aligning the mouse at the same level as the keyboard through purchase of a longer keyboard tray or raising the individual's chair so the desk surface would suffice. The latter solution also required raising the monitor for correct visual line-of-sight and providing forearm support and a raised footrest.

The role of human factors engineers in the development and assessment of medical or learning equipment/technologies is similar to their role in evaluating simulations, discussed previously. The role of the human factors engineer can be to assist in product development using a participatory ergonomic process, as well as conducting iterative usability tests to determine if the device is easy and safe to use by the target audience, if it achieves its intended purpose, and whether it is preferred or liked by users. Preference is subjective and can include information on cost, implementation, system integration, accessibility, maintainability, user engagement, emotional response, user satisfaction, and aesthetic appeal.

Process Design

As is shown in Table 10.1, a number of military training processes could be assessed and improved through applied human factors/ergonomics. Working with the education and training staff, objectives and priorities can undergo process improvement. Support in this area may come from unexpected sources. Interestingly, Lean Six Sigma programs purport to improve processes using terminology, techniques, and goals comparable to human factors/ergonomics (that is, improving efficiency, effectiveness, safety, and cost/benefits). Mentoring and guiding projects initiated by individuals seeking to attain qualifications in Lean Six Sigma projects can accelerate human factors-related goal attainment.

Even the processes of teaching and learning can be informed via research based on human cognitive abilities and performance limitations. For example, class times should be limited, as the average students' attention span is described as lasting for 10 to 20 minutes (Penner 1984). Thus, the information delivery style, mode, or pace should change approximately every 20 minutes, in order to reorient student attention to the lecture. Examples of alternate styles or modes include the traditional classroom format in which the instructor lectures the students, student-led lectures, discussions, question-and-answer sessions, gaming-to-learn exercises, demonstrations, hands-on practice, use of video aids, and storytelling. Information on the physiology involved in learning and memory, and the roles of repetition, exercise, alpha-wave relaxation, and sleep can reinforce current or help introduce new study habits for students (Medina 2008). Knowledge on how to identify one's chronotype, whether one performs better in the morning (lark), or evening (owl), as well as information on zone-napping and which memory tasks are more vulnerable to the effects of sleep deprivation, can help students discover the practices that will work for them and how to self-regulate their work–study patterns accordingly.

Process Design Example 1: Identifying and Mitigating Personal Risk Factors for Academic Attrition

The goal of any military education/training program is for students to learn the presented information and procedures thoroughly enough to pass the course, a passing grade being an indication the soldier should be able to correctly apply learned skills during follow-on assignments. However, this goal is not always attained, particularly in some of the Army's more challenging MOS training programs, where students have difficulty learning the material, or fail the program for a variety of reasons. The Army's 68W AIT program at Fort Sam Houston, Texas, is extremely challenging and their data reveal high attrition (failure) rates. Between 2007 and 2009, the 68W attrition rate peaked as high as 45 percent, but on average was 31 percent (pass rate of 69 percent) (Whittaker and Parsons 2010). In order to make a positive impact on the pass rate and lower the attrition rate, human factors/ergonomics professionals assisted by systematically identifying personal factors contributing to attrition. Once these factors were identified, interventions could be developed and implemented.

To identify personal risk factors Rice and colleagues (2007) used information gained from a large-scale literature review, plus interviews and focus groups with students and faculty, to create a survey, which they administered to 712 MOS 68W students (final analysis used $n = 579$). Risk factors were identified through logistical regression to predict pass/fail status and linear regression to predict final course grades. Table 10.3 shows a rank ordering of the risk factors. This information is subsequently being integrated into a screening tool, the Personal Academic Strategies for Success PASS tool. With PASS, students take a computerized questionnaire at the start of their training and receive personalized feedback on what they can do, based on their strengths and weaknesses, to improve their chances of success in the 68W course.

Table 10.3   68W rank-ordered attrition risk factors

Rank	Variable	Rank	Variable
1	Education level	18	Positive thoughts
2	Study skills	19	ADHD
3	Hardships	20	Parental marital status
4	Continuance commitment	21	Home until enlisted
5	Health status	22	High school GPA
6	Exercise	23	Oppositional defiance disorder
7	Interest in course	24	Achievement
8	Willingness to take course	25	English as a second language
9	Low training importance	26	Marital status (divorced)
10	African American (race)	27	Sleep
11	Self-efficacy	28	Prior medical training
12	High school science grades	29	USAR component
13	Fear of failure	30	Negative thoughts
14	Avoidance coping	31	Responsibility
15	Stress	32	Science orientation
16	Sociability	33	Learning style (visual)
17	Smoking		

A collateral piece of the tool allows instructors to view the composite information for an entire class of 68W students. Based on the ergonomic principle "Know Thy User," this composite information may help instructors alter their teaching style or incorporate specific topics in order to "reach and teach" a particular group of students. Instructors can take the PASS themselves and identify their personal strengths and weaknesses, thus enabling them to identify with and provide support to students with similar profiles as themselves. That is, the instructors have managed to succeed in the military and they may have identified learning and coping techniques that are of particular benefit to themselves, which they can share with students.

Process Design Example 2: Investigation on Interventions for At-Risk Students

A number of "brain-training" technologies and training programs are commercially available that promise to improve concentration, advance reading and retention, speed cognitive processing, and problem-solving, and develop executive level skills. Evaluating such interventions is yet another process area where human factors/ergonomics professionals can assist military education/training. A US

Army Research Laboratory study to investigate the relationship between scores on a neurocognitive timing task (using the Interactive Metronome™, Figure 10.4), and military student performance revealed an association between timing and both grade point average and physical performance (Army Physical Fitness Test scores) (Rice, Butler, and Marra 2007). A follow-on study to examine the utility of the technology to improve the pass rate of Operating Room Technology students who failed the course one time is under way. Similar studies into the effectiveness and usability of such technological interventions are of benefit not only to student programs but also to performance enhancement training programs across the military.

Figure 10.4   Student volunteers participate in a neurocognitive timing task

Process Design Example 3: The Need for Iterative Investigation of the Selection Process

Students   Carefully selecting students into a program of instruction should improve the likelihood that entering students hav the requisite intellectual capacity to succeed. However, if selection procedures are too selective, that is, entrance criteria are extremely difficult to meet, insufficient numbers of personnel may enter the program. Therefore, a selection procedure must find a balance between being difficult enough to reduce attrition, but not so difficult that specialties in shortage are unable to meet manpower requirements. Assessment of student selection procedures is another potential area for involving human factors/ergonomics professionals.

Currently, the military uses the Armed Services Vocational Aptitude Battery (ASVAB) both to qualify individuals for military service using the Armed Forces Qualifying Test (AFQT) and to assign them to specific military occupational specialties or jobs. The ASVAB is made up of 10 sections, combinations of which make up various subscores. Each service has distinct cutoff scores for entrance into the service and classification into specific occupational specialties. Individuals are placed into jobs based on combinations of ASVAB subtest scores, personal preferences, and current military manpower needs at the time of placement. The ASVAB must be continually validated as a selection tool to make sure it is contributing to training success in the contemporary training environments.

Instructors Instructor quality is believed to be related to the achievements of students (Harris and Sass, 2008). The process by which instructors are selected to teach a course is an area in which human factors design can also impact training and education. In addition to having the subject matter expertise to teach a course, an effective instructor also needs to have teaching skills, interest in the course material, commitment to the job, an ability to relate to others, communication and presentation skills, and motivation to teach the course. For illustration, the following aspects of instructor effectiveness have been related to student achievement (Rice 2003):

- greater teaching experience;
- the selectivity or prestige of the institution where the instructor received his or her training;
- advanced degrees in mathematics and science when teaching math and science;
- teacher certification in mathematics when teaching math;
- instructor coursework in the subject area taught;
- instructor coursework in pedagogy (the study of teaching, including instructional strategies and theory);
- higher instructor test scores on literacy levels and verbal abilities.

Areas that are lacking after selection occurs can be supplemented by training. However, training should not be used in place of adequate selection measures, particularly if there is a limited amount of time in which training can be accomplished. The current assumption in military education/training programs is that anyone practicing in a profession or specialty can teach that specialty. This may be so, but their students will probably not achieve at equal levels, as much depends on the abilities of the individual instructor.

Conclusions

This chapter provides an introduction to a few topic areas ripe for human factors involvement in military education and training. It emphasizes how using a team approach integrating the skills of educators (that is, teachers), the perspective of the students, and the expertise of human factors/ergonomics professionals can broaden the scope of analysis and result in an ever-improving quality of the teaching and learning processes.

References

Alfred, P. and Rice, V.J. 2009. Cognitive characteristics of academically "at risk" students attending operating room specialist training. *Proceedings of the 2009 Human Systems Integration Symposium*, Annapolis, MD, March 2009. Available at: http://navalengineers.net/Proceedings/ HSIS2009/Papers/ Alfred_Rice.pdf (accessed: July 19, 2012).

——. 2010. Auditory discrimination and academic performance. *Proceedings of the 3rd AHFE International Conference*, Miami, FL, July 2010.

Alfred, P., DeVilbiss, C., Headley, D., Rice, V., Bazley, C., Jacobs, K. and Vause, N. 2009. Human factors applications in academic settings. *Proceedings of the HFES 53rd Annual Meeting*, San Antonio, TX, October 2009, 1022–1025.

Chapanis, A. 1991. To communicate the human factors message, you have to know what the message is and how to communicate it. *Human Factors Society Bulletin*, 34(11), 1.

DeVilbiss, C., Rice, V.J., Laws, L. and Alfred, P. 2010. If you want to know why students fail, just ask them: Self and peer assessments of factors affecting academic performance. *The Army Medical Department Journal*, October–December, 65–72.

Fisher, T.F., Konkel, S. and Harvey, C. 2004. Musculoskeletal injuries associated with selected university staff and faculty in an office environment. *Work*, 22(3), 195–205.

Harris, D.N. and Sass, T.R. (2008). Teacher training, teacher quality and student achievement. National Center for Analysis of Longitudinal Data In Education Research. Urban Institute.

HFES 2010. *Directory of Human Factors/Ergonomics Graduate Programs in the United States and Canada Human Factors/Ergonomics Undergraduate Programs*. Available at: http://www.hfes.org/web/Students/grad_programs. html (accessed: July 19, 2012).

JISC 2006. *Designing Spaces for Effective Learning: A Guide to 21st Century Learning Space Design*. (Higher Education Funding Council for England). Available at: http://www.jisc.ac.uk/media/documents/publications/learning spaces.pdf (accessed: 19, 2012).

Medina, J. 2008. *Brain Rules*. Seattle, WA: Pear Press.

Miller, G. 1956. The magical number seven, plus or minus two: some limits on our capacity for processing information. *The Psychological Review*, 63, 81–97.

Murphy, R., Penuel, W., Means, B., Korbak, C. and Whaley, A. 2001. *E-DESK: A Review of Recent Evidence on the Effectiveness of Discrete Educational Software*. Menlo Park, CA: SRI International.

O'Dwyer, L.M., Russell, M., Bebell, D. and Tucker-Seeley, K.R. 2005. Examining the relationship between home and school computer use and students' English/language arts test scores. *The Journal of Technology, Learning, and Assessment*, 3(3), January, 1–45.

Penner, J.G. 1984. *Why Many College Teachers Cannot Lecture*. Springfield, IL: Thomas.

Rice, J.K. 2003. *Teacher Quality: Understanding the Effectiveness of Teacher Attributes*. Washington, DC: Economic Policy Institute.

——. 2009. Evaluating and designing education: A collaborative effort between educators and ergonomists. *Proceedings of the Human Factors & Ergonomics Society 53rd Annual Meeting*, San Antonio, TX, October 2009, 1017–1021.

Rice, V.J. and Mays, M.Z. 2003. Work-Related musculoskeletal risk factors and shoulder and neck symptoms in academic personnel. *Proceedings of the Human Factors & Ergonomics Society 46th Annual Meeting*, Baltimore, MD, September–October 2002, 1226–1230.

Rice, V.J.B., Butler, J. and Marra, D. 2007. *Neuro-Cognitive Assessment, Symptoms of Attention Deficit and Hyperactivity Disorder, and Soldier Performance during 68W Advanced Individual Training*. (Technical Report ARL-TR-4292). Aberdeen Proving Ground, MD: US Army Research Laboratory.

Rice, V.J., Butler, J., Marra, D., DeVilbiss, C., Bundy, M., Headley, D., Dixon, M., Patton, D.J. and Rose, P. 2006. *A Prediction Model for the Personal Academic Strategies for Success (PASS) and Academic Class Composite (AC2T) Tool: Supporting Soldiers Attending 68W Advanced Individual Training at Ft. Sam Houston, Texas*. (Customer Report). Fort Sam Houston, TX: US Army Research Laboratory.

Rice, V.J., Butler, J., Marra, D., Vu, T., DeVilbiss, C. and Headley, D. 2007. *Accomplishments and Achievements: Army Medical Department Center and School Advanced Individual Training Attrition*. (Customer Report). Fort Sam Houston, TX: US Army Research Laboratory.

Rice, V.J., Mays, M.Z. and Gable, C. 2009. Self-reported health status of students in-processing into military medical advanced individual training. *Work*, 34(4), 387–400.

Rice, V.J., Pekarek, D., Connolly, V., King, I. and Mickelson, S. 2002. Participatory ergonomics: Determining injury control "buy-in" of US Army cadre. *Work*, 8(2), 191–203.

Rice, V.J., Vu, T., Butler, J., Marra, D., Merullo, D. and Banderet, L. 2009. Fear of failure: Implications for military health care specialist student training, *Work*, 34(4), 465–74.

Schacter, J. 1999. *The Impact of Education Technology on Student Achievement: What the Most Current Research Has to Say.* Santa Monica, CA: Milken Exchange on Education Technology. Available at: http://www.mff.org/pubs/ME161.pdf (accessed: July 19, 2012).

Sivin-Kachala, J. and Bialo, E. 2000. *2000 Research Report on the Effectiveness of Technology in Schools.* 7th Edition. Washington, DC: Software and Information Industry Association.

Tinubu, B.M.S., Mbada, C.E., Oyeyemi, A.L. and Fabunmi, A.A. 2010. Work-related musculoskeletal disorders among nurses in Ibadan, south-west Nigeria: A cross-sectional survey, *BMC Musculoskeletal Disorders.* Available at: http://www.biomed central.com/1471-2474/11/12 (accessed: July 19, 2012).

Whittaker, D. and Parsons, D. 2010. *Chart 68W Attrition Rate 24 Feb 10.* (Internal Report). Fort Sam Houston, TX: Army Medical Department Center & School.

PART III
Assessing and Designing Systems

Chapter 11

The Multi-Aspect Measurement Approach: Rationale, Technologies, Tools, and Challenges for Systems Design

Kelvin S. Oie, Stephen Gordon, and Kaleb McDowell

As scientists and engineers interested in understanding and improving how soldiers interact with technology, we are faced with numerous challenges. Perhaps chief among them is the extremely high dimensionality of the behavioral space within which the science and technology (S&T) community must assess, enhance, and optimize soldier–system performance throughout the technology development cycle. We contend that successful technology design and development will be enhanced with measurement approaches that are able to adequately capture the complex interactions among the many mission, task, environmental, and personnel factors that may have substantial impact on soldier–system performance. Moreover, in the face of this high dimensionality, we suggest that there are no available measurement methods that can, by themselves, provide the information density needed to fully characterize and understand the many aspects of soldier–system performance needed to enable effective systems design and development.

Instead, adequate characterization of soldier–system performance necessitates multiple measurements that are able to simultaneously capture the dynamic changes in mission, task, environmental, and performance variables comprising the behavioral space. This *multi-aspect measurement approach* should seek to integrate state-of-the-art measurement technologies to capture the multiple aspects of soldier–system behavior. Rapid advancements in computing and information technologies have made cutting-edge measurement capabilities more and more available to the S&T community and make the implementation of multi-aspect measurement approaches viable in the short-term future. In this chapter, we present an overview of many of the technologies and tools currently available to support the multi-aspect approach, acknowledging that while the technologies discussed provide remarkable potential to advance this approach, significant scientific and engineering challenges remain. In the final part of the chapter, we identify and briefly discuss several of these key challenges.

A Rationale for the Multi-aspect Measurement Approach

As Army operations continue to move into increasingly dynamic and complex environments, the measures we use to understand how soldier abilities meet the new demands posed by these environments will be more and more critical. The S&T community should of course strive to adopt and integrate the most advanced and effective tools available for measuring, assessing, and understanding soldier performance. However, while there are many technologies and tools to choose from, there is no single measure that alone can fully characterize soldier–system performance across the full spectrum of task and mission contexts. Choosing the right combination of sensor technologies and tools is a critical challenge.

Our focus in this chapter is on measurement technologies and tools that can be used *within operationally relevant environments*. The choice to focus on operationally relevant technologies and tools is grounded in an ecological perspective; many valuable insights about human behavior have come through experimentation within highly controlled laboratory environments. It is a significant concern, however, that the dynamic, complex nature of behavior in the real world, and in particular within Army operational environments, cannot be fully understood, regardless of what measures we use. The tasks and task environments we use are often not representative of those that we perform in real life.

Imposing such constraints in laboratory studies is often justifiable. For example, simplified tasks and environments can help to isolate variables of interest and controls for potentially confounding variables that could affect the interpretation of experimental data and analytic results based upon specific statistical models. In functional neuroimaging studies (for example, in functional magnetic resonance imaging (fMRI) (Oakes et al. 2005)), strict control of the subject's bodily movements is necessary to maximize measurement fidelity.

However, the strict control of tasks and environments in laboratory studies can have fundamental implications for our understanding of human behavior. For example, a long-held belief based upon early laboratory findings in the study of human visual attention has been that centrally located directional cues (for example, an arrowhead) will not induce attentional shifts if they are not predictive of where the target stimulus will appear. This suggests that the attentional shifts associated with such cues are voluntary. In a series of experiments motivated by real-world observations of the influence of observed gaze cues (see Kingstone et al. 2003 for a review), it was demonstrated that (1) centrally displayed, non-predictive eye gaze cues do produce attentional shifts toward the direction of gaze, (2) centrally displayed, non-predictive arrows also produce attentional shifts in the direction of the arrow, and (3) centrally displayed, predictive cues produce attention effects that are greater than predicted by the traditional model of volitional orienting. These findings suggest an automatic/reflexive orienting mechanism. Therefore, based on these findings, as well as similar research results in visual search, this research group has argued that laboratory-based studies isolated from real-world experience may, in fact, generate fundamental misunderstandings of the underlying principles being investigated (Kingstone et al. 2008). In a similar fashion, Foerde, Knowlton, and Poldrack (2006) showed that an individual performing the same memory task multiple times engaged different brain structures within the medial temporal lobe and striatum each time the task was performed; however, no significant differences in performance level were observed. Further, the differences in the brain areas recruited during the performance of this task were shown to be related to the different conditions under which specific items to be remembered were learned.

This suggests that human behavior, and the neurocognitive functions that underlie them, may vary substantially with changes in behavioral context. This makes assessing soldier performance during the performance of real-world tasks in operationally relevant environments critical to our understanding of how soldiers process, integrate, comprehend, and act upon information (that is, ecological validity). Such an understanding is not only vital for generalizing the results of laboratory studies to more naturalistic behaviors and environments (that is, external validity), but also for research results and performance assessments aimed at supporting the design and development of technological capabilities that soldiers will rely upon in the battlespace.

Given this importance, how do we go about measuring soldier performance within operational environments? Generally, there have been four traditional methods that have been used widely in the measurement and assessment of soldier performance and, in particular, soldier cognitive performance: performance measures, subjective ratings, subject matter expert (SME) opinion, and physiological measurement. Performance measures such as reaction time or response time are used extensively in psychological and human factors research on relatively simple tasks. However, as Veltman and Gaillard (1996) argue, performance measures often cannot be used to index cognitive load in complex task settings, especially where behavioral goals require the performance of multiple subtasks, as changing task priorities make it impossible to determine whether such measures accurately reflect task or subtask performance. Operators may also adapt to increasing task loads by "exerting additional effort" (Sarter, Gehring, and Kozak 2006), or by adopting alternative cognitive strategies, which may lead to equivalent assessments of task and cognitive performance when assessed through performance outcome measures alone, even though cognitive load has increased.

This means that within highly complex, dynamic environments, performance-based measures can provide information about cognitive load or state only when some estimate of the operator's effort can also be indexed. Rating scales, which are typically based upon post-hoc, subjective reports of perceived cognitive effort or task difficulty, are often used to provide such estimates. Several instruments, such as the National Aeronautics and Space Administration Task Load Index (NASA-TLX), the Subjective Workload Assessment Technique (SWAT), or the Workload Profile, have been used extensively in previous research (Fréard et al. 2007, Hart and Staveland 1988, Rubio et al. 2004, Scallen, Hancock, and Duley 1995, Tsang and Velazquez 1996, Veltman and Gaillard 1996, Verwey and Veltman 1996, Wu and Liu 2007) and have been shown to be effective in assessing subjective cognitive load associated with routine laboratory tasks and in some operationally relevant contexts (Rubio et al. 2004). However, the resolution of these measures is limited both temporally and in terms of cognitive processing. It has also been argued that individuals do not always accurately report their current psychological, mental, or emotional status (Zak 2004). Veltman and Gaillard (1996) further posit that rating scales are limited by the effects of participants' memory, perception, and biases; for example, participants appear to be unable to discriminate between task demands and the effort invested in task performance. As well, subjective rating scales are not well suited for online estimation of cognitive performance or state, as they often require significant task interruptions, imposing at least some costs to performance due to task switching (Monsell 2003, Speier, Valachich, and Vessey 1999).

In addition to participants' subjective assessments of cognitive performance, SME opinion can be a valuable tool in the assessment of the cognitive demands of soldier–system interaction. SME input may be used to provide a priori or predictive information about the cognitive load an operator may experience under certain conditions (Charlton 2002). For example, Vidulich and Hughes (1991) demonstrated a high correlation (0.94) between the retrospective ratings of F-16 pilots who were presented with six different heads-up-display (HUD) formats during an experimental evaluation and the projective ratings of these HUD formats made by F-16 pilots who were only familiar with the current, operational HUD. Moreover, projective ratings made by college students unfamiliar with HUD displays were found to be uncorrelated with the results of the experimentally derived pilot ratings. These findings provide some support for the incorporation of SME opinion into system design decisions and highlight the importance of relevant expertise in effective projective assessments of the cognitive demands placed on operators. However, SME opinion for use in measurement of performance has practical limitations, as well as limitations inherent to any subjective assessment of human behavioral performance.

The measurement of central and peripheral physiological functions has been used extensively in the assessment of cognitive performance and state, with many different measures found to be reliably associated with different aspects of cognitive performance. Physiological measures provide a more objective, and arguably less intrusive, means of assessment than can be obtained with traditional performance and subjective reports. Unfortunately, physiological measures, while often providing relatively predictable associations with performance, do not seem to have a high degree of sensitivity across different task and environmental conditions, especially when taken in isolation from related measures of central nervous system activity (see below, under "Neurophysiological Measures").

Barandiaran and Moreno (2006) conceptualize this problem from an evolutionary perspective: biological systems are intrinsically purposeful in terms of their self-sustaining nature, a result of their internal, metabolic organization. At the most fundamental level, this is the source of what we call "intentionality." The evolution of the nervous system enabled organisms to actively modify their relationship with the external environment in order to satisfy biologically defined constraints, for example, by enabling organisms to move to different locations in space. In the case of systems that are distinctly cognitive, such as humans, constraint satisfaction and metabolically driven intentionality do not seem able to fully explain the phenomenology of cognition. That is, behavior does not seem to be solely a response to metabolic needs. In such a case, the nervous system can be considered to be "de-coupled" from the metabolic and constructive processes of the organism, such that the functions of the nervous system that support cognitive behavior are no longer explicitly governed by the metabolic organization that supports the nervous system's architecture. A significant implication is that the physiological states of metabolic systems such as the heart, lungs, and skin will not alone be wholly predictive of the dynamic behavior and states of the nervous system. This suggests that measurement approaches based on physiological measures need to take into account the states and function of the nervous system.

Given the limitations of the traditional technologies and tools available for cognitive measurement and assessment, it is unlikely that incremental improvements in our knowledge based upon these approaches alone can, or will, provide the necessary understandings of cognitive function that would be needed to address the challenges of systems design for the current and future operational environment. By contrast, a multi-aspect approach would advocate the simultaneous use and integration of multiple measurement modalities (for example, Halchenko, Hanson, and Pearlmutter 2005). The approach aims to take advantage of both the unique information each measurement modality can bring, as well as the redundant information available across modalities that can provide both validation and robustness to assessments based upon the implemented technologies. In the next part of this chapter, an overview of many of the available measurement technologies and tools is provided, briefly discussing what aspects of soldier performance they assess, as well as their limitations.

Technologies and Tools for Multi-aspect Measurement

In this part of the chapter, technologies and tools that can be implemented in a multi-aspect approach are grouped into four nominal measurement categories: neurophysiological, central and peripheral physiological, body motion and behavioral, and context. Performance measures, subjective rating scales, and SME opinion remain valuable to the characterization of soldier performance, and further discussion of these methods can be found elsewhere in the literature (Gawron 2000, Rubio et al. 2004).

Neurophysiological Measures

Electrophysiological function According to recent estimates, the brain contains approximately 86 billion neurons (Azevedo et al. 2009), which are electrically excitable cells considered to be the primary substrate for processing and transmitting information in the brain. This communication occurs through a combination of both chemical and electrical signaling: Neurons maintain voltage gradients across their cell membranes via metabolically driven cellular mechanisms, generating ion-concentration differences between the intracellular and extracellular spaces. Sufficient changes in membrane voltage driven by chemical changes at the connections between neurons lead to an all-or-none electrochemical pulse called an action potential, which travels rapidly along the cell's length as input signals to other neurons, with each neuron connecting to thousands of other neurons. Understanding the electrophysiological basis of neural function as it relates to cognitive performance, then, has provided researchers with one of the primary tools for understanding human cognition.

Over the past two decades, computational capabilities and sophisticated statistical and modeling approaches have given rise to a wide variety of novel techniques to analyze and interpret electrophysiological data. Such techniques have improved capabilities to localize sources, uncover nonlinear and nonstationary signals, analyze and interpret vast amounts of data, and examine network interdependencies in data. In addition to more traditional approaches to assessing electrophysiological function discussed below, these emerging methods are providing new insights into brain and cognitive function and are giving rise to novel neuroscience-based technologies, including completely new classes of brain-computer interaction technologies (see Chapter 12 in this volume).

Electroencephalogram (EEG)   One of the most common techniques for measuring the electrophysiology of brain function is EEG, which involves the use of small electrodes attached to the scalp (see A in Figure 11.1) that measure the electrical potentials created by the summed activity of populations of neurons in the brain (Andreassi 2000). For research studies, electrodes are typically placed on the scalp according to a standard layout, called the "international 10–20 system," with spacing between electrodes either 10 or 20 percent of the total front-to-back or right-to-left distances across the skull. A typical 10–20 layout has 19 electrodes (see B in Figure 11.1), though additional electrodes can be included by further subdividing the standard distances, with higher-density arrays of 128 (for example, Oostenveld and Praamstra 2001, Sperdin et al. 2009) or even 256 channels (Luu et al. 2010, Nikolaev et al. 2010) in common use today.

Generally speaking, the signals that comprise EEG reflect the spontaneous activity of neural populations (Martin 1985). These continuous signals are typically described as belonging to one of at least five categories based on their frequency content (Andreassi 2000), as shown in Figure 11.2, and changes within these frequency bands have been associated with cognitive performance. For example, as task difficulty increases, activity over the frontal lobes has been shown to induce decreases in alpha-band activity and increases in theta activity (Gevins and Smith 2003). Increased beta activity over the anterior portions of the brain has been shown to

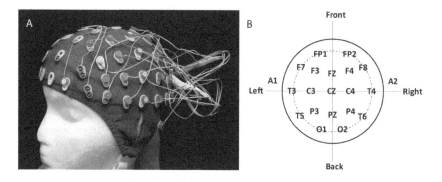

Figure 11.1   **A: Electrode cap for EEG recording. B: International 10–20 electrode montage**

correspond with increased levels of vigilance (Smit, Eling, and Coenen 2004, Valentino, Arruda, and Gold 1993). Increased information processing demands (for example, increased working memory load) is related to theta synchronization (Kahana, Seelig, and Madsen 2001, Klimesch, Freunberger, and Sauseng 2010, Sauseng et al. 2010), while de-synchronization has been previously observed in relation to increasing stimulus complexity (Berlyne and McDonnell 1965) and levels of arousal (Christie et al. 1972).

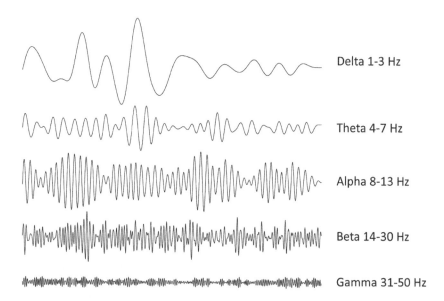

Figure 11.2   **The five primary waveforms or frequency bands in EEG**

EEG provides the highest available temporal resolution (on the order of milliseconds) to measure the dynamic electrical activity generated by cortical neurons and is therefore not confined by the intrinsic time delays inherent to imaging techniques based upon blood flow dynamics (Nunez and Silberstein 2000, see also below). However, because EEG electrodes must measure potentials through the scalp, skull bone, dura mater, and cerebrospinal fluid, the signal strength is often weak, with signals typically in the range of 10–100 µV. In addition, signals from any given neural source are distributed across the scalp surface due to the effects of volume conduction or in some cases not present at the scalp at all (Nunez and Srinivasan 2006, Rutkove 2007). EEG therefore reflects a nonlinear mixture of signal components that may arise from, potentially, all areas of the brain, as well as non-brain sources such as muscular contractions and external electromagnetic interference, which must be taken into account in the analysis and interpretation of EEG data. Moreover, the effects of volume conduction limit the potential spatial resolution possible with EEG; even with high-density electrode arrays ($\geq$ 128 electrodes) and sophisticated mathematical modeling, the spatial accuracy of EEG source estimates is on the order of ~1–2 cm^2 (Liu, Dale, and Belliveau 2002).

Event-related potentials   In addition to the monitoring of more spontaneous activity assessed in EEG, brain activity in response to specific stimuli, or event-related potentials (ERP), has been used extensively to assess human neurocognitive behavior in laboratory studies. ERPs are considered to be brain responses to specific stimuli that manifest themselves as either positive or negative voltage deflections from a baseline value, which occur at different latencies relative to stimulus onset. They are often, therefore, labeled using both the direction (that is, "P" or "N") and latency.

For example, the P300 is an ERP comprising a positive voltage deflection that appears approximately 300 milliseconds after stimulus presentation when a subject is actively engaged in the task of target detection (Picton 1992), and it has been suggested to reflect brain processes of "context updating," wherein an individual's internal model of the environment must be revised (Donchin and Coles 1988). Another ERP, the contingent negative variation (CNV), has been observed between the appearance of successive stimuli when the first is perceived to act as a warning that a response signaled by the second stimulus is required (Walter et al. 1964). The CNV is a long-lasting, negative voltage deflection, differences in which have also been linked to reaction time, levels of attention, and cognitive effort (Andreassi 2000). Motor potentials are ERPs that occur in preparation for and during execution of voluntary movements, even if the subject is only thinking about performing the movement (Pfurtshceller et al. 2003). One specific type of motor potential is the readiness potential, which is a slow-rising negative wave that begins approximately 500–1000 milliseconds before a voluntary movement and peaks at the time of response (Kutas and Donchin 1974).

While ERPs have been linked to many aspects of task and cognitive performance, their use in real-time applications and operational environments may be limited. ERPs are generally obtained by averaging over multiple time-locked trials, as the magnitude of the brain-related ERP signal is often low relative to the combination of unrelated activity produced by other neural and physiological processes (for example, muscular contractions) and noise from external sources. Under the assumption that these contaminating sources are randomly distributed, averaging over a sufficient number of stimulus presentations should reveal only brain responses associated with stimulus processing. Averaging over multiple time-locked events is clearly limited for the use of such signals as inputs to real-time systems. This may be exacerbated by increased levels of signal contamination in operational environments that are more dynamic and complex than those typically seen in laboratory experimentation.

Hemodynamic function   Measuring the electrical activity generated within the brain is not the only means of analyzing brain function. Another approach involves measuring changes in blood flow, blood volume, and blood-oxygen saturation levels in different regions of the brain as a result of cognitive performance. Like all of the body's organs, the brain consumes nutrients (glucose) and oxygen during its metabolic processes. A local reduction in glucose and oxygen stimulates an increase in local arteriolar vasodilation, increasing local blood flow and blood volume and, therefore, oxygen transport to the area. These processes, termed the "hemodynamic response," occur over the course of several seconds following the initiation of neural activity.

However, there is no simple relationship between the magnitude of hemodynamic response and any single physiological parameter, as the coupling of neural dynamics and hemodynamics is both indirect and only approximately linear within a certain range of neural activation (Lauritzen 2001). Devor et al. (2005) present data that show increases in hemodynamic response beyond saturation of a local site of neural activation (Fox et al. 1988), indicating that the hemodynamic signal observed at a given location in the brain is driven, at least in part, by electrophysiological activity over a relatively broad neighboring region. Thus, the complex spatiotemporal integration of hemodynamic response should be carefully considered when interpreting these changes as related to brain function.

Functional near-infrared spectroscopy (fNIRs)   While the use of hemodynamics-based functional neuroimaging measurement modalities such as fMRI and positron emission tomography (PET) have been essential to the explosive growth observed in cognitive neuroscience, their deployment in operationally relevant environments is clearly infeasible. By contrast, fNIRs provides a non-invasive, minimally intrusive, relatively inexpensive, and portable method for hemodynamic imaging in real-world environments. fNIRs works by irradiating the scalp with near-infrared light in the range of 700–900 nm. Most biological tissues are relatively transparent to light in this range. However, oxygenated hemoglobin (oxy-Hb) and deoxygenated hemoglobin (deoxy-Hb) in the blood are not, and differentially absorb light in this functional range. Observed changes in the concentrations of oxy-Hb and deoxy-Hb can then be used to measure the hemodynamic response, providing an index of brain activity (Aslin and Mehler 2005, Izzetoglu et al. 2007).

As is indicated above, along with other measures based upon the hemodynamic response, fNIRs signals associated with brain activations are relatively slow, with temporal resolution on the order of seconds versus the millisecond timing available with electrophysiological measures. fNIRs does have good spatial accuracy on the order of ~1 cm^2, without necessitating the complex mathematical modeling needed to provide such spatial resolution for source localization in EEG, even when applied to very-high-density fields. Finally, fNIRs is generally limited to measurements of the outer cortical mantle, with the signal sensitive to hemodynamic changes within the outer 2–4 mm of the cortex (Bunce et al. 2006).

Central and Peripheral Physiological Measures

Cardiovascular function One of the most frequently measured aspects of physiological function is cardiovascular activity. Changes in heart rate and heart rate variability have been linked to differences in cognitive task demand or difficulty (Boutcher et al. 1998, Veltman and Gaillard 1998), emotional activity (Ikehara and Crosby 2005, Sinha, Lovallo, and Parsons 1992, Wagner, Kim, and Andre 2005), and cognitive load (Noel, Bauer, and Lanning 2005, Reeves, Schmorrow, and Stanney 2007, Veltman and Gaillard 1998). As well, frequency domain or spectral analysis can be applied to heart rate variability data to identify frequency components known as sinus arrhythmia, which have been linked to mental effort (Allanson and Fairclough 2004).

Blood pressure, blood volume, and changes in blood oxygen saturation levels have also been associated with states of mental and physical stress (Ikehara and Crosby 2005, Ohsuga, Shimono, and Genno 2001). As well, blood pressure has been linked to changes in emotional state (Sinha, Lovallo, and Parsons 1992) and is a useful method for differentiating between subjects in a state of challenge versus those in a state of threat (Blascovich et al. 1999). In addition, regional blood flow and levels of perfusion have been used to identify arousal and emotion, as well as frustration and mental load (Or and Duffy 2007).

Electrocardiogram (ECG) One of the most common methods for measuring heart rate and related measures is known as ECG, which uses anywhere from two to several dozen electrodes placed on the skin.

The electrodes register the electrical potentials generated by the muscles of the heart, which are typically in the range of 100–5,000 μV. Normal ECG waveforms are composed of a series of deflections denoted as P, Q, R, S, or T waves (Figure 11.3). The time between successive R-waves is known as the R–R interval or interbeat interval.

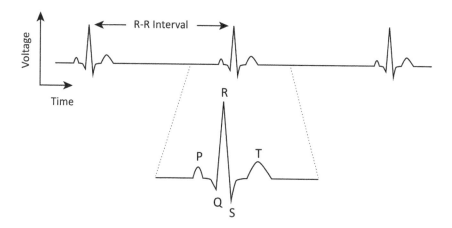

Figure 11.3   The ECG waveform

ECG electrode placement is most common across the chest or on the limbs; a sample three-electrode layout is shown in illustration A, Figure 11.4. Because the electrical potentials in ECG are relatively large (for example, compared to signals of interest in electroencephalography, see above), these signals are less susceptible to contamination by external noise sources, though muscle contractions related to body movements can introduce undesirable artifacts into the recorded signal (Andreassi 2000), as well as induce changes in heart rate (Jennings et al. 1981).

Impedance cardiogram (ICG) ICG is performed by passing a low amplitude, high-frequency current (~3–5 mA at 75–100 Hz) across the chest by attaching stimulating electrodes to the neck and lower abdomen (see illustration B, Figure 11.4). Additional electrodes are attached near the stimulating electrodes in order to measure the voltage difference across the chest, and impedance can be computed according to Ohm's law. As blood moves through the cardiovascular system, impedance levels change and several quantities can be computed, such as heart rate and blood pressure.

While ICG is able to provide more information than ECG, it does so at the cost of increased complexity for both the measurement process and the computation of signals of interest. ICG must also pass current

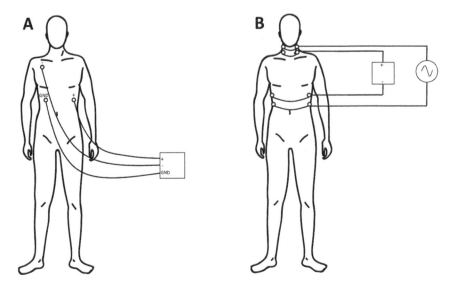

Figure 11.4 Sample electrode configurations for A: three-electrode ECG; B: an impedance cardiogram

through the body and, like ECG, is susceptible to signal artifacts related to body movements.

Pulse oximetry   Pulse oximetry is used to determine the relative saturation levels of oxygenated hemoglobin in the blood. In a manner similar to fNIRs, red and infrared light sources are used to pass light through the skin and sensors record the light reflected back. Because the different wavelengths of light are absorbed at different rates by oxygenated versus deoxygenated hemoglobin, the ratio of red versus infrared light can be used to determine the level of oxygen saturation in the blood (SpO_2). Physiological processes, including cognitive processes, use oxygen in the blood and can therefore be indexed by changes in SpO_2. Moreover, since the absorption of red and infrared light is also affected by blood volume within the peripheral tissues, pulse oximetry can be used to detect heart rate measures by identifying spikes in the absorption levels. However, depending on the location of the sensors, these measures of heart rate will exhibit different latencies compared to measures such as ECG due to the pulse transit time required for the blood to reach the sensor site.

Because the technology relies on relatively low light levels that must pass through the body's tissues, SpO_2 measurements can only be performed on locations on the body at which the tissues are thin enough to allow light to pass through and be detected. These areas include the fingers, toes, and earlobes. A similar technique, photoplethysmography, uses infrared light alone to calculate measures such as heart rate and blood volume. Both pulse oximetry and photoplethysmography are sensitive to body movements, shifts in sensor location, and decreases in blood flow.

Respiratory function   Measures of respiratory function are also commonly used physiological indices of cognitive performance. Respiratory rate, inter-breath interval, and inspiratory time and flow have all been used to identify states of high and low cognitive load (Noel, Bauer, and Lanning 2005, Ohsuga, Shimono, and Genno 2001), task demand and difficulty (Backs and Selijos 2005, Veltman and Gaillard 1998), and emotional states (Boiten 1998, Wagner, Kim, and Andre 2005). The detection of respiration rate also helps to identify respiratory components of heart rate variability, known as respiratory sinus arrhythmia, which has been used as an indicator of cognitive load (Ohsuga, Shimono, and Genno 2001).

Several technologies are available for the measurement of respiratory function, including the use of strain gauges or potentiometers attached to bands around a subject's chest. The sensors record the expansion and compression of the rib cage. From these recordings, the breathing cycle can be evaluated, and calibration of these measurements allows the extraction of relative values of inspiratory and expiratory flow volume as a function of chest expansion.

Respiratory analysis has also been performed using thermal measurements. By analyzing the temperature changes that occur around the nose and mouth, basic indices of respiration rate and inter-breath can be extracted. For example, Veltman and Gaillard (1998) used a thermistor attached to the nose, while Or and Duffy (2007) used thermal imaging techniques. In both cases, the temperature drop around the nasal passageways was used to indicate the movement of air and, thus, respiration.

Skin temperature and conductance   Subtle differences in facial skin temperature can indicate changes in emotional states, for example, during blushing (Ekman 1992), as well as indexing cognitive effort and task difficulty (Genno et al. 1995, Genno 1997). Further, Ishikawa et al. (1998) and also Ohsuga, Shimon, and Genno (2001) have shown that nose skin temperature is sensitive to changes in cognitive effort, while forehead skin is not. However, as skin temperature changes with changes in environmental temperature, it is suggested that the difference between nose and forehead skin temperatures could be used as an index of physiological function (Ishikawa et al. 1998). Recent research efforts are using thermal imaging along with automated image processing to detect changes in regional, facial blood flow patterns indicative of psychophysiological processes, such as auditory startle response (Levine, Pavlidis, and Cooper 2001) or deception (Pavlidis, Eberhardt, and Levine 2002).

Electrodermal activity   Electrodermal activity (EDA), also commonly known as galvanic skin response, is the result of increased sweat production and is often measured on the palmar surface of the hands or plantar surface of the feet in response to increased arousal (Ikehara and Crosby 2005, Mandryk, Inkpen, and Calvert 2006, Reeves, Schmorrow, and Stanney 2007), emotional stress (Perala and Sterling 2007, Wagner, Kim, and Andre 2005), and cognitive load (Allanson and Fairclough 2004). EDA recordings have been linked to levels of attention, vigilance (Andreassi 2000), and stimulus significance (Wingard and Maltzman 1980).

Because of the increased density of (eccrine) sweat glands on the hands and feet, they are the preferred locations for EDA measurement. There are two main approaches for EDA measures. The first approach uses two electrodes to measure skin conductance. A constant, low-amplitude current or voltage is applied to the skin, which is used to determine conductance. The second approach measures skin potential levels. Most often, electrode placements for both techniques have one electrode on the hand and the other on the forearm, or one electrode on each medial phalanx of the middle and index fingers of the same hand (illustration A, Figure 11.5) (Andreassi 2000). If foot placement is desired, for example to keep the subject's hands free of obstructions (illustration B, Figure 11.5), then the electrodes can be placed on the medial plantar surface (Boucsein 1992).

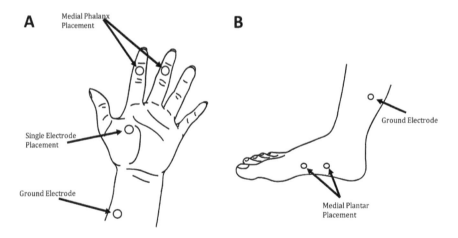

Figure 11.5 Electrode placements for EDA monitoring for A: the hand; B: the foot

EDA is a relatively simple measure of physiological function; however, the EDA signal is relatively slow to change in response to events, with a time course on the order of 1–2 seconds. As well, it can be sensitive to artifacts resulting from body movements or other thermal influences that increase sweat production.

Body Motion and Behavioral Measures

The results of the functioning of neurophysiological and physiological processes are often manifested in intentional behavior, making measurements of such behaviors an invaluable part of the multi-aspect measurement approach. The outcomes, speed, and precision of bodily movements can reflect underlying processes and state. Furthermore, involuntary motor responses (for example, micro-expressions (Ekman, 2009)) give insight into cognitive and affective states of an operator. In this part of the chapter, we overview a number of technologies that provide researchers with the capability of recording and characterizing full-body motion, as well as other behavioral measures reflective of cognitive performance and behavior.

Electromyogram The electrical activity related to muscular contraction, using methods similar to EEG, ECG, and EDA, is recorded through a technique known as electromyography (EMG). In EMG, electrodes are attached to the surface of the skin to measure the electrical potentials arising from muscular activity, which are typically in the range of 100–2,000 µV. There is no singular, standard configuration for performing EMG, but in cognitive assessment it is common to record activity from the muscles around the forehead, face, and neck, as this information can be used to identify facial expressions and muscle tension, as well as artifacts that might affect simultaneous recording of EEG (Jung et al. 2000, Makeig et al. 2009).

Increases in EMG activity prior to the execution of a task have been implicated in faster response times (Davis 1940), and this activity can be used as an indication that the subject was preparing for movement (Malmo and Malmo 2000). Facial EMG has been observed in response to the pleasantness of particular stimuli (Cacioppo, Bush, and Tassinary 1990). In particular, activity of the frontalis, zygomaticus, and corrugator muscles of the face has been used to differentiate between stressful and non-stressful reactions (Carlson, Singelis, and Chemtob 1997) and to identify expressions of anger (Jancke 1996). Differences in tonic muscle tension have also been shown to be useful in identifying stress (Andreassi 2000).

However, EMG is generally not a good indicator of behavioral complexity and, like many other electrophysiological techniques, EMG can be subject to movement artifacts that originate from muscles that are not of interest in a given assessment, as well as noise artifacts from external sources.

Actigraphy An alternative method to monitoring motor activity and behavior, called actigraphy, employs an array of accelerometers placed at different locations on the subject's body to obtain orientation and acceleration measurements, which can then be used to characterize motor behaviors. Actigraphy has been used to provide basic measurements of vigilance and reaction time (Russo et al. 2005), as well as to identify wake–sleep cycles (RTO 2004). Additionally, depending upon sensor sensitivity, actigraphy can be used to measure heart rate by detecting small vibrations that occur as blood pulses through the wrists. However, these measurements can be easily corrupted by movement artifacts under high-motion or vibration conditions (for example, in a moving vehicle). As well, it can be difficult to reconstruct complex behaviors from actigraphic information without a detailed model of the user.

Electrooculogram   The tracking of eye movements, including gaze and fixation behavior, blink rate, and pupillary responses can provide valuable measurements of attention allocation (Fairclough and Venables 2006) and cognitive and fatigue state (Stern, Boyer, and Schroeder 1994). The electrophysiological technique for recording eye activity, electrooculography (EOG), is based upon the existence of a consistent electrical potential between the cornea and the retina of the eye (Andreassi 2000). As the eye rotates about its axes, the voltage difference causes measurable changes in the potential across each axis. EOG measurements, therefore, are made by placing electrodes across the eyes in both the horizontal and vertical directions. Figure 11.6 illustrates a simple layout for recording horizontal eye movements. Placing an additional electrode between the eyes is helpful in differentiating the independent movements of the eyes that occur during vergence motions.

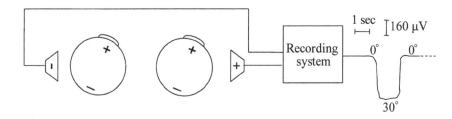

Figure 11.6   Schematic representation of the horizontal EOG (adapted from Andreassi 2000)

EOG signals can also be used to determine saccades and fixations. Fixations are periods where the eyes remain focused on a particular position in visual space, and saccades are rapid simultaneous movements of both eyes that are observed from one fixation point to another, typically lasting less than 200 milliseconds. If the position of the head in 3D space is known, then the EOG signal can be used to calculate absolute point-of-regard, that is, the location in space to which the eyes are directed. EOG can also be used to extract information about eye blinks and blink rate, which have been used as indicators of task difficulty and demand (Veltman and Gaillard 1998, Yamada 1998), levels of fatigue (Stern, Boyer, and Schroeder 1994), and task engagement (Fairclough and Venables 2006).

Videooculogram Because it is difficult to extract absolute point-of-regard information from EOG data due to uncertainties about absolute head position and orientation, video-based eye-tracking techniques have become a prevalent method for monitoring the allocation of visual attention based on gaze behavior (Smith, Shah, and da Vitorio Lobo 2003, Duchowski 2002). Automated video eye-tracking, or videooculography (VOG), uses cameras and automated image analysis algorithms to capture, compute, and record eye motion and activity. In a desktop arrangement, one or more cameras are placed facing the subject, and a volume in 3D space is calibrated, allowing the subject to move freely within a small range (~1–2 ft³). Head-mounted VOG systems operate under the same basic principles as desktop systems, but allow more freedom of movement and may even allow the subject to walk freely around their environment. The trade-off to this added mobility is that most head-mounted systems will track only a single eye, require head-tracking sensors to provide absolute point-of-regard information, and can potentially impede the subject's field of view.

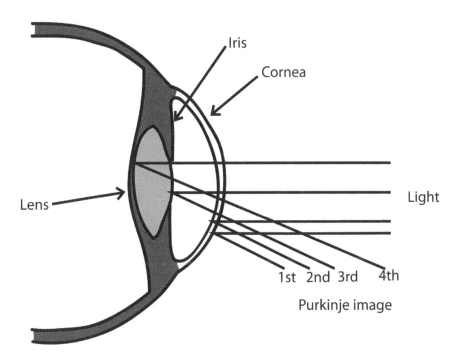

Figure 11.7   The four Purkinje images (adapted from Glenstrup and Engell-Nielsen 1995)

Many VOG systems rely on the properties of the pupil and cornea to reflect infrared light, and in most of these applications an infrared light source is used to ensure stable lighting conditions. Tracking is typically based on distance measures between the corneal reflection, called the "1st Purkinje image" (Figure 11.7), and the center of the pupil. Dual-Purkinje tracking systems use reflections from both the 1st and 4th Purkinje images, but because the back of the lens only provides a weak reflection, this technique requires that ambient light conditions be carefully controlled (Glenstrup and Engell-Nielsen 1995).

When using Purkinje images there are two methods for illuminating the eye. In the method known as "bright pupil," the infrared (IR) source is co-axial with the eye. This produces the common effect known as "red eye." It is difficult to maintain this co-axial condition though, and in these systems, the head is often restricted from movement. Dark pupil methods, on the other hand, use IR sources that are offset from the eye. The dark pupil technique is used by most commercial systems and allows the head to move within a confined 3D space.

However, not all VOG techniques use IR illumination and Purkinje images. The main tools behind VOG are the image-processing algorithms that allow the eyes to be located and tracked; Purkinje images merely ease the computational demands of these algorithms. Commercial eye-tracking systems are available that do not use IR illumination and rely solely on image-processing techniques.

VOG methods also can be used to determine pupil dilation. Pupil dilation is, among other things, an autonomic reaction to changes in lighting conditions, with pupil diameter increasing with decreasing light levels. However, pupil dilation also occurs in response to cognitive and affective states. For example, Partala and Surakka (2003) have demonstrated pupillary responses to arousing and exciting visual stimuli. Task-evoked pupillary responses have also been demonstrated in response to increased cognitive load (Beatty 1982), or increased sensory, perceptual, or attentional demands (Beatty and Lucero-Wagoner 2000).

Motion capture and video analysis   More complete behavioral recordings can be made through the use of motion capture and video analysis technologies. Motion capture systems are typically composed of an array of motion-based sensors, such as accelerometers, gyroscopes, or strain gauges (Arminian and Najafi 2004), or optic (Moeslund and Granum 2001), magnetic (Lange and Lappe 2007), or ultrasound technologies (Warren et al., 2001). Motion capture systems have been used successfully to characterize emotional behaviors (Kapur et al. 2005), as well as providing contextual information for the interpretation of electrophysiological data and other cognitive measures (Makeig 2009, Makeig et al. 2009).

Often, motion capture methods require the subject to wear a specialized array of passive or active markers that may be distracting to the subject. In lieu of such techniques, automatic video analysis using computer vision algorithms can be used. A common technique employs advanced pattern recognition and image-processing tools to identify facial expressions, which can be used to identify emotional states such as frustration, tension, and stress (Cohen et al. 2002, Susskind et al. 2007). Cheng and Trivedi (2006) have also demonstrated the use of video-based motion capture for the prediction of a subject's intent to produce a specific action, that is, making a turn while driving a vehicle.

Speech recognition and analysis Simple microphones, in combination with advanced language and signal-processing techniques, can also be used to measure and assess soldier–system interactions. For example, current speech recognition software can be used to detect and process simple commands from spoken words. Moreover, while systems capable of fully processing natural language have not yet been developed, statistical analyses can be performed to evaluate comprehension and to detect emotional states, such as surprise (Lau et al. 2006, Osterhout and Holcomb 1992). Recently, Wood, Torkkola, and Kundalkar (2004) presented results showing differences in a number of speech parameters (for example, pause duration and variability) when a driver is engaged in neutral versus intense conversations, suggesting the potential to use speech recognition and analysis technologies to index cognitive load. Similarly, Ball and Breese (1999) have developed a Bayesian classifier to estimate a user's emotional state during interaction with a speech-based computer interface, finding model parameters that could be mapped to changes in emotional and arousal levels that are consistent with previous empirical findings, such as increased pitch and speech rate with increased arousal.

Context Measures

External conditions While monitoring environmental and task conditions does not directly offer a window into the soldier cognitive state, such information can provide the researcher operating in the real world with invaluable information to interpret physiological and neurophysiological data. For example, in laboratory environments, external conditions (for example, temperature, humidity, and air quality) are kept relatively constant to limit their impact on task performance and cognitive state. In real-world environments, such control may not be possible, and previous results have shown associations among ambient temperature, heat stress, and cognitive state (Hancock and Vasmatzidis 2003). Because of the various effects that the environment may have on soldier cognitive state and performance, it is important, if a specific environmental variable changes during the course of an experiment, that this variable is measured. Table 11.1 lists a number of measurable external conditions, along with typical measurement technologies, that may be used as contextual measures to improve the accuracy of cognitive assessment in operational environments.

Table 11.1 **External and task condition measures and representative measurement technologies**

Variable	Measurement technology
External conditions	
Ambient temperature	Thermistor, thermocouple
Humidity	Polymer-based, hygrometer
Air pressure	Barometer
Air quality	Solid-state semiconductor devices, electrochemical
Vehicle vibration/shock	Accelerometer, gyroscope
Background sound/noise levels	Microphone
Task conditions	
Task Type	Operator interface input, task models
Command information	Operator input
Vehicle command state	Software monitoring of vehicle controls/system states
Geospatial position	Global positioning system devices

Task conditions Especially in the assessment of soldier–system performance in operationally relevant environments, information about task conditions is critical to the interpretation of neurophysiological, physiological, and behavioral data. Unlike in laboratory-based experiments, behavior in the operational world will likely involve multiple simultaneous task demands and changing task priorities, and methods for automatic activity recognition would provide invaluable information for the validity of cognitive metrics that may be highly task dependent (Berka et al. 2007).

Monitoring and recording information from vehicle controls and user interface systems can provide a method for more directly assessing how attention is allocated during task performance by indicating what tasks are being attended to by the operator. Moreover, Fogarty et al. (2005) illustrate how information from low-level operator interactions with computer interfaces during a realistic computer-programming task can be used to develop novel sensors of cognitive or functional state, such as the operator's level of task engagement.

Summary

We have discussed several sensor technologies and tools that can be used in a multi-aspect approach to characterizing soldier–system performance. Where measurement capabilities of relevant soldier–system attributes are lacking, novel sensor and measurement systems will need development. Innovative sensor development is continuous, and implementation and integration of other measurement capabilities such as event-related optical signaling (Gratton and Fabiani 2001) could provide new, enabling capabilities to the S&T community. Moreover, sensors should ideally be as user-acceptable and non-intrusive to soldier–system behavior as possible, and novel approaches to such development, for example, using simple web-camera-based imaging for non-contact measurement of heart rate (Poh, McDuff, and Picard 2010), should be an active research arena.

Multi-aspect Measurement Integration

Integrating these different sensors into a multi-aspect measurement system is non-trivial. Data streaming from multiple sensors must be sampled and acquired, synchronized when the temporal relationship among variables is of interest, and recorded and stored in specified data formats by data acquisition and data-handling hardware–software systems. Such data-handling across measures reflect processes whose dynamics can range widely, from the millisecond changes in brain activity to changes in ambient temperature that occur over minutes and hours to aspects of culture that change very slowly, but all of which may impact soldier–system performance in a given mission context. Synchronization of data streams from different measurement systems therefore remains a significant challenge for implementing any complex data acquisition paradigm, including those that would embody the multi-aspect approach. Perhaps an even more difficult issue, the scaling of variables from dramatically different sources (for example, brain electrical signals, digital scene processing, networked linguistic analysis) need to be considered. Several researchers have offered solutions (for example, Schalk et al. 2004, Vankov, Bigdely-Shamlo, and Makeig 2010); however, overcoming these challenges remains an open area for research and technology development.

Perhaps the most challenging areas of research are within the data analysis approaches needed to discover patterns within very high-dimensional data sets that robustly reflect ongoing behavioral dynamics. Computational algorithms used during different processing steps, including pre-processing and/or filtering for the discrimination or removal of signal artifacts and noise, continue to be challenged by the nonlinearity and nonstationarity of real-world data. Ultimately, data analyses must provide some form of metric that can be used to characterize an attribute of interest. Contextual variables, including information such as the task being performed (that is, activity recognition), mission status, or even information about individual differences can be used to inform the interpretation of computed metrics to provide estimates about an individual's cognitive, attentional, or functional state or performance, either post-hoc or in real time. Researchers have taken numerous approaches to multi-aspect measurement, and in some sense the concept is integral to an ecological perspective where interaction with the environment is critical to understanding behavior. Still, previous approaches were generally geared toward characterizing relatively constrained situations or isolated aspects of more complex, dynamic scenarios. The concept proposed here calls for measuring and developing computational and modeling approaches that are able to capture and provide more complete understandings of the very high-dimensional behavioral spaces that describe soldier–system performance in real-world environments.

Challenges for Systems Design

Designing systems that are consistent with human capabilities and limitations is not, of course, a trivial problem when one considers the innumerable factors that influence human behavior and the many aspects of systems design and development that need to be considered. Figure 11.8 provides a schematic representation of a systems design approach, identifying some of the many issues of the systems design problem space that warrant consideration and some of the interactions among them. While the approach presented here is not necessarily specific to a multi-aspect measurement approach, in this part of the chapter, we present a high-level analysis of this approach and some illustrative examples as a useful exercise in considering some of the many challenges for systems design.

At the center are the soldier, the system, and their interactions. The soldier–system is considered to be situated within both an environment and a mission context, where different environmental and mission constraints, such as weather, terrain, and tactical and strategic objectives, affect the soldier–system. Conversely, actions taken by the soldier–system affect both mission (for example, the priorities of different mission objectives or whether certain mission objectives have or have not been met) and environment (for example, proximity to hostile or friendly forces).

Developing technology solutions to provide mission-enabling capabilities will take an understanding of the relevant attributes and interactions of the soldier, system, environment, and mission. Identification of variables that are able to capture those aspects of the behavioral space relevant to successful soldier–system performance is a critical first step in the design process, and will largely drive the choice of sensor combinations needed to provide an adequate understanding of soldier–system performance related to the targeted capability needs.

As has been discussed in this chapter, the development, integration, analysis, and interpretation of such sensor technologies and tools is non-trivial. In cases where measurement capabilities of relevant soldier–system attributes are inadequate or too heavy, bulky, or power-hungry, novel sensor and measurement systems will need development. Over the past decade tremendous advances across a wide range of sensor systems have made multi-aspect sensing in realistic soldier–system testing scenarios a current possibility. With the current pace of technological advancement, multi-aspect sensing capabilities in fieldable situations are expected in the near future. Similarly, tools and techniques for the integration, analysis, and interpretation of multi-aspect measurement systems have rapidly advanced over the past decade. Preliminary examples of this approach are evidenced in modern exercise equipment, smartphones, and the transportation industry.

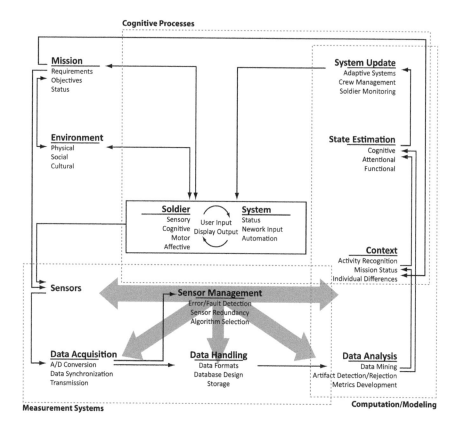

Figure 11.8 A soldier–system-centered framework for systems design

Further, data-mining approaches being used in advertising and Internet technologies are showing great promise for interpreting very high-dimensional behavioral spaces.

Finally, individual and group characterizations based upon interpretations of these metrics can be used to update the system. Such system updates can be provided in different forms, for example, through fundamentally influencing how we understand human capabilities and limitations. System updates may then be accomplished through the integration of such knowledge into the design of soldier–system interactions that reflect our best understanding of human capabilities and limitations. System updating can also be accomplished by integrating measures of soldier state and/or behavior as system inputs. For example, real-time soldier monitoring could be used to provide commanders and decision-makers with information about a soldier's current state to inform them of potential performance risks. These estimates of state could also be used in real-time adaptive systems that can automatically update soldier–system interactions by invoking automated systems to alleviate task load (Kaber et al. 2001, Parasuraman et al. 1992; for further discussion related to brain–computer systems designs, see Chapter 12 in this volume).

Conclusion

In this chapter, we have argued for a multi-aspect approach to the measurement and assessment of soldier–system performance to support systems design and development. This conclusion is based upon the increasing demands placed on our soldiers' cognitive abilities within an increasingly complex and information-intensive battlespace. These demands make understanding soldier cognitive behavior and performance critical to the design of systems that can meet the increasing challenges of the current and future operational environment. Thus, taking advantage of the joint and unique sources of information that exist within the broad range of available measurement and assessment technologies offers a path forward in providing the scientific knowledge and understandings needed to support effective systems design for the future.

References

Allanson, J. and Fairclough, S.H. 2004. A research agenda for physiological computing. *Interacting with Computers*, 16, 857–78.

Andreassi, J.L. 2000. *Psychophysiology: Human Behavior and Physiological Response*. Mahwah, NJ: Lawrence Erlbaum.

Arminian, K. and Najafi, B. 2004. Capturing human motion using body-fixed sensors: Outdoor measurement and clinical applications. *Journal of Visualization and Computer Animation*, 15(2), 79–94.

Aslin, R.N. and Mehler, J. 2005. Near-infrared spectroscopy for functional studies of brain activity in human infants: Promise, prospects, and challenges. *Journal of Biomedical Optics*, 10(1), 011009: 1–3.

Azevedo, F.A.C., Carvalho, L.R.B., Grinberg, L.T., Farfel, J.M., Ferretti, R.E.K., Leite, R.E.P., Filho, W.J., Lent, R. and Herculano-Houzel, S. 2009. Equal numbers of neuronal and nonneuronal cells make the human brain an isometrically scaled-up primate brain. *Journal of Comparative Neurology*, 513, 532–41.

Backs, R.W. and Selijos, K.A. 1994. Metabolic and cardiorespiratory measures of mental effort: The effects of level of difficulty in a working memory task. *International Journal of Psychophysiology*, 16, 57–68.

Ball, G. and Breese, J. 1999. *Modeling the Emotional State of Computer Users*. Redmond, WA: Microsoft Corporation.

Barandiaran, X. and Moreno, A. 2006. On what makes certain dynamical systems cognitive: A minimally cognitive organization program. *Adaptive Behavior*, 14, 171–85.

Beatty, J. 1982. Task-evoked pupillary response, processing load, and the structure of processing resources. *Psychological Bulletin*, 91(2), 276–92.

Beatty, J. and Lucero-Wagoner, B. 2000. The pupillary system, in *Handbook of Psychophysiology*, edited by J.T. Cacioppo, G. Berntson and L.G. Tasskinary. 2nd Edition. Cambridge: Cambridge University Press, 142–62.

Berka, C., Levendowski, D.J., Lumicao, M.N., Yau, A., Davis, G., Zikovic, V.T., Olmstead, R.E., Tremoulet, P.D. and Craven, P.L. 2007. EEG correlates of task engagement and mental workload in vigilance, learning, and memory tasks. *Aviation, Space, and Environmental Medicine*, 78(5), B231–B244.

Berlyne, D.E. and McDonnell P. 1965. Effects of stimulus complexity and incongruity on duration of EEG desynchronization. *Electroencephalography and Clinical Neurophysiology*, 18(2), 156–61.

Blascovich, J., Berry-Mendes, W., Hunter, S.B. and Salomon, K. 1999. Social "facilitation" as challenge and threat. *Journal of Personality and Social Psychology*, 77(1), 216–33.

Boiten, F. 1998. The effects of emotional behavior on components of respiratory cycle. *Biological Psychology*, 49, 29–51.

Boucsein, W. 1992. *Electrodermal Activity*. New York: Plenum Press.

Boutcher, S.H., Nugent, F.W., McLaren, P.F. and Weltman, A.L. 1998. Heart period variability of trained and untrained men at rest and during mental challenge. *Psychophysiology*, 35, 16–22.

Bunce, S.C., Izzetoglu, M., Izzetoglu, K., Onaral, B. and Pourrezaei, K. 2006. Functional near-infrared spectroscopy: An emerging neuroimaging modality. *IEEE Engineering in Medicine and Biology Magazine*, 25(4), 54–62.

Cacioppo, J.T., Bush, L.K. and Tassinary, L.G. 1990. Microexpressive facial actions as a function of affective stimuli: Replication and extension. *Personality and Social Psychology Bulletin*, 18, 515–26.

Carlson, B., Singelis, T.M. and Chemtob, C.M. 1997. Facial EMG response to combat-related visual stimuli in veterans with and without post traumatic stress disorder. *Applied Psychophysiology and Biofeedback*, 22, 247–59.

Charlton, S.G. 2002. Measurement of cognitive states in test and evaluation, in *Handbook of Human Factors Testing and Evaluation*, edited by S.G. Charlton and T.G. O'Brien. Mahwah, NJ: Lawrence Erlbaum Associates, 97–126.

Cheng, S.Y. and Trivedi, M.M. 2006. Turn–intent analysis using body pose for intelligent driver assistance. *Pervasive Computing*, 5(4), 28–37.

Christie, B., Delafield, G., Lucas, B., Winwood, M. and Gale, A. 1972. Stimulus complexity and the EEG: Differential effects of the number and the variety of display elements. *Canadian Journal of Psychology*, 26(2), 155–70.

Cohen, I., Sebe, N., Garg, A., Lew, M.S. and Huang, T.S. 2002. Facial expression recognition from video sequences. *Proceedings of ICME'02, International Conference on Multimedia and Expo*, Lausanne, Switzerland, August 2002, 121–24.

Davis, R.C. 1940. *Set and Muscular Tension*. Bloomington, IN: University of Indiana Press.

Devor, A., Ulbert, I., Dunn, A.K., Narayanan, S.N., Jones, S.R., Andermann, M.L. Boas, D.A. and Dale, A.M. 2005. Coupling of the cortical hemodynamic response to cortical and thalamic neuronal activity. *Proceedings of the National Academy of Sciences*, 102(10), 3822–3827.

Donchin, E. and Coles, M.G.H. 1988. Is the P300 component a manifestation of context updating? *Behavioral and Brain Sciences*, 11, 357–74.

Duchowski, A.T. 2002. A breadth-first search of eye-tracking applications. *Behavior Research Methods, Instruments and Computers*, 34(4), 455–70.

Ekman, P. 1992. An argument for basic emotions. *Cognition and Emotion*, 6, 169–200.

——. 2009. Lie catching and micro expression, in *The Philosophy of Deception*, edited by C. Martin. New York: Oxford University Press.

Fairclough, S.H. and Venables, L. 2006. Prediction of subjective states from psychophysiology: A multivariate approach. *Biological Psychology*, 71, 100–110.

Foerde, K., Knowlton, B.J. and Poldrack, R.A. 2006. Modulation of competing memory systems by distraction. *Proceedings of the National Academy of Sciences*, 103(31), 11778–11783.

Fogarty, J., Ko, A.J., Aung, H.H., Golden, E., Tang, K.P. and Hudson, S.E. 2005. Examining task engagement in sensor-based statistical models of human interruptibility. *Proceedings of the SIGCHI Conference on Human Factors in Computing Systems*, Portland, OR, April 2005, 331–40.

Fox, P.T., Raichle, M.E., Mintun, M.A. and Dence, C. 1988. Nonoxidative glucose consumption during focal physiologic neural activity. *Science*, 241, 462–64.

Fréard, D., Jamet, E., Le Bohec, O., Poulain, G. and Botherel, V. 2007. Subjective measurement of workload related to a multimodal interaction task: NASA-TLX vs. workload profile. *Human Computer Interaction*, 4552, 60–69.

Gawron, V.J. 2000. *Human Performance Measures Handbook*. Mahwah, NJ: Lawrence Erlbaum.

Genno, H. 1997. Non-contact method for measuring facial skin temperature. *International Journal of Individual Ergonomics*, 19, 147–59.

Genno, H., Kanbara, O., Matsumoto, K., Suzuki, R., Osumi, M. and Mabuchi, K. 1995. Relationship between thermal sensation and the temperature of skin and deep tissues in various environments. *Journal of the Japanese Society of Thermology*, 15(4), 2–7.

Gevins, A. and Smith, M.E. 2003. Neurophysiological measures of cognitive workload during human–computer interaction. *Theoretical Issues in Ergonomic Science*, 4, 113–21.

Glenstrup, A.J. and Engell-Nielsen, T. 1995. *Eye controlled media: Present and future state*. Copenhagen: University of Copenhagen.

Gratton, G. and Fabiani, M. 2001. Shedding light on brain function: The event-related optical signal. *Trends in Cognitive Sciences*, 5(8), 357–63.

Halchenko, Y.O., Hanson, S.J. and Pearlmutter, B.A. 2005. Multimodal integration: fMRI, MRI, EEG, MEG, in *Advanced Image Processing in Magnetic*

Resonance Imaging, edited by L. Landini, V. Positano and M.F. Santarelii. Boca Raton, FL: CRC Press, 223–65.

Hancock, P.A. and Vasmatzidis, I. 2003. Effects of heat stress on cognitive performance: The current state of knowledge. *International Journal of Hypothermia*, 19(3), 355–72.

Hart, S.G. and Staveland, L.E. 1988. Development of NASA-TLX (task load index): Results of empirical and theoretical research, in *Human Mental Workload*, edited by P.A. Hancock and N. Mechkati. Amsterdam: North Holland Press, 239–50.

Ikehara, C.S. and Crosby, M.E. 2005. Assessing cognitive load with physiological sensors. *Proceedings of the 38th Hawaii Conference on System Sciences*, Big Island HI, January 2005, 295.

Ishikawa, K., Genno, H., Kanbara, O., Yasuda, M. and Osumi, M. 1998. Evaluation of stress using facial skin temperature. *Proceedings of PIE*, 158–59.

Izzetoglu, M., Bunce, S., Izzetoglu, K., Onaral, B. and Pourrezaei, K. 2007. Functional brain imaging using near-infrared technology: Assessing cognitive activity in real-life situations. *IEEE Engineering in Medicine and Biology Magazine*, 26(4), 38–46.

Jancke, L. 1996. EMG in an anger-provoking situation: Individual differences in directing anger outwards or inwards. *International Journal of Psychophysiology*, 23, 207–14.

Jennings, J.R., Berg, W.K., Hutchinson, J.S., Obrist, P., Porges, S. and Turpin, G. 1981. Publication guidelines for heart rate studies in man. *Psychophysiology*, 18(3), 226–31.

Jung, T-P., Makeig, S., Humphries, C., Lee, T-W., McKeown, M.J., Iragui, V. and Sejnowski, T.J. 2000. Removing electroencephalographic artifacts by blind source separation. *Psychophysiology*, 37, 162–78.

Kaber, D.B., Riley, J.M., Tan, K-W. and Endsley, M.R. 2001. On the design of adaptive automation for complex systems. *International Journal of Cognitive Ergonomics*, 5(1), 37–57.

Kahana, M.J., Seelig, D. and Madsen, J.R. 2001. Theta returns. *Current Opinion in Neurobiology*, 11, 739–44.

Kapur, A., Kapur, A., Virji-Babul, N., Tzanetakis, G. and Driessen, P.F. 2005. Gesture-based affective computing on motion capture data, in *Affective Computing II 2005*, edited by J. Tao, T. Tan and R.W. Picard. Berlin/Heidelberg: Springer-Verlag, 1–7.

Kingstone, A., Smilek, D. and Eastwood, J.D. 2008. Cognitive ethology: A new approach for studying human cognition. *British Journal of Psychology*, 99, 317–40.

Kingstone, A., Smilke, D., Ristic, J., Friesen, C.K. and Eastwood, J.D. 2003. Attention, researchers! It is time to take a look at the real world. *Current Directions in Psychological Science*, 12, 176–80.

Klimesch, W., Freunberger, R. and Sauseng, P. 2010. Oscillatory mechanisms of process binding in memory. *Neuroscience and Biobehavioral Reviews*, 34, 1002–1014.

Kutas, M. and Donchin, E. 1974. Studies of squeezing: Handedness, responding hand, response force, and asymmetry of readiness potential. *Science*, 186, 545–48.

Lange, J. and Lappe, M. 2007. The role of spatial and temporal information in biological motion perception. *Advances in Cognitive Psychology*, 3(4), 419–28.

Lau, E., Stroud, C., Plesch, S. and Phillips, C. 2006. The role of structural prediction in rapid syntactic analysis. *Brain and Language*, 98(1), 74–88.

Lauritzen, M. 2001. Relationship of spikes, synaptic activity, and local changes of cerebral blood flow. *Journal of Cerebral Blood Flow and Metabolism*, 21, 1367–1383.

Levine, J.A., Pavlidis, I. and Cooper, M. (2001). The face of fear. *The Lancet*, 357(9270), 1757.

Liu, A.K., Dale, A.M. and Belliveau, J.W. 2002. Monte carlo simulation studies of EEG and MEG localization accuracy. *Human Brain Mapping*, 16, 47–62.

Luu, P., Geyer A., Fidopiastis, C., Campbell, G., Wheeler, T., Cohn, J. and Tucker, D.M. 2010. Reentrant processing in intuitive processing. *PLoS One*, 5(3), e9523.

Makeig, S. 2009. Mind monitoring via mobile brain–body imaging, in *Foundations of Augmented Cognition: Neuroergonomics and Operational Neuroscience*, edited by D. Schmorrow, I.V. Estabrooke and M. Gootjen. Berlin/Heidelberg: Springer, 749–58.

Makeig, S., Gramann, K., Jung, T-P., Sejnowski, T.J. and Poizner, H. 2009. Linking brain, mind, and behavior: The promise of mobile brain/body imaging (MOBI). *International Journal of Psychophysiology*, 73(2), 95–100.

Malmo, R.B. and Malmo, H.P. 2000. On electromyographic (EMG) gradients and movement-related brain activity: Significance for motor control, cognitive functions, and certain psychopathologies. *International Journal of Psychophysiology*, 38, 143–207.

Mandryk, R.L., Inkpen, K.M. and Calvert, T.W. 2006. Using psychophysiological techniques to measure user experience with entertainment technologies. *Journal of Behavior and Information Technology*, 25(2), 141–58.

Martin, J.H. 1985. Cortical neurons, the EEG, and the mechanisms of epilepsy, in *Principles of Neural* Science, edited by E.R. Kandel and J.H. Schwartz. 2nd Edition. Amsterdam: Elsevier, 636–47.

Moeslund, T.B. and Granum, E. 2001. A survey of computer vision-based human motion capture. *Computer Vision and Image Understanding*, 81, 231–68.

Monsell, S. 2003. Task switching. *TRENDS in Cognitive Science*, 7(3), 134–40.

Nikoleav, A.R., Gepshtein, S., Gong, P. and van Leeuwen, C. 2010. Duration of coherence intervals in electrical brain activity in perceptual organization. *Cerbral Cortex*, 20, 365–82.

Noel, J.B., Bauer, K.W. and Lanning, J.W. 2005. Improving pilot mental workload classification through feature exploitation and combination: A feasibility study. *Computers and Operations Research*, 32, 2713–2730.

Nunez, P.L. and Silberstein, R.B. 2000. On the relationship of synaptic activity to macroscopic measurements: Does co-registration of EEG with fMRI make sense? *Brain Topography*, 13, 79–96.

Nunez, P.L. and Srinivasan, R. 2006. *Electric Fields of the Brain: The Neurophysics of EEG*. Oxford: Oxford University Press.

Oakes, T.R., Johnstone, J., Ores Walsh, K.S., Greischar, L.L., Alexander, A.L. Fox, A.S. and Davidson, R.J. 2005. Comparison of fMRI motion correction software tools. *NeuroImage*, 28, 529–43.

Ohsuga, M., Shimono, F. and Genno, H. 2001. Assessment of phasic work stress using autonomic indices. *International Journal of Psychophysiology*, 40, 211–20.

Oostenveld, R. and Praamstra, P. 2001. The five percent electrode system for high-resolution EEG and ERP measurements. *Clinical Neurophysiology*, 112, 713–19.

Or, C.K.L. and Duffy, V.G. 2007. Development of a facial skin temperature-based methodology for non-intrusive mental workload measurement. *Occupational Ergonomics*, 7, 82–94.

Osterhout, L. and Holcomb, P.J. 1992. Event-related potentials elicited by syntactic anomaly. *Journal of Memory and Language*, 31(6), 785–806.

Parasuraman, R., Bahri, T., Deaton, J.E., Morrison, J.G. and Barnes, M. 1992. *Theory and Design of Adaptive Automation in Aviation Systems*. (Report No. NAWCADWAR-92033-60). Warminster, PA: US Department of the Navy, Naval Air Warfare Center—Aircraft Division.

Partala, T. and Surakka, V. 2003. Pupil size variation as an indication of affective processing. *International Journal of Human–Computer Studies*, 59, 185–98.

Pavlidis, I., Eberhardt, N.L. and Levine, J.A. 2002. Seeing through the face of deception. *Nature*, 415, 35–37.

Perala, C.H. and Sterling, B.S. 2007. *Galvanic Skin Response as a Measure of Soldier Stress*. (Technical Report ARL-TR-4114). Aberdeen Proving Ground, MD: US Army Research Laboratory.

Pfurtshceller, G., Muller, G., Pfurtshceller, J., Gerner, H.J. and Rupp, R. 2003. "Thought"-control of functional electrical stimulation to restore handgrip in a patient with tetraplegia. *Neuroscience Letters*, 351, 33–36.

Picton, T.W. 1992. The P300 wave of the human event-related potential. *Journal of Clinical Neurophysiology*, 9(4), 456–79.

Poh, M-Z., McDuff, D.J. and Picard, R.W. 2010. Non-contact, automated cardiac pulse measurements using video imaging and blind source separation. *Optics Express*, 18(10), 10762–10774.

Reeves, L.M., Schmorrow, D.D. and Stanney, K.M. 2007. Augmented cognition and cognitive state assessment technology—near-term, mid-term, and long-term research objectives, in *Foundations of Augmented Cognition*, edited by D. Schmorrow and L.M. Reeves. Berlin/Heidelberg: Springer, 220–28.

RTO 2004. *Operator Functional Assessment*. (Report No. RTO-TR-HFM-104). Neuilly-sur-Seine Cedex, France: North Atlantic Treaty Organization, Research and Technology Organisation.

Rubio, S., Diaz, E., Martin, J. and Puente, J.M. 2004. Evaluation of subjective mental workload: A comparison of SWAT, NASA-TLX, and workload profile methods. *Applied Psychology*, 53(1), 61–86.

Russo, M.B., Vo, A., Labutta, R., Black, I., Campbell, W., Greene, J., McGhee, J. and Redmond, D. 2005. Human biovibrations: Assessment of human life signs, motor activity, and cognitive performance using wrist-mounted actigraphy. *Aviation, Space, and Environmental Medicine*, 76(7), C64–C74.

Rutkove, S.B. 2007. Introduction to volume conduction, in *The Clinical Neurophysiology Primer*, edited by A.S. Blum and S.B. Rutkove. Totowa, NJ: Humana Press.

Sarter, M., Gehring, W.J. and Kozak, R. 2006. More attention must be paid: The neurobiology of attentional effort. *Brain Research Reviews*, 51, 145–60.

Sauseng P., Griesmayr, B., Freunberger, R. and Klimesch, W. 2010. Control mechanisms in working memory: A possible function of EEG theta oscillations. *Neuroscience and Biobehavioral Reviews*, 34, 1015–1022.

Scallen, S.F., Hancock, P.A. and Duley, J.A. 1995. Pilot performance and preference for short cycles of automation in adaptive function allocation. *Applied Ergonomics*, 26(6), 387–403.

Schalk, G., McFarland, D.J., Hinterberger, T., Birbaumer, N. and Wolpaw, J.R. 2004. BCI2000: A general-purpose brain–computer interface (BCI) system. *IEEE Transactions on Biomedical Engineering*, 51(6), 1034–1043.

Sinha, R., Lovallo, W.R. and Parsons O.A. 1992. Cardiovascular differentiation of emotions. *Psychosomatic Medicine*, 54, 422–35.

Smit, A.S., Eling, P.A.T.M. and Coenen, A.M.L. 2004. Mental effort affects vigilance enduringly: After-effects in EEG and behavior. *International Journal of Psychophysiology*, 53, 239–43.

Smith, P., Shah, M. and da Vitorio Lobo, N. 2003. Determining driver visual attention with one camera. *IEEE Transactions on Intelligent Transportation Systems*, 4(4), 205–18.

Speier, C., Valacich, J.S. and Vessey, I. 1999. The influence of task interruption on individual decision making: An information overload perspective. *Decision Sciences*, 30(2), 337–60.

Sperdin, H.F., Cappe, C., Foxe, J.J. and Murray, M.M. 2009. Early, low-level auditory–somatosensory multisensory interactions impact reaction time speed. *Frontiers in Integrative Neuroscience*, 3(2), 1–10.

Stern, J.A., Boyer, D. and Schroeder, D. 1994. Blink rate: A possible measure of fatigue. *Human Factors*, 36(2), 285–97.

Susskind, J.M., Littlewort, G., Bartlett, M.S., Movellan, J. and Anderson, A.K. 2007. Human and computer recognition of facial expressions of emotion. *Neuropsychologica*, 45, 152–62.

Tsang, P.S. and Velazquez, V.L. 1996. Diagnosticity and multidimensional subjective workload ratings. *Ergonomics*, 39(3), 358–81.

Valentino, D.A., Arruda, J.E. and Gold, S.M. 1993. Comparison of QEEG and response accuracy in good vs poorer performers during a vigilance task. *International Journal of Psychophysiology*, 15(2), 123–33.

Vankov, A., Bigdely-Shamlo, N. and Makeig, S. 2010. Data River—A software platform for real-time management of multiple data streams. *Fourth International BCI Meeting*, Asilomar, CA, May–June 2010.

Veltman, H.J.A. and Gaillard, A.W.K. 1996. Physiological indices of workload in a simulated flight task. *Biological Psychology*, 42, 323–42.

———. 1998. Physiological workload reactions to increased levels of task difficulty. *Ergonomics*, 41(5), 656–69.

Verwey, W.B. and Veltman, H.J.A. 1996. Detecting short periods of elevated workload: A comparison of nine workload assessment techniques. *Journal of Experimental Psychology: Applied*, 2(3), 270–85.

Vidulich, M. and Hughes, E.B. 1991. Testing a subjective metric of situation awareness. *Proceedings of the 35th Human Factors Society Annual Meeting*, San Francisco, CA, September 1991, 1307–1311.

Wagner, J., Kim, J. and Andre, E. 2005. From physiological signals to emotions: Implementing and comparing selected methods for feature extractions and classification. *Proceedings of the IEEE International Conference on Multimedia and Expo*, Amsterdam, July 2005, 940–43.

Walter, W.G., Cooper, R., Aldridge, V.J., McCallum, W.C. and Winter, A.L. 1964. Contingent negative variation: An electrical sign of sensory–motor association and expectancy in the human brain. *Nature*, 203, 380–84.

Warren, W.H., Jr., Kay, B.A., Zosh, W.D., Duchon, A.P. and Sahuc, S. 2001. Optic flow is used to control human walking. *Nature Neuroscience*, 4(2), 213–16.

Wingard, J.A. and Maltzman, I. 1980. Interest as a predeterminer of the GSR index of the orienting reflex. *Acta Psychologica*, 40, 153–60.

Wood, C., Torkkola, K. and Kundalkar, S. 2004. Using driver's speech to detect cognitive workload. *Proceedings of SPECOM 2004, 9th Conference on Speech and Computer*, St. Petersburg, Russia, September 2004.

Wu, C. and Liu, Y. 2007. Queueing network model of driver workload and performance. *IEEE Transactions on Intelligent Transportation Systems*, 8(3), 528–37.

Yamada, F. 1998. Frontal midline theta rhythm and eyeblinking activity during a VDT task and a video game: Useful tools of psychophysiology in ergonomics. *Ergonomics*, 41(5), 678–88.

Zak, P.J. 2004. Neuroeconomics. *Philosophical Transactions of the Royal Society of London, Series B: Biological Sciences*, 359(1451), 1737–1748.

Chapter 12

Future Soldier-System Design Concepts: Brain–Computer Interaction Technologies

Brent Lance, Jorge Capo, and Kaleb McDowell

A major emphasis in the ongoing transformation of Army operations is a focus on information. Superiority in the generation, manipulation, and use of information—that is, information dominance—is widely believed to be critical to enabling military dominance on the current and future battlefield. However, one of the consequences of such a paradigm is an increase in the cognitive demands placed on soldiers. In particular, these cognitive demands are imposed by the need to effectively use the vast amount of information that can be obtained, processed, and presented by advanced, network-centric information systems to make correct decisions and take appropriate actions. System designs that overwhelm the operator's capability to make sense of information, especially under the high-stress conditions of the battlefield, will hamper soldier performance and negatively impact mission success and soldier survivability (Klein 1996). One of the major goals of technology developers, then, should be to design systems that can function naturally with soldiers' abilities to effectively perceive and process information, taking advantage of their capabilities and accounting for their limitations, in order to maximize soldier-system performance. Thus, one must ask the question: How should this goal be achieved?

One potential approach to reaching the goal of designing systems that integrate the soldiers' natural information-processing abilities is to leverage the rapidly growing area of neurotechnology, which we define here as any technology that (1) fundamentally influences how people understand the nervous system, (2) integrates knowledge of nervous system function, or (3) integrates measures of nervous system activity. Ultimately, such a neurotechnological approach may lead to enhanced system development through the creation of technologies that exploit the capabilities of nervous system structure and function, optimize across both soldier and system capabilities, and account for differences both within and between soldiers. In this chapter, we examine the potential impact of one specific area of neurotechnology on the future of soldier-system design: brain–computer interaction technologies (BCITs). Specifically, we describe and distinguish the term "BCIT" from the standard term "brain–computer interface" (BCI). We then discuss critical components of BCITs, which include brain signal detection, with an emphasis on those most relevant to BCIT design; and brain signal analysis and interpretation. Finally, we close with three examples of novel BCITs that target themes within this book: increases in complexity induced by technology; soldier-system performance under battlefield conditions; and enhancement of the system design and evaluation process.

Overview of Brain–Computer Interaction Technologies

A BCIT is a system that adapts to or integrates dynamic changes in an individual's brain state, as determined from neural signals, into its function. Although such systems, by definition, must incorporate brain signals, this does not preclude the integration of alternative non-neural physiological, behavioral, and contextual information. Quite the opposite, as is discussed elsewhere in this volume (see Chapter 11); we foresee the BCITs that integrate these multiple sources of information as likely to provide increased performance over those that do not. Future BCITs will be geared to both improving soldier–systems and augmenting the system design process. For example, BCITs may address some of the current limitations in sensing and decision-making in autonomous systems by leveraging the capabilities of the human brain to augment the capabilities of the autonomous system. However, it is a nontrivial process to measure and analyze signals from the brain, which results in BCITs having a high overhead measurement cost. To ultimately be accepted by the development community, BCIT systems must outperform both traditional and newly developing techniques for system design. That is, a BCIT must demonstrate in performance that its benefits are worth the logistical, technological, and scientific difficulty of collecting and processing brain and related data.

We distinguish the term "BCIT" from the standard term "BCI," which some have argued can only be used to describe systems meeting a very strict definition. As an example, a BCI has been defined as a system that meets four requirements: a BCI must (1) rely on brain signals, (2) be driven by the intentional decisions of the user, (3) perform in real time, and (4) provide feedback to the user about their performance (Pfurtscheller, Allison et al. 2010). This definition arises from the origins of BCI research, which was done with the goal of supporting specific patient populations, in particular patients who are paralyzed, or even "locked in," with no motor control whatsoever, including no control over the face or eyes (Wolpaw et al. 2002). The goal of many BCI researchers was, therefore, to replace the function of a failed aspect of the nervous system. Thus, the performance of the system only had to provide some capability in cases where capabilities did not exist. In contrast, when designing systems for the healthy population, the goal must be to provide capabilities that augment existing and intact physical and mental capabilities. To enable achievement of this different goal, our definition of a BCIT is more general, requiring only that a BCIT adapt to or integrate information derived from brain signals into its function. This definition does not exclude BCIs; rather, under this terminology BCIs are considered a specific type of BCIT.

Brain Signal Detection

Throughout this chapter, we will focus on electroencepthalogram (EEG)-based brain imaging technologies, which have undergone tremendous advances over the past ten years. EEG systems detect brain signals by reading the current change across the scalp caused by the simultaneous action potential discharge of millions or billions of neurons (Nunez and Srinivasan 2006). Relative to other brain-imaging modalities, EEG signals have the following characteristics: they are highly temporally accurate, detecting changes in scalp current immediately, but are also spatially inaccurate.

There are many alternative methods to detect brain function, each with its own set of trade-offs; yet we consider EEG as the most likely to function as a platform for BCITs that support near-term soldier–system design. This is because EEG is the least expensive and most mature portable noninvasive technology for detecting brain signals. Invasive techniques, such as electrocorticography provide a higher signal quality than EEG, but these techniques require electrodes to be surgically implanted under the skull (Shenoy et al. 2007), making electrocorticography impractical for a healthy population. Immobile techniques such as magnetoencephalography (MEG) can provide higher signal quality than EEG (Hämäläinen et al. 1993), while functional magnetic resonance imaging (fMRI) can provide higher spatial resolution (Belliveau et al. 1991), but both technologies require extremely large, room-sized mechanisms that can cost several million dollars, are limited in availability, have operational costs that severely limit their applicability to BCITs, and, due to their immobility, require the separation of brain activity and body behavior. However, while these immobile techniques are impractical for the majority of BCIT research and application, there exists a role for these approaches in the broader development process, that is, using fMRI to inform EEG processing for BCIT development (Hinterberger et al. 2004). Finally, competing mobile, noninvasive technologies such as functional near-infrared spectroscopy (fNIRs), which detect hemodynamic response, an indirect measure of brain activity (Chance et al. 1998), have the potential to be used successfully in future BCITs (Girouard, Solovey, and Jacob 2010), but these technologies lack the strong research base into processing techniques and system development that currently exists for EEG. Integrating EEG and fNIRs may provide increased performance for future BCITs (Izzetoglu et al. 2007).

The high temporal accuracy and low spatial accuracy inherent in EEG recording occurs because the brain signals being detected originate primarily within the brain's cortical tissues, while the EEG system records the signals at the scalp (Nunez and Srinivasan 2006). During the transfer of the electrical current from the cortex to the scalp, it passes through several different layers of fluid and tissue, including the skull, resulting in a mixing of electrical signals from various sources in the brain being detected at the scalp, obscuring the original source of the signal. Modern techniques (Onton and Makeig 2006) have made significant improvements in the spatial accuracy of EEG signals. However, these techniques tend to take significant processing time and resources, making real-time BCIT development difficult. EEG signals are also highly susceptible to artifacts in the signal. Both biological artifacts, caused by eye or muscle movement; and environmental artifacts, caused by other sources of electrical or radio frequencies in the local environment, appear in EEG recordings. Fortunately, many of these artifacts can be cleaned out of EEG signals through the use of mathematical techniques (Fatourechi et al. 2007, Iwasaki et al. 2005, McDowell, Kerick, and Oie 2010) or can be used as additional information for the BCIT.

Except for a few extremely novel prototype sensors (see, for example, Sullivan, Deiss, and Cauwenberghs 2007), EEG signals are recorded through electrodes attached to the scalp. Current state-of-the-art laboratory systems consist of a high-density array of 64–256 electrodes and require electroconductive gel to be applied for each electrode; the electrodes are connected to a signal amplifier through wires. While these EEG systems are critical for BCIT development, due to the large set-up times, the expertise needed to operate them, and a general lack of mobility, the BCITs that could employ these laboratory systems are limited in scope and capability. However, recent advances in electrode, wireless, and circuit technologies have resulted in mobile EEG systems that are substantially more usable for BCITs (Figure 12.1).

EEG systems have progressed considerably since their invention. The earliest EEG systems had relatively few wired electrodes, which had to be attached to the scalp with electroconductive gels or pastes. Laboratory systems have advanced both sensor design (for example, on-sensor amplifiers to improve the signal-to-noise ratio) and increased sensor coverage (that is, high density). In the past 10 years, significant advancements are making EEG technologies mobile and usable to the point of them being commercially available in toys and games (such as the Star Wars Force Trainer or the Emotiv EPOC). Advancements include micro-electromechanical electrodes (illustration A in Figure 12.1; Lin, Ko et al. 2008) or fabric-based electrodes (illustration B in Figure 12.1; Lin, Liao et al. 2011) that are intended for safe, unobtrusive, and comfortable long-term wear; and wireless mobile system designs (for example, the Mindo-16 system, (illustration C in Figure 12.1;Lin, Wang, Wang, Chen, Tseng, and Ko 2011), which will enable the recording of high-quality signals in the real world. As these technologies progress, we foresee high-quality, high-density wireless mobile systems that are commercially available and integrated into commonly available headwear such as hats and helmets.

Brain Signal Analysis and Interpretation

Signal analysis and interpretation methods are as critical to successful BCIT development as brain measurement. Over its long history, EEG has been primarily processed in two ways: through analyzing spectral content and analyzing event-related potentials (ERPs). Because of this history, many of the early BCIs leveraged these spectral and ERP measures. Study of the spectral content of the

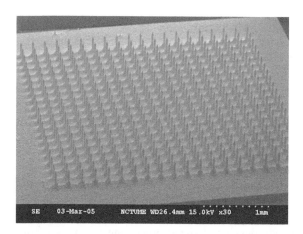

Figure 12.1   EEG systems

EEG signal and how it changes over time is one of the earliest forms of processing EEG data, first occurring in 1932 (Swartz and Goldensohn 1998). For example, by performing a Fourier transformation and power spectral analysis, the EEG signal can be broken down into frequency bands (for example, the alpha, beta, and theta bands) that can provide meaningful information to a system (Basar et al. 1999). ERPs are measurable deflections in the electrical potential recorded in the EEG signal in response to specific events (for example, visual or auditory stimuli). While there are methods for studying single-trial ERPs (Jung et al. 2001), ERPs are often very difficult to detect after a single event, so ERPs are typically obtained by time-locking the EEG signal to the time of the events' occurrence and then averaging the time-locked signal over multiple trials until a stereotypical ERP waveform appears (Rugg and Coles 1996).

The combination of recent increases in computational capabilities with existing and newly developed signal processing techniques from areas outside neuroscience has led to the development and application of many new signal analysis approaches for EEG analysis, resulting in new potential approaches for implementing future BCITs. For example, existing BCIs have used event-related synchronizations (ERS) or event-related desynchronizations (ERD) (Pfurtscheller and Lopes da Silva 1999). Similar to an ERP, an ERS is an increase in spectral power in a frequency band in response to specific visual environmental stimuli, while an ERD is a corresponding decrease in spectral power. Since ERD/ERS are highly specific to individual frequency bands, an EEG recording from a specific area of the scalp may display both an ERD and an ERS to a given event.

Among the hundreds of other EEG analysis approaches that could support BCIT development are the capabilities to spatially filter the EEG signal (for example, common spatial patterns; Dornhege, Blankertz, and Curio 2003, Müller-Gerking, Pfurtscheller, and Flyvbjerg 1999) to examine neural activity at precise spatial locations (Makeig et al. 1996, Pascual-Marqui, Michel, and Lehmann 1994), pull out specific internally generated neural events (for example, epileptic spikes; Mormann et al. 2007), or uncover the functional connectivity between neural regions (for example, granger causality or phase lag index; Stam et al. 2007, Hesse et al. 2003).

The modeling and interpretation of brain signals, regardless of modality, in relation to human behavior is a central challenge in the future development of BCITs. Importantly, and as is exemplified in the example BCITs at the end of this chapter, successful BCITs may not require a complete model of the nervous system, but rather will require the identification of some meaningful aspect of the relationship between neural activity and human behavior. There are several approaches that could be taken to uncover such aspects. Below, two common, but very different, approaches are discussed.

There is a long history of attempting to relate neural signals to psychological/ cognitive processes. In relation to BCITs, this long history has been extremely useful to provide interpretable starting points from which technologies can be developed. However, psychological/cognitive frameworks are only attempts to describe behavioral phenomena that arise from what is extremely complex neural activity. Critically, as BCITs are developed by incorporating measures of neural activity, it is necessary to optimize them based on neural activity as opposed to associated psychological processes. An example reflecting this need arises from the original attempts to use cortical signals to control the moment-to-moment actions of specific BCIs (as described by Wolpaw et al. 2002). However, these devices have seen limited success, in part, because the cortical signals being used to control them were more reflective of goal selection than they were of the moment-to-moment control. Interestingly, recent research efforts have begun to move toward goal-selection models of BCIs as opposed to moment-to-moment models (Wolpaw 2007). While this example is still framed in reference to psychological constructs (i.e., goal selection), it highlights the difficulty in interpreting how to use neural signals pragmatically in application. As an important caveat, there is also the expectation that additional neural processes will also not directly map onto intended control constructs (such as emotion, attention, or deception) to a degree that will be sufficient for implementing systems such as a BCIT. Instead, these neural processes may map to processes that are, at best, similar in nature to psychological conceptions or, more drastically, to concepts which have not yet been conceived. Use of a priori psychological or cognitive constructs as opposed to actual neural activity may, in fact, lead researchers down misguided paths.

An alternative to this psychologically driven approach for developing BCITs is to use a data-driven approach, where pattern classification and, potentially, machine-learning methods are used to uncover the relationship between neural activity and brain states or mental commands. Prior to the application of these data-driven approaches to BCI research, early BCI systems were not capable of adapting to the user. Instead, the user would have to adapt their mental states in order to produce the brain signals that the BCI could reliably incorporate, a process which could take months of arduous trial-and-error skill acquisition. However, by using data-driven approaches, the system becomes capable of learning about the user's brain signals, allowing for this calibration period to be drastically reduced (Müller et al. 2004).

Classification approaches segment sets of data into discrete categories, and similar to standard statistical methods such as an analysis of variance or *t*-test, are concerned with the differences between groups of data. However, classification techniques have more in common with statistical regression. Statistical techniques such as a *t*-test can be used to demonstrate that two sets of data are statistically different to some selected level of significance, while a classifier will find a best-fit segmentation that separates a set of data into two or more discrete classes. This segmentation can then be used to segment new data points, beyond those originally used to develop the segmentation, into one of these discrete classes. Classification approaches can be either "supervised" or "unsupervised," depending on whether a data set containing ground-truth information about the class to which the data belongs can be obtained (Hastie, Tibshirani, and Friedman 2009). As most BCIs tend to use supervised approaches (Lotte et al. 2007), we will only cover these approaches in this chapter.

To make use of a supervised classification algorithm, an initial data set that is annotated with the class to which it belongs must be provided to the algorithm. The data must be separated into a training set (used to develop the class segmentation) and a test set (used to test how well the segmentation performs on new data that the classifier has not previously seen). Often this is done through "cross-validation," where the data are broken up into test and training sets in multiple different ways and an instance of the classifier is developed for each test/training set pair in order to ensure that the approach performs well across the entire data set (Hastie, Tibshirani, and Friedman 2009). While classification can be performed simply by providing the system with raw sensor data, performance can be markedly enhanced by using features extracted from the EEG data as inputs to the classification algorithm instead of the raw data. This is where existing methods of EEG processing can be used to inform the classification process by supplying the needed features to the classification algorithms. By using a priori knowledge about what is important in the EEG signal, input features can be derived that represent the data in a way that may not be readily available to the classifier through the raw data or that relieve the classifier of the burden of feature extraction. For example, spectral properties, ERPs, ERS, and Common Spatial Patterns (CSP) have all been used as feature inputs to classifiers (Müller et al. 2008, Garcia-Molina, Tsoneva, and Nijholt 2009).

There are many classification algorithms that have been used in BCIs, each with its own set of capabilities that leads to specific advantages and disadvantages, and there are several reviews of classifiers, machine-learning, and BCI that describe these capabilities (Lotte et al. 2007, Müller et al. 2004, Müller et al. 2008, Ochoa 2002). In this chapter, we very briefly discuss two methods that have been successful in BCI applications (Müller et al. 2004, Müller et al. 2008): Linear Discriminant Analysis (LDA) and Support Vector Machines (SVM). In their fundamental forms, LDA and SVM are two-class linear classification algorithms that function by drawing a linear hyperplane between the two classes of data, although they use different mathematical criteria to draw this line. LDA works by drawing a linear hyperplane that defines a normal vector on which to project the data for a maximal separation between the two classes' means, and minimizes the variance within each class. SVM works by drawing a line that provides the "maximum margin" between the two classes, that is, it maximizes the distance between the points in each class that are closest to the other class. While both methods work well with linearly separable data, these methods can fail to classify complex, nonlinear data. However, SVMs are capable of projecting the data into a higher-dimensional space, where it becomes easier to linearly separate the data. SVMs can also use what are called "kernel functions" to model nonlinearity in the data. Commonly used kernel functions include polynomial functions for Gaussian radial basis functions (Müller et al. 2004, Hastie, Tibshirani, and Friedman 2009) or for an intuitive explanation, see Daumé III (2004).

BCITs at their core require very pragmatic models and interpretations of nervous system activity. Presented here are two of several approaches toward solving this very difficult problem. These approaches are discussed as they exemplify very different perspectives; however, we believe that even these two approaches are not mutually exclusive. In particular, considering the complexity of the human brain and the vast measurement and analysis approaches that are used in understanding neural processing, we argue that psychological/cognitive approaches should guide the identification and selection of features and classification algorithms for data-driven approaches. Likewise data-driven approaches, when performed in a principled fashion and subject to extensive post-classification analysis, will offer neurally grounded approaches to refining theories about how humans generate behavior.

Existing BCIs

BCI research started in the 1970s and focused on using technology to replace a failed aspect of the nervous system in patient populations. In particular, much of the original focus was on technologies intended for completely paralyzed, or "locked in," patients who have no real ability to interact with the outside world. Much of the early BCI work therefore focused on communication and control.

Communication-based BCIs included applications for dialing telephones (for example, calling 911) and spelling words. In fact, several research groups (Cheng et al. 2002, Wang and Jung 2011) have developed phone dialing systems based on a steady-state visually evoked potential (SSVEP). The SSVEP is a brain signal that is induced by a steadily flashing visual stimulus that is composed of frequency components appearing at the harmonic frequencies of the flashing stimulus, and that is detectable via EEG over the occipital lobe. By displaying the buttons on a phone as a set of visual stimuli on the screen and blinking them at different frequencies, the SSVEP can be used to determine which of the stimuli is being attended to, providing the system with enough information to dial a phone for the user. It is foreseeable that this same type of approach could be applied to operating a system interface or for controls such as sending email.

There are also systems that provide the ability for a locked-in patient to write by detecting the intent of the user to produce specific letters in order. Two example spelling systems are the P300 speller (Guan, Thulasidas, and Wu 2005) and the Hex-o-spell (Blankertz et al. 2006). The P300 speller is based on the P300 ERP, a brain signal related to contextual relevance (Donchin and Coles 1988). The system works by displaying each of the letters in the alphabet in a grid. Then, each of the columns and rows in the grid will flash in a sporadic random order. By concentrating on the desired letter, the brain will produce a P300 signal whenever the desired letter flashes. The system will detect this signal and correlate it with the column and row that are flashing, providing it with the grid coordinates of the letter that the user intends to occur next. By placing the letters into a word processer, the system functions as a spelling aid for the user.

In contrast, the Hex-o-spell system arrays the letters in a hexagonal fashion and uses a motor imagery paradigm. Motor imagery signals are based on the difference in brain signals between imagining different parts of the body moving, such as imagining moving a right hand versus imagining moving a left foot. Specifically, the Hex-o-Spell system differentiates between imagining right hand movement and right foot movement. By imagining right hand movement, the user can rotate a centrally positioned arrow clockwise, while imagining right foot movement will stop rotation and increase the length of the arrow. This allows the user to select a hexagon containing a group of letters or signals, and then select one of the letters or signals within that group. Unfortunately, while potentially extremely useful for "locked-in" patients, these communication BCIs are extremely slow when compared to a member of a healthy population using a keypad/keyboard.

A second area of BCI focus has been on "direct control," that is, systems that use brain signals to directly control the movement of a system. One example of a direct-control application is using a BCI to control a mouse cursor (Li et al. 2008, McFarland et al. 2008, 2010, Wolpaw, McFarland, Neat, and Forneris, 1991). McFarland et al. (2008) developed an EEG-based BCI system that allows two-dimensional cursor control with enough accuracy to select one of four targets on a screen. Similar to Hex-o-Spell, this system uses motor imagery-induced localized changes in EEG frequency power called "sensorimotor rhythms," which are recorded from the central scalp over the junction of the frontal and parietal lobes in order to produce the control signal for the cursor.

Direct-control BCIs have been used to control a wide variety of applications, including controlling a virtual character (Bayliss 2003, Lalor et al. 2005, Pfurtscheller, Leeb et al. 2006), a flight simulator (Middendorf et al. 2002), an unmanned aerial vehicle (Akce, Johnson, and Bretl 2010), a prosthesis (Guger et al. 1999), and a wheelchair (Tanaka, Matsunaga, and Wang 2005). However, while there are many different methods for obtaining the brain signals used to drive these direct-control systems, they tend to share similar rates of information transfer, about 20–24 bits per minute (bpm) (Bell et al. 2008). Similar to the communications systems, this relatively slow information transfer rate will severely limit potential direct-control applications for healthy populations.

The specific choice of applications relevant to a locked-in population allows the traditional BCI to be held to a different standard than that for applications relevant to a healthy population. Fundamentally, instead of improving on a person's existing capability, the system for a locked-in patient only has to provide a capability that the patient completely lacks. For a healthy individual, none of the described technologies provides performance similar to that which can be done manually. Phones can be dialed and documents typed much more quickly and easily than using a BCI with an information transfer rate of 20–24 bpm, while direct-control BCIs are associated with "ataxic" (irregular and jerky) movement (Wolpaw 2007).

An illustration of the challenge faced in developing neurotechnologies in healthy populations can be seen in the comparison of a "normal" mouse movement to a BCI-controlled mouse movement (Figure 12.2; Hochberg et al. 2006).

Clearly, direct-control BCIs must be improved to be relevant to future soldier systems. One critical insight into improving direct-control BCIs is the observation that the cortical brain features were being used for the early BCIs in a way that was very different from how humans perform motor control (Wolpaw et al. 2002). In particular, the human motor system has a highly evolved, efficient system between the cortical signals and end-effector muscles to facilitate the control of motor movements. However, early BCIs attempted to use the higher cortical function as a moment-to-moment signal to the end effector, thereby removing the downstream processing that the brain normally relies upon to perform motor movements (Wolpaw 2007). Some current research attempts to replace this moment-to-moment control model of BCI with a "goal selection" model, where the downstream motor pathways are compensated for with an efficient autonomous system (Wolpaw 2007). Examples of this type of research have been recently demonstrated by using brain signals to control robots (Bell et al. 2008, Iánez et al. 2010) and wheelchairs (Galán et al. 2008, Vanacker et al. 2007) with artificially intelligent systems that allow the autonomous robot or wheelchair to execute goals provided by the BCI technology.

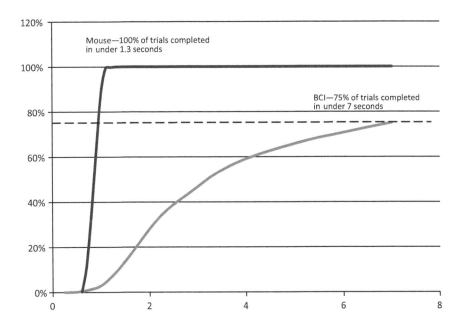

Figure 12.2   Comparison of BCI versus mouse control when selecting targets with a cursor (adapted from Hochberg et al. 2006)

These highlights reflect just a portion of the research in BCI, an area that has seen an exponential growth rate in the past decade (Hamadicharef 2010). While this chapter provides only a brief overview of BCI research, there are several excellent reviews of BCIs that can provide additional information (Millán et al. 2007, Nijholt and Tan 2008, Wolpaw 2007, Wolpaw et al. 2002).

Developing BCITs for Soldier-System Design

The combination of several factors, including the rapid advancement of mobile brain signal detection technologies; the enhancement of signal processing and interpretation methods and techniques; the continued growth of small, low-power, lightweight computational capabilities; and the exponential growth of BCI research with notable successful applications for specific patient populations leads to an expectation of further success in BCITs. However, to enable application to soldier–system designs, BCIT developers and integrators must have clear conceptions of how such technologies will be able to make the quantum leap to outperforming healthy individuals using alternative state-of-the-art technologies. We posit here not only that the basis for such concepts exists in the literature but also that a clear developmental path exists for prototype technologies. Consistent with the theme of this book, we discuss three very specific BCIT examples for (1) mitigating the impact on human performance of complexity introduced by technology, (2) mitigating the decrements in soldier-system performance induced by operational conditions on the battlefield, and (3) enhancing the system design and evaluation process.

For each example, we describe the basic concept for the benefit of using a brain signal, evidence that the brain signal can currently be detected; an existing, related system that has been developed, and a potential soldier–system application that could be developed based on the described existing technology.

Enhanced Aided Target Recognition

As technology advances, it is becoming increasingly commonplace to overwhelm the human operator with massive amounts of information, thereby severely decreasing soldier–system performance. There are several potential approaches toward mitigating this situation, for example, training or pharmaceuticals could be used in an attempt to enhance the capabilities of the human, intelligent algorithms could be used to triage the information transmitted to the operator, or more manpower could be applied to the situation. However, each of these solutions has obvious limitations. In this first BCIT example, we discuss leveraging the inherent capabilities of the human brain to (1) increase the speed in which the operator can process data and (2) develop a rapidly reconfigurable aided target recognition technology that adapts to targets the operator perceives as important.

As compared to computer systems, the human brain has formidable pattern recognition capabilities (Kapoor, Shenoy, and Tan 2008). Recently, it has been shown that by examining neural signatures in a rapid serial visual presentation (RSVP) paradigm, these pattern recognition capabilities can be exploited to increase the speed of information-processing in a target recognition task (Gerson, Parra, and Sajda 2006). In an RSVP paradigm, images are displayed rapidly (between 5 and 10 images per second) to the system user, which is on the order of a tenfold increase in speed of image presentation as compared to a self-paced task. If the human user is looking for a target among the images, when the target is identified the brain will produce a distinctive ERP called a P300, which is a measurable positive deflection in the voltage recorded at the scalp occurring 300–600 millisecond post-target presentation (Donchin and Coles 1988). Importantly, this ERP can be detected as soon as the user is consciously aware that they saw the target. By measuring the neural signal, it is possible for a system to determine approximately when the user saw a target image, thus narrowing the potential target to one of a small number of images (Mell, Bach, and Heinrich 2008).

Several researchers have recently developed systems to use this paradigm to leverage the human brain to process data. One such system is the Cortically Coupled Computer Vision system (Sajda, Pohlmeyer, Wang, Parra et al. 2010). This system uses the RSVP paradigm in conjunction with a computer vision system and provides three separate methods for doing so: a "computer-first" implementation, where computer vision algorithms are used to preprocess the image data to improve human performance at the RSVP task; a "human-first" implementation, where EEG data from the RSVP task are used to provide a small set of training images that are used to improve performance of the computer vision system; and a "tightly coupled" implementation, where the computer vision and RSVP operate in parallel with integrated results. Experiments have shown that such a system can significantly decrease the amount of time needed to search satellite imagery (Figure 12.3) and improve the percentage of targets detected (Sajda, Pohlmeyer, Wang, Parra et al. 2010).

There are many potential Army-relevant applications of similar BCITs that can increase the speed of processing visual information. Future force capabilities are likely to include extensive sensor coverage over a highly dynamic battlespace. These sensors could include manned and unmanned vehicles, and unattended ground sensors. This will lead to the collection of a vastly increased quantity of multimodal data, including imagery and video that will need to be processed. By leveraging a BCIT such as Cortically Coupled Computer Vision, it should be possible to develop technologies such as rapidly reconfigurable aided target recognition systems. As a specific example, there are known instances of insurgents hiding improvised explosive devices (IEDs) in holes in walls and covering the IEDs with posters and flags (Wheeler 2011). The posters were often intended to attract US soldiers by being either very pro- or very anti-American. Looking for posters that fit into these categories is an extremely difficult problem for computer vision systems, and developing and fielding a custom system to identify them

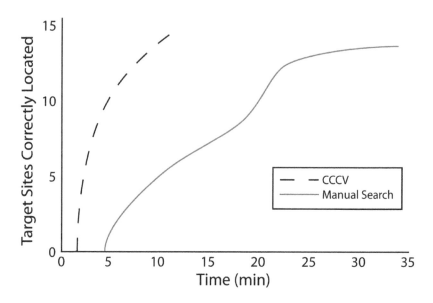

Figure 12.3 **Manual search of satellite imagery for targets compared to a search of the imagery after being triaged by the cortically coupled computer vision system (Adapted from Sajda, Pohlmeyer, Wang, Hanna et al. 2010)**

would be time- and cost-prohibitive. However, the human brain can quickly and immediately identify these posters. By using an iterative version of a computer vision system that is tightly coupled to an RSVP paradigm (or even a standard target presentation paradigm), it should be possible to train the BCIT to identify posters that have low-level visual similarities (for example, containing red, white, and blue; manmade objects; the letters U, S, and A; and so on) to the target posters. This would allow the computer vision portion of the system to rapidly process multiple data streams to identify potential IEDs.

The research to date suggests that such a system is possible for particular target types. The success of these BCITs will depend on both the human capability to perceive the target (that is, camouflage may increase the difficulty of perception) and the computer vision capability to extract and integrate salient target features within a naturalistic scene. There are also concepts that could be used to enhance the effectiveness and generalizability of these BCITs. For example, multiple soldiers or subject matter experts (SMEs) could use the system in parallel (Wang and Jung 2011) to augment the strength of the neural signal-to-noise ratio. Further, using different types of sensor systems such as hyperspectral imaging could increase both the detection rate for the human and the feature space for the computer vision, and large neural and scene reference databases could increase the speed and accuracy of the overall BCIT technology. With these types of advancements,

BCITs for the repurposing of aided target recognition systems becomes a distinct possibility.

Improving Performance under Stress and Fatigue

Perhaps one of the most critical issues facing system development is that operational conditions vary dramatically and influence both soldiers and their use of the systems. While some of these conditions (temperature, environmental particulates, sleep deprivation, and time pressures) are feasibly included in research and systems development, other conditions (extremely long-term fatigue, psychological stressors associated with family separations, or life-threatening situations) are extremely difficult to account for and can have strong negative impacts on performance. Accounting for all such performance-impacting conditions and their interactions during system design is unlikely. An alternative approach is to attempt to monitor the soldier's nervous system and relate changes in nervous system processing to changes in performance. In this second BCIT example, we discuss monitoring the soldier nervous system to (1) increase the knowledge about the individual soldier's changing capabilities, with a particular emphasis on fatigue, stress, and affective state and how they relate to performance; and (2) develop systems that adapt to these changes for enhanced performance.

There has been considerable research examining neural processing with regard to fatigue, stress, and affective state. For example, multiple researchers have demonstrated the relationship between theta, alpha, delta, and beta frequency band neural activity and fatigue (Lal and Craig 2001). A long history of research for over a half century has demonstrated relationships between different forms of stress and EEG signals (Ulett and Gleser 1952, Shackman et al. 2009). Finally, current research shows that it is possible for neural measures to provide unbiased information about the affective state of an individual, and that it may be possible to detect affective information that is subconscious or deliberately hidden from external behavior (Garcia-Molina, Tsoneva, and Nijholt 2009).

This extant literature suggests not only that the nervous system contains information that would be potentially very useful to system development but also that methods and technologies exist to retrieve that information. However, translating this research to real-world battlefield conditions will be nontrivial. For example, there is a large body of research on EEG-based affect measurement (Li and Lu 2009, Lin, Huang et al. 2010, Garcia-Molina, Tsoneva, and Nijholt 2009, Murugappan, Ramachandra, and Sazali 2010, Nijboer et al. 2009, Schaaff 2008, Zander and Jatzev 2009). However, each of the many methods of inducing, measuring, and verifying emotional and affective states provide information on different aspects of an individual's emotional state (Harmon-Jones, Amodio, and Zinner 2007, Mauss and Robinson 2009). This implies that if the battlefield conditions induce sufficiently varied affective states (which they are highly likely to do), it could be very difficult to extend the laboratory research on mental state detection to reliable BCITs. However, there are several potential approaches to address this problem. First, the focus must be on extending the most robust findings, such as are revealed in the fatigue literature (Lal and Craig 2001), to realistic conditions. The second focus must be on what the nervous system is actually doing, not on higher-level constructs that may not be revealed through the neural data. For example, much of the research on detecting affect directly from brain and other physiological signals focuses on the detection of affective dimensions such as valence or arousal, due in large part to a lack of success at categorizing stereotypical emotions such as anger or fear (Mauss and Robinson 2009). Because of this, it has been argued that psychological constructs such as attention, boredom, and anger may not directly reflect the underlying neural states that can be detected from brain signals. The final focus must be on the need to leverage future mobile neurotechnological systems to understand more about operational stressors through collecting neural data in stressful training environments, and eventually through operationally fielding data collection systems, and analyzing the resulting data.

One relevant and promising BCIT system is that developed by Lin, Ko, et al. (2008), which uses the frequency power spectra of neural signals collected through EEG to predict performance at a simulated driving task based on the expected fatigue of the driver. The driving simulator consists of a real car that is mounted on a six degree-of-freedom motion platform, which is surrounded by screens upon which seven projectors display the simulated driving scene. Performance at the driving task is defined by the amount of time it takes for the subject to react to a force that perturbs the vehicle sideways. The system predicts this reaction time using a multiple regression based on the principal components analysis (PCA) of the power spectra of the EEG signal. The regression output, when compared with the actual reaction time of the subject, has an average r correlation value of 0.9, demonstrating a high level of performance prediction on the driving task. Further, the researchers have demonstrated that low driver performance can be somewhat mitigated with a loud tone, and that this mitigation is also represented in the EEG power spectrum (Lin, Huang et al. 2010, Jung et al. 2010). The extensive literature and these developmental efforts (Lin, Ko, et al. 2008) suggest that it may be possible to expand such technologies to also be sensitive to neural activity associated with stress and affect. That is, one could develop a performance prediction technology based on multidimensional neural correlates of fatigue, stress, and affect that, in theory, would have broader generalizability across environments.

There are many potential Army-relevant applications for these kinds of technologies, which can monitor brain signals relevant to soldier performance and may indicate the effects of battlefield-relevant conditions. For example, it may be possible to develop fatigue/stress/affect-based BCITs geared toward supporting driving performance or robotic asset control in an Army-relevant secure mobility task (McDowell et al. 2008). In a wide range of applications, systems are envisioned that contrast the neural signatures to a range of expected responses, including neural or behavioral responses, based on current conditions, and use that contrast to support enhanced resource appropriations (for example, who should do what task), ultimately impacting team dynamics and communications. Further, the Army has a long and continued interest in using advanced virtual environments, serious games, and intelligent tutoring technologies to provide improved training, cognitive assessment, and cognitive rehabilitation to soldiers (Smith 2010). Monitors of fatigue/stress/affect could be used to inform training and tutoring systems that can adapt to the detection of states indicative of a poor learning performance; such systems would use existing pedagogical strategies to attempt to improve learning. Finally, many skills performed by soldiers must be performed well under realistic conditions. Systems used for training or evaluating these skills could use neurally based monitoring capabilities to modify computer-generated environments to ensure that they are evoking the intended responses.

Enhancing the System Design and Evaluation Process

SMEs are critical components of the system design and evaluation process, providing feedback and guidance along the entire developmental process. For both soldier–systems (a vehicle crew station) and autonomous systems (unmanned robots), a significant role of the SME is to observe system use and behaviorally report various types of errors. BCITs have the potential to increase the quality of information available over behavioral or self-report methodologies. Specifically, we discuss potential BCITs that aim toward enhancing the resolution (timing and weighting) of the error-reporting process, which could then be used to speed up the system design process.

The human brain has been shown to be sensitive to errors, in a manner akin to the sensitivity to target stimuli described previously. When an individual perceives an error, whether they observe it or perform it, their brain will produce a type of ERP called an "error-related negativity" (ERN) (Falkenstein et al. 2000) or "error-related potential" (ErrP) (Ferrez and Millán 2005). An ERN is a negative deflection in the electrical voltage recorded from the scalp, which occurs about 100 milliseconds after the error is perceived, with amplitude related to the perceived importance of the error (Dehaene, Posner, and Tucker 1994). An ErrP can be either a negative or positive deflection, occurring 100–200 milliseconds after the error was observed (Falkenstein et al. 2000). We use the term "error-related potential" to describe both concepts in this chapter. Error potentials can be caused either by the individual making an error or by a system making the error, and thereby failing to execute the user's directions. One paradigm for inducing and detecting system-mistake error potentials involves manipulating a cursor on a screen to move it into contact with a target. A small percentage of the time, the cursor will move in an incorrect direction, inducing the error potential (Ferrez and Millán 2008a). By analyzing these error potentials, it is possible to obtain information about the severity of the error, which is related to the amplitude of this signal (Gehring et al. 1993), as well as the precise time of the error (Buttfield, Ferrez, and Millán 2006). Several researchers have demonstrated the ability to reliably detect single-trial error potentials in real time, a development of importance for real-time BCIT design (Ferrez and Millán 2008b, Parra et al. 2003).

Researchers have integrated error-related brain signal detection capabilities into several systems, including a system that uses brain signals to spell words (Visconti et al. 2008) and a system that controls a robotic arm using brain signals (Kreilinge, Neuper, and Mueller-Putz 2010). The system developed by Kreilinger, Neuper, and Mueller-Putz (2010) detects an ERD based on motor imagery to control the robot's movement or send a stop command to the robot. However, the robot also blinks a red light-emitting diode when it has received a stop command and is ceasing movement within 1 second. If the system detected an error potential in response to this blink, then it would relay the stop command, continuing movement based on the motor imagery signal. Integrating the ability to detect error potentials into these systems provides them with the capability to quickly respond to and correct the errors that they make, leading to improved performance of the overall system.

Potentially, by integrating error-related potentials with other information, such as SME behavioral responses, the quality of the information regarding the reported errors can be improved, opening the door to numerous BCIT applications. There are many Army-relevant system design tools that could be developed or enhanced by this capability to have high-resolution error detection. For example, a tool could be developed to enhance the performance (or even the user acceptability) of semi-autonomous navigation systems. That is, SMEs familiar with a system can, through observation of system behavior, often predict that the system will make an error well before the system actually performs the erroneous behavior. Through behavioral responses, SMEs will often note a problem and provide approximate timing and location information. However, by monitoring the error potentials and behavioral responses of SMEs, very precise timing and "quality of error" information could be obtained when SMEs detect the conditions giving rise to an error, not just when the mistake is made. This information would give developers additional insights into the time course of errors, and should make it much easier to alter the behavior of the system to address emergent erroneous behavior. In the long term, human–robot interaction could be improved by using error potentials to notify the system of mistakes it is making in real time or be used to improve the user acceptance issues associated with placing semi-autonomous navigation systems on manned vehicles (McDowell et al. 2008).

Similarly, error detection-based tools would also be of value in the human factors analysis of soldier–systems. By monitoring the error potentials of the human factors SMEs as they observe soldiers interacting with the equipment being evaluated, the precise timing of errors, whether caused by equipment malfunctions or performed by the soldiers, could be recorded for later incorporation into the analysis. Ideally, both of these applications would be performed in real time with the SME observing the performance. However, nearer-term solutions would likely involve sets of SMEs observing videos of the autonomous system or human factors experiment. This would still provide the required additional information, allowing analysis and improvements to be made, while not requiring in-field sensors with real-time processing.

Conclusion

We argue that a potentially valuable approach to system development is to design systems that maximize soldier–system performance by functioning naturally with soldiers' abilities to effectively perceive and process information, taking advantage of their capabilities and accounting for their limitations; the use of neurotechnology and particularly the integration of BCITs offers such an approach. However, for this approach to succeed, the focus of BCITs needs to move away from building technologies for something healthy humans can already perform extremely well (that is, reproducing the control of a limb) or for something that the human nervous system is not evolved to accomplish (that is, controlling more than two arms). By achieving these goals, we can move past the current BCI research that has so far had a limited impact outside of specific patient populations and make strong contributions to future system design.

We illustrate three examples of BCITs that work in harmony with the human's natural neural processing: 1) an enhanced automated target recognition system that uses human pattern-recognition capabilities to improve its performance, 2) a performance enhancement technology that mitigates fatigue-based decrements in human performance at a driving task, and 3) system development tools that use precise knowledge of when SMEs detect mistakes. In addition, none of these examples requires a deep and complete understanding of the relationship between brain activity and behavior, and can be developed with current technologies and understandings of neural signals. In examples 1 and 3, the system relies on the superior ability of humans to naturally detect objects and errors in their environments, while in example 2, we are opening the bandwidth between the human and the computer, allowing the computer to adjust to fluctuations in the human's performance.

As neurotechnologies begin to leverage existing soldier capabilities and broaden the communication bandwidth between the soldier and the system, we foresee dramatic new human–system interactions arising that lead to a revolution in the effectiveness of soldier–system capabilities.

References

Akce, A., Johnson, M. and Bretl, T. 2010. Remote teleoperation of an unmanned aircraft with a brain–machine interface: Theory and preliminary results. *IEEE International Conference on Robotics and Automation*, Anchorage, Alaska, May 2010, 5322–5327.

Basar, E., Basar-Eroglu, C., Karakas, S. and Schürmann, M. 1999. Are cognitive processes manifested in event-related gamma, alpha, theta and delta oscillations in the EEG? *Neuroscience Letters*, 259(3), 165–68.

Bayliss, J.D. 2003. Use of the evoked potential P3 component for control in a virtual apartment. *IEEE Transactions on Neural Systems and Rehabilitation Engineering*, 11(2), 113–16.

Bell, C.J., Shenoy, P., Chalodhorn, R. and Rao, R.P.N. 2008. Control of a humanoid robot by a noninvasive brain–computer interface in humans. *Journal of Neural Engineering*, 5, 214.

Belliveau, J.W., Kennedy Jr., D.N., McKinstry, R.C., Buchbinder, B.R., Weisskoff, R.M., Cohen, M.S., Vevea, J.M., Brady, T.J. and Rosen, B.R. 1991. Functional mapping of the human visual cortex by magnetic resonance imaging. *Science*, 254(5032), 716.

Blankertz, B., Dornhege, G., Krauledat, M., Schröder, M., Williamson, J., Murray-Smith, R. and Müller, K.R. 2006. The Berlin brain–computer interface presents the novel mental typewriter Hex-o-spell. *Proceedings of the 3rd International Brain–Computer Interface Workshop and Training Course*. Graz, Austria, September 2006, 108–109.

Buttfield, A., Ferrez, P.W. and Millán, J.R. 2006. Towards a robust BCI: Error potentials and online learning. *IEEE Transactions on Neural Systems and Rehabilitation Engineering*, 14(2), 164–68.

Chance, B., Anday, E., Nioka, S., Zhou, S., Hong, L., Worden, K., Li, C., Murray, T., Ovetsky, Y., Pidikiti, D. and Thomas, R. 1998. A novel method for fast imaging of brain function, non-invasively, with light. *Optics Express*, 2(10), 411–23.

Cheng, M., Gao, X., Gao, S. and Xu, D. 2002. Design and implementation of a brain–computer interface with high transfer rates. *IEEE Transactions on Biomedical Engineering*. 49(10), 1181–1186.

Daumé III, H. 2004. *Support Vector Machines for Natural Language Processing*. [Online.] Available at: http://hal3.name/docs/cs544_svms.pdf (accessed: July 12, 2012).

Dehaene, S., Posner, M.I. and Tucker, D.M. 1994. Localization of a neural system for error detection and compensation. *Psychological Science*, 5(5), 303–305.

Donchin, E. and Coles, M.G.H. 1988. Is the P300 component a manifestation of context updating? *Behavioral and Brain Sciences*, 11(03), 357–74.

Dornhege, G., Blankertz, B. and Curio, G. 2003. Speeding up classification of multi-channel brain–computer interfaces: Common spatial patterns for slow cortical potentials. *Proceedings of the First International IEEE EMBS Conference on Neural Engineering*, Capri, Italy. March 2003, 595–98.

Falkenstein, M., Hoormann, J., Christ, S. and Hohnsbein, J. 2000. ERP components on reaction errors and their functional significance: A tutorial. *Biological Psychology*, 51(2–3), 87–107.

Fatourechi, M., Bashashati, A., Ward, R.K. and Birch, G.E. 2007. EMG and EOG artifacts in brain–computer interface systems: A survey. *Clinical Neurophysiology*, 118(3), 480–94.

Ferrez, P.W. and Millán, J.R. 2005. You are wrong!—automatic detection of interaction errors from brain waves. *Proceedings of the Nineteenth International Joint Conference on Artificial Intelligence*, Edinburgh, Scotland, July–August 2005, 1413–1418.

Ferrez, P.W. and Millán, J.R. 2008a. Error-related EEG potentials generated during simulated brain–computer interaction. *IEEE Transactions on Biomedical Engineering*, 55(3), 923–29.

———. 2008b. Simultaneous real-time detection of motor imagery and error-related potentials for improved BCI accuracy. *Proceedings of the 4th International Brain–Computer Interface Workshop and Training Course*, Graz, Austria, September 2008, 197–202.

Galán, F., Nuttin, M., Lew, E., Ferrez, P.W., Vanacker, G., Philips, J. and Millán, J.R. 2008. A brain-actuated wheelchair: Asynchronous and non-invasive brain–computer interfaces for continuous control of robots. *Clinical Neurophysiology*, 119(9), 2159–2169.

Garcia-Molina, G., Tsoneva, T. and Nijholt, A. 2009. Emotional brain–computer interfaces. *Proceedings of the 3rd International Conference on Affective Computing and Intelligent Interaction (ACII) and Workshops*, Amsterdam, September 2009, 1–9.

Gehring, W.J., Goss, B., Coles, M.G.H., Meyer, D.E. and Donchin, E. 1993. A neural system for error detection and compensation. *Psychological Science*, 4(6), 385–90.

Gerson, A.D., Parra, L.C. and Sajda, P. 2006. Cortically coupled computer vision for rapid image search. *IEEE Transactions on Neural Systems and Rehabilitation Engineering*, 14(2), 174–79.

Girouard, A., Solovey, E.T. and Jacob, R.J.K. 2010. Designing a passive brain computer interface using real time classification of functional near-infrared spectroscopy. *International Journal of Autonomous and Adaptive Communications Systems*.

Guan, C., Thulasidas, M. and Wu, J. 2005. High performance P300 speller for brain–computer interface. *IEEE International Workshop on Biomedical Circuits and Systems*, Singapore, December 2004.

Guger, C., Harkam, W., Hertnaes, C. and Pfurtscheller, G. 1999. Prosthetic control by an EEG-based brain–computer interface (BCI). *Proceedings of the 5th European Conference for the Advancement of Assistive Technology*, Düsseldorf, Germany, November 1999.

Hamadicharef, B. 2010. Brain–computer interface (BCI) literature—a bibliometric study. *10th International Conference on Information Sciences Signal Processing and their Applications (ISSPA)*, Kuala Lumpur, Malaysia, May 2010, 626–29.

Hämäläinen, M., Hari, R., Ilmoniemi, R.J., Knuutila, J. and Lounasmaa, O.V. 1993. Magnetoencephalography—theory, instrumentation, and applications to noninvasive studies of the working human brain. *Reviews of Modern Physics*, 65(2), 413–97.

Harmon-Jones, E., Amodio, D.M. and Zinner, L.R. 2007. Social psychological methods in emotion elicitation, in *Handbook of Emotion Elicitation and Assessment*. edited by J.A. Coan and J. J. B. Allen. Series in Affective Science. New York: Oxford University Press, 91–105.

Hastie, T., Tibshirani, R. and Friedman, J.H. 2009. *The Elements of Statistical Learning: Data Mining, Inference, and Prediction*. New York: Springer Verlag.

Hesse, W., Möller, E., Arnold, M. and Schack, B. 2003. The use of time-variant EEG Granger causality for inspecting directed interdependencies of neural assemblies. *Journal of Neuroscience Methods*, 124(2), 27–44.

Hinterberger, T., Weiskopf, N., Veit, R., Wilhelm, B., Betta, E. and Birbaumer, N. 2004. An EEG-driven brain–computer interface combined with functional magnetic resonance imaging (fMRI). *IEEE Transactions on Biomedical Engineering*, 51(6), 971–74.

Hochberg, L.R., Serruya, M.D., Friehs, G.M., Mukand, J.A., Saleh, M., Caplan, A.H., Branner, A., Chen, D., Penn, R.D. and Donoghue, J.P. 2006. Neuronal ensemble control of prosthetic devices by a human with tetraplegia. *Nature*, 442(7099), 164–71.

Iánez, E., Azorín, J.M., Úbeda, A., Ferrández, J. and Fernández, E. 2010. Mental tasks-based brain–robot interface. *Robotics and Autonomous Systems*, 58(12), 1238–1245.

Iwasaki, M., Kellinghaus, C., Alexopoulos, A.V., Burgess, R.C., Kumar, A.N., Han, Y.H., Lüders, H.O. and Leigh, R.J. 2005. Effects of eyelid closure, blinks, and eye movements on the electroencephalogram. *Clinical Neurophysiology*, 116(4), 878–85.

Izzetoglu, M., Bunce, S.C., Izzetoglu, K., Onaral, B. and Pourrezaei, K. 2007. Functional brain imaging using near-infrared technology. *IEEE Engineering in Medicine and Biology Magazine*, 26(4), 38–46.

Jung, T.P., Huang, K.C., Chuang, C.H., Chen, J.A., Ko, L.W., Chiu, T.W. and Lin, C.T. 2010. Arousing feedback rectifies lapse in performance and corresponding EEG power spectrum. *32nd Annual International Conference of the IEEE Engineering in Medicine and Biology Society (EMBC)*, Buenos Aires, August–September 1010, 1792–1795.

Jung, T.P., Makeig, S., Westerfield, M., Townsend, J., Courchesne, E. and Sejnowski, T.J. 2001. Analysis and visualization of single-trial event-related potentials. *Human Brain Mapping*, 14(3), 166–85.

Kapoor, A., Shenoy, P. and Tan, D. 2008. Combining brain–computer interfaces with vision for object categorization. *Proceedings of the 2008 IEEE Conference on Computer Vision and Pattern Recognition*, Anchorage, Alaska, June, 2008.

Klein, G. 1996. The effect of acute stressors on decision making, in *Stress and Human Performance*, edited by J.E. Driskell and E. Salas. Mahwah, NJ: Lawrence Erlbaum, 49–88.

Kreilinger, A., Neuper, C. and Mueller-Putz, G. 2010. Control of an artificial arm with time coded motor imagery and error potential detection. *Fourth*

International BCI Meeting, Asilomar, CA, May–June 2010. Available at: http:// bcimeeting.org/2010/poster_abstracts.shtml#R1. (accessed: July 27, 2012).

Lal, S.K.L. and Craig, A. 2001. A critical review of the psychophysiology of driver fatigue. *Biological Psychology*, 55(3), 173–94.

Lalor, E.C., Kelly, S.P., Finucane, C., Burke, R., Smith, R., Reilly, R.B. and Mcdarby, G. 2005. Steady-state VEP-based brain–computer interface control in an immersive 3D gaming environment. *EURASIP Journal on Applied Signal Processing*, 2005(19), 3156–3164.

Li, M. and Lu, B.L. 2009. Emotion classification based on gamma-band EEG. *Annual International Conference of the IEEE Engineering in Medicine and Biology Society (EMBC) 2009*, Minneapolis, MI, September 2009, 1223–1226.

Li, Y., Wang, C., Zhang, H. and Guan, C. 2008. An EEG-based BCI system for 2D cursor control. *IEEE International Joint Conference on Neural Networks*, Hong Kong, 2008, 2214–2219.

Lin, C.T., Huang, K.C., Chao, C.F., Chen, J.A., Chiu, T.W., Ko, L.W. and Jung, T.P. 2010. Tonic and phasic EEG and behavioral changes induced by arousing feedback. *NeuroImage*, 52(2), 633–42.

Lin, C.T., Ko, L.W., Chiou, J.C., Duann, J.R., Huang, R.S., Liang, S.F., Chiu, T.W. and Jung, T.P. 2008. Noninvasive neural prostheses using mobile and wireless EEG. *Proceedings of the IEEE*, 96(7), 1167–1183.

Lin, C.T., Liao, L.D., Liu, Y.H., Wang, I.J., Lin, B.S. and Chang, J.Y. 2011. Novel Dry polymer foam electrodes for long-term EEG measurement. *IEEE Transactions on Biomedical Engineering*, 58(5), 1200–1207.

Lin, C.T., Wang, W.R., Wang, I.J., Chen, S.F., Tseng, K.C. and Ko, L.W. 2011. Development of a miniaturized, mobile, and wireless 16-channel EEG system. *Proceedings of the 17th Annual Meeting of the Organization for Human Brain Mapping*. Quebec City, Canada, June, 2011.

Lotte, F., Congedo, M., Lécuyer, A., Lamarche, F. and Arnaldi, B. 2007. A review of classification algorithms for EEG-based brain–computer interfaces. *Journal of Neural Engineering*, 4, R1–R13.

Makeig, S., Bell, A.J., Jung, T.P. and Sejnowski, T.J. 1996. Independent component analysis of electroencephalographic data. *Advances in Neural Information Processing Systems*, 8, 145–51.

Mauss, I.B. and Robinson, M.D. 2009. Measures of emotion: A review. *Cognition and Emotion*, 23(2), 209–237.

McDowell, K., Kerick, S.E. and Oie, K. 2010. Non-linear brain activity in real-world settings: Movement artifact and the phase lag index. *Proceedings of the 27th Army Science Conference*, Orlando, FL, November 2010.

McDowell, K., Nunez, P., Hutchins, S. and Metcalfe, J.S. 2008. Secure mobility and the autonomous driver. *IEEE Transactions on Robotics*, 24(3), 688–97.

McFarland, D.J., Krusienski, D.J., Sarnacki, W.A. and Wolpaw, J.R. 2008. Emulation of computer mouse control with a noninvasive brain–computer interface. *Journal of Neural Engineering*, 5(2), 101–110.

McFarland, D.J., Sarnacki, W. and Wolpaw, J. 2010. Electroencephalographic (EEG) control of three-dimensional movement. *Journal of Neural Engineering*, 7(3), 36007.

Mell, D., Bach, M. and Heinrich, S.P. 2008. Fast stimulus sequences improve the efficiency of event-related potential P300 recordings. *Journal of Neuroscience Methods*, 174(2), 259–64.

Middendorf, M., McMillan, G., Calhoun, G. and Jones, K.S. 2002. Brain–computer interfaces based on the steady-state visual-evoked response. *IEEE Transactions on Rehabilitation Engineering*, 8(2), 211–14.

Millán, J., Ferrez, P., Galán, F., Lew, E. and Chavarriaga, R. 2007. Non-invasive brain–machine interaction. *International Journal of Pattern Recognition and Artificial Intelligence*, 20(10), 1–13.

Mormann, F., Andrzejak, R.G., Elger, C.E. and Lehnertz, K. 2007. Seizure prediction: The long and winding road. *Brain*, 130(2), 314–33.

Müller, K.R., Krauledat, M., Dornhege, G., Curio, G. and Blankertz, B. 2004. Machine learning techniques for brain–computer interfaces. *Biomedical Engineering*, 49(1), 11–22.

Müller, K.R., Tangermann, M., Dornhege, G., Krauledat, M., Curio, G. and Blankertz, B. 2008. Machine learning for real-time single-trial EEG analysis: From brain–computer interfacing to mental state monitoring. *Journal of Neuroscience Methods*, 167(1), 82–90.

Müller-Gerking, J., Pfurtscheller, G. and Flyvbjerg, H. 1999. Designing optimal spatial filters for single-trial EEG classification in a movement task. *Clinical Neurophysiology*, 110(5), 787–98.

Murugappan, M., Ramachandran, N. and Sazali, Y. 2010. Classification of human emotion from EEG using discrete wavelet transform. *Journal of Biomedical Science*, 3(4), 390–96.

Nijboer, F., Carmien, S.P., Leon, E., Morin, F.O., Koene, R.A. and Hoffmann, U. 2009. Affective brain–computer interfaces: Psychophysiological markers of emotion in healthy persons and in persons with amyotrophic lateral sclerosis. *3rd International Conference on Affective Computing and Intelligent Interaction and Workshops*, Amsterdam, September 2009, 1–11.

Nijholt, A. and Tan, D. 2008. Brain–computer interfacing for intelligent systems. *IEEE Intelligent Systems*, 23(3), 72–79.

Nunez, P.L. and Srinivasan, R. 2006. *Electric Fields of the Brain: The Neurophysics of EEG*. Oxford and New York: Oxford University Press.

Ochoa, J.B. 2002. *EEG Signal Classification for Brain–Computer Interface Applications*. Thesis. Lausanne, Switzerland: École Polytechnique Fédérale de Lausanne.

Onton, J. and Makeig, S. 2006. Information-based modeling of event-related brain dynamics. *Progress in Brain Research*, 159, 99–120.

Parra, L.C., Spence, C.D., Gerson, A.D. and Sajda, P. 2003. Response error correction—a demonstration of improved human–machine performance using real-time EEG monitoring. *IEEE Transactions on Neural Systems and Rehabilitation Engineering*, 11(2), 173–77.

Pascual-Marqui, R.D., Michel, C.M. and Lehmann, D. 1994. Low-resolution electromagnetic tomography: A new method for localizing electrical activity in the brain. *International Journal of Psychophysiology*, 18(1), 49–65.

Pfurtscheller, G. and Lopes da Silva, F.H. 1999. Event-related EEG/MEG synchronization and desynchronization: Basic principles. *Clinical Neurophysiology*, 110(11), 1842–1857.

Pfurtscheller, G., Allison, B.Z., Brunner, C., Bauernfeind, G., Solis-Escalante, T., Scherer, R., Zander, T.O., Mueller-Putz, G., Neuper, C. and Birbaumer, N. 2010. The hybrid BCI. *Frontiers in Neuroscience*, 2(3), 1–11.

Pfurtscheller, G., Leeb, R., Keinrath, C., Friedman, D., Neuper, C., Guger, C. and Slater, M. 2006. Walking from thought. *Brain Research*, 1071(1), 145–52.

Rugg, M.D. and Coles, M.G.H. 1996. *Electrophysiology of Mind: Event-Related Brain Potentials and Cognition*. Oxford and New York: Oxford University Press.

Sajda, P., Pohlmeyer, E., Wang, J., Hanna, B., Parra, L.C. and Chang, S.F. 2010. Cortically-coupled computer vision. *Brain–Computer Interfaces*. Springer, 133–48.

Sajda, P., Pohlmeyer, E., Wang, J., Parra, L.C., Christoforou, C., Dmochowski, J., Hanna, B., Bahlmann, C., Singh, M.K. and Chang, S.F. 2010. In a blink of an eye and a switch of a transistor: Cortically coupled computer vision. *Proceedings of the IEEE*, 98(3), 462–78.

Schaaff, K. (2008). *EEG-Based Emotion Recognition*. Thesis. Karlsruhe, Germany: Universitat Karlsruhe.

Shackman, A.J., McMenamin, B.W., Maxwell, J.S., Greischar, L.L. and Davidson, R.J. 2009. Right dorsolateral prefrontal cortical activity and behavioral inhibition. *Psychological Science: A Journal of the American Psychological Society*, 20(12), 1500–1506.

Shenoy, P., Miller, K.J., Ojemann, J.G. and Rao, R.P.N. 2007. Generalized features for electrocorticographic BCIs. *IEEE Transactions on Biomedical Engineering*, 55(1), 273–80.

Smith, R. 2010. The long history of gaming in military training. *Simulation and Gaming*, 41(1), 1–12.

Stam, C.J., Nolte, G. and Daffertshofer, A. 2007. Phase lag index: Assessment of functional connectivity from multi-channel EEG and MEG with diminished bias from common sources. *Human Brain Mapping*, 28(11), 1178–1193.

Sullivan, T.J., Deiss, S.R. and Cauwenberghs, G. 2007. A low-noise, non-contact EEG/ECG sensor. *IEEE Proceedings Biomedical Circuits and Systems Conference*, Montreal, Canada, November 2007, 154–57.

Swartz, B.E. and Goldensohn, E.S. 1998. Timeline of the history of EEG and associated fields. *Electroencephalography and Clinical Neurophysiology*, 106(2), 173–76.

Tanaka, K., Matsunaga, K. and Wang, H.O. 2005. Electroencephalogram-based control of an electric wheelchair. *IEEE Transactions on Robotics*, 21(4), 762–66.

Ulett, G.A. and Gleser, G. 1952. The effect of experimental stress upon the photically activated EEG. *Science*, 115(2999), 678–82.

Vanacker, G., Millán, J., Lew, E., Ferrez, P., Moles, F., Philips, J., Van Brussel, H. and Nuttin, M. 2007. Context-based filtering for assisted brain-actuated wheelchair driving. *Computational Intelligence and Neuroscience*, 1–13.

Visconti, G., Dal Seno, B., Matteucci, M. and Mainardi, L. 2008. Automatic recognition of error potentials in a P300-based brain–computer interface. *Proceedings of the 4th International Brain–Computer Interface Workshop and Training Course*, Graz, Austria, September 2008, 238–43.

Wang, Y. and Jung, T.P. (2011). A collaborative brain–computer interface for improving human performance. *PLoS One*, 6(5). [Online.] Available at www.plosone.org (accessed: July 12, 2012).

Wang, Y.-T., Wang, Y. and Jung, T.-P. 2011. A cell-phone-based brain–computer interface for communication in daily life. *Journal of Neural Engineering*, 8(2), 025018.

Wheeler, V. 2011. Inside the bomber's cauldron: 100 hurt locker heroes blitz village. *The Sun*. Available at: http://www.thesun.co.uk/sol/homepage/news/campaigns/our_boys/3609056/100-heroes-blitz-bomb-Cauldron.html (accessed: July 27, 2012).

Wolpaw, J.R. 2007. Brain–computer interfaces as new brain output pathways. *The Journal of Physiology*, 579(3), 613–19.

Wolpaw, J.R., Birbaumer, N., McFarland, D.J., Pfurtscheller, G. and Vaughan, T.M. 2002. Brain–computer interfaces for communication and control. *Clinical Neurophysiology*, 113(6), 767–91.

Wolpaw, J.R., McFarland, D.J., Neat, G.W. and Forneris, C.A. 1991. An EEG-based brain–computer interface for cursor control. *Electroencephalography and Clinical Neurophysiology*, 78(3), 252–59.

Zander, T.O. and Jatzev, S. 2009. Detecting affective covert user states with passive brain–computer interfaces. *3rd International Conference on Affective Computing and Intelligent Interaction and Workshops*. Amsterdam, September 2009, 1–9.

Chapter 13

Soldier-centered Design and Evaluation Techniques

Pamela A. Savage-Knepshield

We don't want to figure out what all those buttons do or why they are set up the way they are. We just want to get on with our lives and do our jobs well. When we make use of technology, we want to focus on achieving our goals, not on deciphering the technology. The design should be in the background of our attention.

(Vicente 2004)

Introduction

Soldier-centered design places the soldiers, who will ultimately use a system, squarely in the center of the design process and ensures that their needs are the foremost consideration when making design trade-offs and decisions. Common soldier-centered design methods include focus groups (more commonly called "user juries" in the Army), usability testing, card sorting, participatory design, questionnaires, and interviews. Including active involvement of target end-users early in the design process ensures that a system's design will be informed by the needs of the people who ultimately will be using it. Without an understanding of a system's role in the mission and the constraints in which it will be used, there is absolutely no guarantee that a system will meet its users' needs. Meeting soldier and unit needs also requires an understanding of their physical and cognitive characteristics, the work that they do, the environments in which they work, the resources that they have available to them, and the toll that their work and the environment takes on them mentally, physically, and psychologically (e.g., developing trust in their equipment). ISO 13407 (ISO 1999) identifies the following four activities as essential to a successful user-centered design process: (1) requirements gathering to understand the users' context of use; (2) specifying the user, organizational, and technological requirements; (3) producing concepts, architectures, visual designs, and prototypes; and (4) carrying out user-based assessments of the system.

Usability engineering, which entails closely integrating design and evaluation through incremental development and iterative refinement of a system based on soldier input and feedback, is another critical component of a soldier-centered design process. During the early design phases, low-fidelity paper prototypes are typically used. As the system concept evolves and becomes more concrete, so too does the fidelity of the test stimuli and highly interactive software simulations are used to collect feedback. This is followed by prototype equipment or technology demonstration units that have limited functionality, but hardware that is generally representative of the final product. When the final system becomes available, limited user-testing is conducted by the Army Test and Evaluation Command. Inherent in this process is the notion that soldier involvement continues throughout design and development, and even after the system has been fielded (as discussed in Chapter 14). It is critical that issues identified post-fielding are captured and documented so they can be used to inform their mitigation efforts as well as influence future system design efforts to prevent past mistakes from recurring.

The idea of user-centered design, whether conceptualized as soldier-, customer-, or human-centered, is not new. It seeks to optimize the user interface (UI) around how people work, rather than force users to change how they work to accommodate the system. Don Norman is credited with coining the term "user-centered design" in his research laboratory at the University of California San Diego in the 1980s (Abras, Maloney-Krichmar, and Preece 2004). Use of the term became widespread after the publication of Norman and Draper's (1986) *User-Centered System Design: New Perspectives on Human–Computer Interaction* and Norman's (1988) *The Psychology of Everyday Things* (Abras et al. 2004). The practice of user-centered design flourished with the growth of the PC industry and found renewed momentum when online businesses found a direct relationship between a website's usability and sales revenue (Wroblewski 2008). The body of literature demonstrating that user-centered design is good business as well as good for business continues to grow. Among the proven benefits in the commercial sector are reduced training costs, better operational efficiency, fewer errors, reduced frustration, less need to consult documentation, and more time to focus on other tasks (Donahue, Weinschenk, and Nowicki 1999).

One of the basic tenets of soldier-centered design is to "know thy user." By adhering to this tenet and having the design process focus on accommodating soldiers and the work that they perform, military systems can be designed to better support their users. Never before has the utility, usefulness, and usability of military systems received more media attention. One poignant example is an article written by Colonel John Charlton in which he describes his battlefield experience using the Force XXI Battlefield Command Brigade and Below (FBCB2) system crossing the Line of Departure into Iraq (Charlton 2003). In the article, he describes his experience as a 3rd Infantry Division commander who was forced to use digital maps displayed on a computer screen in his Bradley when his paper maps did not provide the level of detail needed during movement in an unfamiliar terrain and country. It was not until a sandstorm created a no-visibility situation, forcing him to rely upon an FBCB2 to maneuver his tank, much as pilots use instrumentation in bad weather conditions, that his conversion to digital battle command was complete. Not only does he espouse the battle-changing virtues of the FBCB2, he also points out its limitations, or, in his words, "What needs fixing." Excerpts from his article describing the limitations of the UI follow:

Everyone I talked with about FBCB2 complained about the operating system and graphical user interface (GUI). It is about the most non-intuitive operating system and GUI I have ever used. Even the simplest task took multiple steps to accomplish and some of the procedures simply did not make sense [...]. The FBCB2 developers really need to work on making the GUI more intuitive and user-friendly [...]. The messages are so cumbersome that nobody used them ... The FBCB2 system is physically too large for use in combat vehicles. (Charlton 2003: 17)

Usability issues are not unique to this system. In general, military systems are plagued by cumbersome data entry procedures, meaningless error messages, cryptic abbreviations, and confusing navigation steps among a multitude of cluttered, inconsistent screens that do not follow the conventions that people have grown accustomed to when using PCs. There is a failure to adequately address usability issues such as these when target end-users and human factors engineers are not included early in the system acquisition process and when contractual documents lack human performance requirements. Once defense contractors have secured a contract, there is little impetus or guidance upon which to assess the usability, utility, or usefulness of a system and its UI. Efforts are currently under way to resolve these issues and are discussed later in this chapter.

Human System Integration in the Acquisition Process

Manpower and Personnel Integration (MANPRINT) is the Army's Human Systems Integration (HSI) program. It was established in 1984 in response to concerns that human error was causing equipment failures and that systems were not being designed to maximize soldier–system performance. Through systematic consideration of the impact of materiel design on soldiers throughout the system development process, it seeks to optimize total system performance, reduce life-cycle costs, and minimize the risk of soldier loss or injury. Accomplishing this requires not only a thorough understanding of the needs of the soldier and unit but also the identification and mitigation of risks associated with use of the system. Managing risk requires knowledge of a risk's impact on human–system performance, probability of occurrence, severity, and priority (relative to other risks present). The following seven key design areas, or domains, are the focus areas for MANPRINT risk analyses (Department of the Army 2001):

- *Manpower:* The numbers of military and civilian personnel required and available to operate, maintain, sustain, and provide training for a system.
- *Personnel:* The cognitive and physical capabilities required to learn to operate, maintain, and sustain a system.

- *Training:* The instruction or education and on-the-job or unit training required to provide personnel with the essential job skills, knowledge, and attitudes to effectively use a system to accomplish their goals.
- *Human Factors Engineering:* The integration of human characteristics into system definition, design, development, and evaluation to optimize human–machine performance under operational conditions.
- *System Safety:* The design features and operating characteristics of a system that serve to minimize the potential for human or machine errors or failure that cause injurious accidents.
- *Health Hazards:* Consideration of the design features and operating characteristics of a system that create significant risks of bodily injury or death; prominent sources of health hazards include acoustics energy, chemical substances, biological substances, temperature extremes, radiation energy, oxygen deficiency, shock (not electrical), trauma, and vibration.
- *Soldier Survivability:* The characteristics of a system that can reduce fratricide, detectability, and probability of being attacked, and minimize system damage, soldier injury, and fatigue (both cognitive and physical).

Systematic assessment of human–system interaction is required in order to identify, assess, and mitigate risks present in each domain. Although the MANPRINT program specifies policy and provides oversight to ensure that government and defense contractor MANPRINT practitioners are included in materiel development, their level of involvement varies across Army programs. This variance is a function of several factors, including the point in the acquisition process at which the practitioner is engaged in the design process, the amount of funding allocated for the effort, and the willingness of the defense contractor and program manager to engage in soldier-centered design activities. The incorporation of MANPRINT in the Army acquisition process during the mid-1980s forced a radical change in the way that contractors did business with the Army, requiring them to focus on the human element and design systems that fit soldiers' needs and capabilities (Skelton 1997). Ideally, human performance parameters should be specified in requirements and contractual documents in the same manner as any other component of the system. However, in actual practice, this has not fully materialized (Booher and Minninger 2003). Efforts to address this shortcoming within the Army have been recently initiated by the Office of the Deputy Chief of Staff, G-1 MANPRINT Directorate, the US Army Research Laboratory (ARL) Human Research and Engineering Directorate (HRED), and the Training and Doctrine Command (TRADOC), which is the organization responsible for leading Army requirements determination (Department of the Army 2007). This initiative, called "Moving MANPRINT to the Left," is directed at ensuring that MANPRINT considerations are included in early system engineering planning efforts and during analyses of alternatives prior to Milestone A.[1] This is significant

---

1   Milestones are major decision-points used to oversee and manage acquisition programs (Schwartz 2010). There are currently three major milestones: Milestone A, when

Figure 13.1 An early design of the manpack radio and its HMI (display and keypad)

because it is at this point in the acquisition process that human–system interaction is first considered through the development of key performance parameters (KPPs) and system requirements; it is also the appropriate time to begin paying early attention to interfaces and interface complexity (NRC 2008). The following case study drawn from the Joint Tactical Radio System (JTRS) Program illustrates the importance of early human factors engineering (HFE) involvement in the acquisition process.

TRADOC regulation 350–70 contains guidelines for creating task performance measures or KPPs. One example, drawn from the JTRS program, illustrates a KPP that adheres to these guidelines: "the design shall allow trained operators and maintainers to perform all critical tasks required to install, operate and maintain the radio correctly on the first attempt 90 percent of the time" (Savage-Knepshield 2009). To facilitate alphanumeric entry, another requirement specified that the human–machine interface (HMI) should consist of a 3 × 4-inch keypad. It was believed that this keypad layout (Figure 13.1) coupled with the use of the multitap method prevalent on cell phones for text messaging, would minimize the amount of training necessary. However, unlike cell phones, the HMI did not

approval is sought to enter the Technology Development phase; Milestone B, when entry into the Engineering and Manufacturing Development phase is approved; and Milestone C, when entry into the Production and Deployment phase is approved.

have dedicated navigation and delete buttons. To provide this functionality, two dual-function buttons and two modes of operation were implemented. One mode was used for entering alphanumeric data and the other for navigating and deleting characters in a text field. Users switched between the two modes by pressing a shift button. When the shift button was active, an "up" arrow symbol appeared in the status region of the display (Figure 13.1).

Early usability study results (run with software simulations of the initially proposed UI) found that soldiers were unable to meet the KPP when performing the critical task of logging into the manpack radio due to the excessive cognitive workload imposed by the HMI's dual-function buttons and modal functionality. It was too difficult for soldiers to keep track of which mode was currently active. The display's mode indicator provided little help since a soldier's attention was focused on the keypad when entering data, not the display. Inability to meet this KPP provided the impetus to conduct a series of user assessment studies focused on redesigning the radio's HMI (Savage-Knepshield 2009). The redesign (Figure 13.2) facilitated the ability of soldiers to log into the radio and to navigate its menu hierarchy, which was necessary in order to successfully execute other critical tasks. However, changing the design resulted in additional programmatic costs since it necessitated changing requirements which had already been formalized under contract. Conducting these studies earlier in the design process, prior to Milestone B, would have facilitated the ability to incorporate UI and human performance requirements in the initial capabilities document, thereby avoiding rework and increased cost later in the acquisition process. These studies are required in order to set clear, comprehensive key performance parameters and system requirements; and they pay early attention to interfaces and interface complexity as recommended by the National Research Council (NRC) (NRC 2008). For additional MANPRINT case studies, refer to Booher 2003. For search engine access to a compendium of ARL technical reports, which include those documenting HFE and MANPRINT experimentation and soldier assessments, visit the ARL website.[2]

When one considers the well-documented benefits of including end-users in the design process, it is difficult to understand why it is not a well-used and accepted design practice. Why would a program not want to include soldier-centered design practices early and often? Common objections based on powerful myths include the following arguments: designing usable systems is just common sense, conducting user studies slows down the design process and add significant cost, and an effective design can be obtained by following UI design guidelines and check lists. There is also a commonly held belief that it is too difficult to locate the appropriate soldiers for participation in user studies and that too many participants are needed to obtain significant results. None of these objectives are valid.

2   Available at: www.arl.army.mil/www/default.cfm?page=239 (accessed: July 19, 2012).

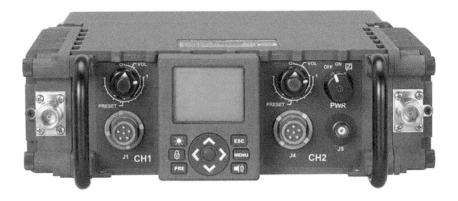

Figure 13.2   The redesigned manpack radio and HMI

The purpose of this chapter is twofold. First, it provides evidence to debunk myths held by those in program management and development who are resistant to incorporating soldier-centered design techniques. Secondly, it provides concrete examples and artifacts from soldier-centered design and evaluation techniques (for example, usability testing and card sorting for defining software menus, and symbology selection for hardware component labeling) that have had a positive impact on soldier performance. When used early and iteratively, these techniques among others have reduced design risk, improved soldier–system interaction and, total system performance.

Debunking the Myths

In this part of the chapter, a dozen of the most commonly held myths related to soldier-centered design are presented along with reality that refutes them. Where available, evidence refuting the myths has been drawn from recent design efforts conducted and/or supported by ARL. Among these myths are those that relate to usability-testing (Chrusch 2000, Becker 2011), measuring usability return-on-investment (ROI) (Rosenberg 2004), and usability as it relates to the software engineering process (Seffah and Metzker 2004).

Myth 1: Designing usable systems is just common sense.

Everything is common sense—in retrospect. Since all developers have common sense, software should be easy to use and require minimal training. The truth is that designing usable systems is part art and part science, and requires a background in behavioral or social sciences (Rubin and Chisnell 2008). The mistaken belief that it is merely common sense promotes the delusion that usability can be attained

Designing Soldier Systems

at no cost and that software developers are capable UI designers (Purba 2002). A good example of this fallacy in practice is taken from a manpack radio design effort. Unlike previous military radios, this radio is "software-defined" with components that are controlled by means of software rather than hardware (for example, filters, amplifiers, and modulators/demodulators). Therefore, it has many parameters that must be selected and set up in order to operate on a network. This requires, among other things, three pieces of information: an Internet Protocol (IP) address, a subnet mask, and a gateway. In an attempt to make it easier for soldiers to fill in this information, the developer defined each octet[3] as a separate field without using any periods (for example, 255 255 255 0). This made perfect sense since the development tool required this information parsed in this particular format, but would it cause problems for our users? A few team members suspected that it might not be an issue for addresses that conformed to the optimal case in which there were three non-zero digits, but were concerned that it might cause problems for cases that were not optimal (Figure 13.3). We would need concrete evidence on which to recommend a design change if our suspicions were correct.

IP address	255.255.255.232
Subnet mask	255.255.255.255
Gateway	255.255.255.255
DHCP*	Enabled

IP address	10.250.14.10
Subnet mask	255.255.255.0
Gateway	10.250.14.1
DHCP*	Enabled

Optimal entries   **Suboptimal entries**

Figure 13.3 IP, subnet mask, and gateway addresses
*   DHCP: Dynamic Host Configuration Protocol[4]

Prior to conducting the usability test, we decided to test optimal addresses (those with three digits in each octet) during the first week of the two-week test and those that did not conform to the optimal case during the second week. Having comparative data would allow us to understand the impact that the design would have on the user in "best" and "worst" case scenarios.

Participants were asked to judge the extent to which they agreed with a usability statement (that is, mistakes were easy to correct) using a five-point rating scale in which "1" meant that they "strongly disagreed" with the statement and "5" meant that they "strongly agreed." Usability was judged acceptable when 85 percent of

---

3   In computing, an octet is a unit of digital information that consists of 8-bit numbers or "octets" separated by a period. They are made up of numbers ranging from 0 to 255 (Ou 2006).
4   DHCP allows network devices to automatically obtain a valid Internet protocol address from a server.

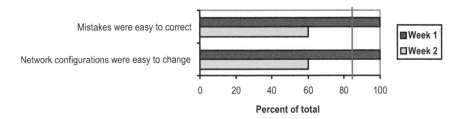

Figure 13.4   Usability test results

the participants judged an item acceptable (a rating of "4" or "5") as indicated by the vertical line in Figure 13.4. We found that the usability target was exceeded when we tested the best-case scenario, but fell short when we tested the worst-case scenario, as shown in Figure 13.4.

As we had expected, participants experienced a great deal of difficulty entering the data during the second week. Typical reactions included the following:

- "When you put in your IP address and subnet mask and all that stuff, why did you break it up? Minimize the number of button clicks. Just like when you are putting in the frequency, instead of breaking it up just put it across and arrow down (to the next entry)."
- "Changing each octet in the IP—that's insane. I usually don't change the IP on my radio because my radio doesn't have an IP. I just change the IP on my computer and that is usually all one field and I type it out consecutively."

According to participants, common sense dictated that leading zeros and trailing zeros in fixed fields should be automatically filled in by the system, that fixed decimal points should be automatically filled in for fields in which they are mandatory, and that the system should prevent them from making obvious errors, especially in cases in which there is a known valid range or set of acceptable values. Although they had never experienced entering an IP address using a radio, they had previous computer experience with IP addresses that influenced their expectations on how it should work. This is a critical concept when it comes to system design. After years of working with computers and other electronic devices, most people have formed mental models about how they work. At a conceptual level, these models are used to understand the mental representations that people form when they interact with a physical device (Gentner and Stevens 1983). Here, mental models are considered to be representations of objects, events, and processes that people construct through interaction with their environments. These models generally describe physical objects and how they work in terms of a device's internal structure and processes (Halasz and Moran 1983, Norman 1983). They provide a mechanism for people to understand and predict the effect that their actions will have on their environment. Designing systems that capitalize on existing knowledge facilitates or augments interaction with that system. Designing a system that is incongruent with one's mental model will result in negative training transfer that will hinder its usability. Usability problems ensue when there is a mismatch between a user's mental model and how a system actually works. This may seem like common sense, but it is not. If it were, all systems would work as we expect them to, which is certainly not the case.

To mitigate the usability issues identified during user testing, the following design changes were recommended:

- Edit boxes containing fixed length fields shall insert leading zeroes (for example, IP addresses) in whole digit fields, and following zeroes in decimal position fields.
- Edit boxes containing numerical fields shall align right (digits moved to right side of the field) in whole digit fields, and align left in decimal digit fields.
- Fixed use of decimal points shall appear appropriately in fields (for example, receive/transmit frequency and IP address).
- Data validation shall be incorporated when their value ranges are known.

Myth 2: Relying upon design guidance provided by "former" soldiers, who are considered subject matter experts (SMEs), is just as good as obtaining guidance from "active-duty" soldiers who have recently returned from theater.

Relying upon the opinion of former soldiers, who are SMEs, will most likely result in a system design that will not meet target end-user needs. As discussed in the opening paragraph of this chapter, designs must be grounded in information (size, shape, strength, fitness, skills, and abilities) for the people who will ultimately use them. For example, Iraq veterans who are now serving in Afghanistan will quickly explain that environmental changes (terrain and elevation) are placing different demands on them and their equipment. No longer are they rolling down asphalted highways at breakneck speeds in urban environments, suffering from excessive summer heat while conducting dismounted patrols under the heavy weight of their body armor and gear, and chatting with university graduates (Gray 2009). Now, they painstakingly travel long distances along treacherous dirt roads through bitter cold at high altitudes encountering illiterate farmers who may have never spoken to an American (Gray 2009). What worked in Iraq, may not work in Afghanistan. Tactics, techniques, and procedures (TTPs), which guide how soldiers conduct their missions, are constantly changing in response to new environments and new threats; it is difficult for those not actively engaged in the fight to be intimately aware of these changes. In order to design a system that accommodates rapidly evolving TTPs, it is necessary to recruit the participation of recent combat veterans in soldier-centered design activities. Only by gathering empirical data from actual users can one assess how well the UI for a system will meet their needs and expectations. This is particularly relevant for systems that are undergoing rapid design in order to meet soldiers' urgent needs in theater. Given this, it is not surprising that SMEs and developers alike were astonished to learn that a system design based solely upon their input was met unfavorably by soldiers participating in a usability study. This system's PM had rejected early participation of an ARL HRED human factors engineer in the design process. By the time ARL became involved and conducted a usability test, it was already too late to incorporate many of the necessary design changes due to cost and schedule constraints.

Myth 3: Conducting soldier-centered design activities slows down the design process.

Actually, this is true when applied late in the development process. Design changes will be more time-consuming to implement and may interact with other system components in unforeseen ways. The more complete a system is, the more that will have to be retested; and at later stages in the development process, this will require time that may not be available. However, with upfront planning and knowledge of the development schedule, soldier-centered design activities can be integrated within the overall design process. According to Bosert (1991), "Usability engineering has demonstrated reductions in the product-development cycle by 33 to 50 percent." To realize this benefit, the most critical information needed by the design team is the date(s) by which the developers must have guidance so it can be implemented in the upcoming software or hardware build. Armed with this knowledge, the design team can plan the timing of their user assessments to adequately accommodate software and hardware schedules. Furthermore, given the nature of an incremental design process, it makes good sense to plan user studies early in the process to ensure that the appropriate data is available when it is needed to make informed design trade-offs within an iterative or spiral design process.

Myth 4: Conducting soldier-centered design activities is costly.

The cost to conduct usability tests depends upon a variety of factors, including how much functionality is being designed and tested; the size of the test team; and the costs associated with conducting user studies, which includes the number of participants, participant fees, number of location(s), locations of test sites, how results will be presented (lengthy report or a briefing/slide presentation), how quickly study results are needed, and the cost of rental fees for facilities and test equipment.

The Nielsen Norman Group's rates for testing website usability range from $22,000 (testing a few home pages) to $45,000 (conducting comparative usability tests) (Nielsen Norman Group n.d.). Nielsen (2008), a long-time proponent of discount usability test techniques, has come to realize that field research is required to supplement lab-based testing and that more in-depth testing is required if researchers truly expect to understand the next generation of user needs.

Simple Usability, a British research and usability testing firm in Leeds, provides a research and usability cost estimator on their website. Using this estimator, the projected fee for a three-day website usability test using 18 participants with live observation of research as it occurs and results prepared in slide presentation format within 48 hours of the last day of research is US$20,791 (Simple Usability n.d.).

How does this compare to research conducted for military systems? Two examples are presented from studies conducted by ARL HRED. At one extreme is a study conducted by one human factors engineer who, in a single day, traveled to a local NJ National Guard facility to conduct a usability study with 12 soldiers. During a previous usability study, the durability of the latch that connects the HMI to the chassis of the radio had been identified as an HFE issue. The HMI is removable so a soldier can connect it to a cable, which is then connected to the radio. This allows a soldier to bury the radio in a rucksack and still have access to the radio's controls via the cabled HMI. The contractor failed to mitigate the issue and this study sought to determine the extent to which this would impact the usability of the tethered HMI feature. Would a plastic latch used to secure a removable component be durable enough to withstand repeated use in a harsh military environment? The usability task entailed having soldiers remove and reattach the component three times as quickly as possible while being timed in a group setting (Figure 13.5).

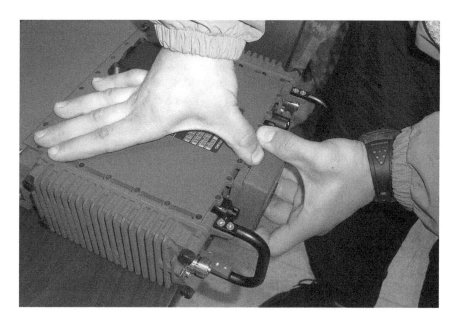

Figure 13.5   A soldier attempts to remove the HMI from the manpack radio

Soldiers practiced the task until they felt comfortable enough to begin the timed trials. The competitive aspect of the study was not lost on soldiers. They strove to beat each other's times, and in doing so, replicated, to a certain extent, the harsh treatment that the radio would be subjected to in its intended environment. The following day results were presented to the PM with the recommendation that the plastic latch be replaced with one that was more durable. This recommendation was based on the observation that during the study, a piece of the plastic latch had chipped off. This study provided the PM with the data necessary to support an earlier request for a sturdier latch (that is, one not made out of plastic), which had not been fulfilled. In this example, no additional cost was incurred by the PM to conduct this study since the human factors engineer was already 100 percent funded by this program. However, using a fully loaded (mid-range) General Schedule (GS) Pay Scale for a GS-13, the cost of this study was calculated at under $1,500 (two days of work).

At the other extreme is a study spanning one month that entailed one week of preparation, two weeks of execution (42 hours of actual testing with participants), and one week for analysis and report generation. The intent of this study was to validate a list of mission-critical radio functions and evaluate the usability of a radio's UI software and hardware with a focus on: software menu content and navigation, alphanumeric entry on a 4 × 3-button keypad, ergonomic design of its HMI, intuitiveness of HMI and status bar labeling, and its quick reference card and menu tree navigation aid. This study was conducted at 6 different military installations (Fort Gordon, GA; Fort Benning, GA; Naval Amphibious Base, Little Creek, VA; Camp Lejeune, NC; Pope Air Force Base, NC; and Fort Bragg, NC) with 18 warfighters, 2 moderators (one from ARL and another from the defense contractor's company), and 4 observers. Included in this cost estimation are air and local travel expenses, and lodging and meals at the Federal Government per diem rates for each locality visited by the ARL study moderator for a total of $3,125. Add to this, $14,000 for the ARL moderator (4 weeks of a fully loaded mid-range GS-13 salary) for a grand total of $17,125. By comparison, these costs are substantially less than those charged by either of the commercial firms mentioned above, especially given the study's scope and depth. The ARL study was considerably more extensive in scope than simply testing the usability of a few web pages or a website, and data was gathered over a longer period of time (42 vs. 24 hours). The cost for observers was not included in the cost comparison nor was the cost incurred by the other moderator who was a defense contractor. However, we found it invaluable to have these design members present during the study since it enabled us to develop a common understanding of the soldiers' context of use, problem space, and mission-critical needs, as well as the usability problems that they encountered and what needed to be done in order to resolve them (Savage-Knepshield 2009).

Resolving issues identified during early MANPRINT interventions such as these will result in a cost savings because it is a lot less expensive to make these changes at this stage in development than to retrofit a whole fleet of fielded systems with a change that is determined to be required post-fielding. HFE is an investment—spend a little now, save a lot later.

How much should be spent on usability engineering? The general rule of thumb is that 10 percent of the total development budget should be allocated to usability engineering (First Insights 2011, Gupta 2004, Nielsen 2003, 2008). Using this rule of thumb as a guideline, a program such as the JTRS Handheld, Manpack, and Small (HMS) Form Fit, which was estimated to be a $37 billion effort (GAO 2006), would need to allocate almost $4 billion to usability engineering. Obviously, this guideline is not appropriate for Major Defense Acquisition Programs such as this one. In actuality, JTRS has invested approximately .0037 percent of its total budget in ARL soldier-centered design and evaluation activities. Although the amount of funding invested by JTRS HMS defense contractors is not available, it is noteworthy that none has provided full-time HFE support to design and development and at least two of the contractors did not have this expertise in-house and relied upon the expertise provided by ARL. Given that this rule of thumb is inappropriate for government use, this remains an open question requiring additional thought and investigation.

Myth 5: Design guidelines and check lists are sufficient to create a good UI.

Guidelines and check lists are best used as a starting point or as a memory aid to ensure that some aspect of design or test and evaluation is not overlooked. They are inadequate for making design decisions because they are either too specific or not specific enough. Consider the following guideline for a push-to-talk button that is not specific enough, "If a manual press-to-talk button is used, it shall be convenient to both left- and right-handed people and designed to avoid inadvertent actuation" (MIL-STD-1472F 2001: 51). A push-to-talk button that was approximately 4 inches in diameter adhered to this guideline, but feedback obtained from soldiers during an early user assessment indicated that it was too large to be practical during tactical use. Since this dissatisfier was identified fairly early in the acquisition process, the PM had sufficient time to locate an acceptable replacement. Additional issues that may be encountered when solely relying upon design guidelines or check lists include the following:

- Guidelines and check lists may be too specific. An inter-button separation distance of 0.25 in, which is specified in MIL-STD-1472F may be impossible to achieve on a small, handheld device.
- Guidelines and check lists may conflict with one another. According to MIL-STD-1472F, "The user should not be required to learn mnemonics, codes, special or long sequences, or special instructions." However, this is exactly what is required of soldiers by devices that require user IDs and passwords.
- Guidelines and check lists may be incomplete. MIL-STD-2525C (2008) is the Department of Defense Interface Standard for Common Warfighting

Symbology. Absent from this standard are symbols for backlighting, keypad lock, and volume, which were needed to label buttons on a manpack radio. (A soldier-centered design technique for selecting button labels is presented in the next part of this chapter.)

Myth 6: Designing a consistent UI will ensure a safe, usable design.

Consistency in the surface properties of an interface will not ensure usability (Grudin 1989). According to Grudin, when UI consistency becomes the primary design driver, it averts design attention away from supporting the system's users and their work. Striving for internal UI consistency, the developers of one military global positioning system (GPS) receiver made all top-level screen layouts consistent across their design. Without being able to view the two screens side by side, it was not always readily apparent which screen was being viewed (Figure 13.6).

Figure 13.6   Present position and fire support screens

This resulted in a fatal error when the Fire support screen was confused with the Present position screen (Savage-Knepshield and Martin 2005). A usability study was conducted with 16 soldiers who reenacted a scenario in which they were preparing to call in coordinates for an air strike. After entering information about the target's location, but before they could complete their task, the GPS batteries ran out. After replacing the batteries, the soldiers were asked to continue their task. Six out of the 16 soldiers (38 percent) erroneously called in their present position as the target. As is the case with most usability issues that have catastrophic consequences, there were multiple contributing factors including: (1) a misheld belief that the GPS would return to the Fire Support screen, (2) the similarity (consistency) between the Fire support and Present position screens, and (3) the human perceptual system's weakness for detecting unexpected visual changes (change blindness). This example demonstrates that consistency for consistency's sake is not good design practice and that it is critical for designers to understand a system's context of use in order to design an effective and safe UI.

Myth 7: It is too difficult to locate soldiers to participate in soldier-centered design activities.

There are formal processes in place to request soldier participation, but it must be realized that all Army assets have specific military missions and training requirements. To request SMEs from the Armed Forces, one must prepare a written request using DD Form 2536,[5] between 30 and 90 days prior to an event and submit it to the Commander of the Military Installation or the appropriate Military Service's Public Affairs Office as indicated on the form, keeping in mind that sponsors are required to pay, when necessary, the standard military allowance for quarters and meals. Internally, ARL has access to soldiers through its parent organization the US Army Research, Development and Engineering Command (RDECOM). In addition to this, about 36 active-duty soldiers are assigned to ARL to provide its scientists, engineers, and researchers with the subject matter expertise that they require when developing technologies to meet soldier needs (Maxwell 2011). It is critical that designers and developers expend the extra effort necessary to obtain the participation of active-duty soldiers who are representative of the intended user population in order to truly judge the extent to which specific designs and technology implementations will meet end-user needs. This is achieved by drawing participants from the relevant military occupational specialty(ies) (MOS) as identified in the Target Audience Description. This document provides information about the personnel who will use, operate, maintain, train, and repair a system.

Myth 8: Soldiers do not know what they need.

Soldiers understand their role in the feedback process and, based on personal experience, they are able to predict the extent to which a system will meet their mission needs; they understand that one day their lives may depend upon it. When using survey instruments to collect feedback, it is important to ensure that (1) questions/probes are clearly worded, concise, relevant, unbiased, and unambiguous; (2) questions flow in a logical order; (3) the physical appearance of the survey is aesthetically pleasing (for example, well organized with a good use of white space); and (4) the questionnaire items are limited to absolutely essential items. For further guidelines for effective questionnaire development, refer to Azzara (2010), Bradburn, Sudman, and Wansink (2004), Sudman, Bradburn, and Schwartz, (2010), and Willis (2005).

---

5   Available at: www.dtic.mil/whs/directives/infomgt/forms/eforms/dd2536.pdf (accessed: July 19, 2012).

Myth 9: The "usability people" can make the system user-friendly after the software developers are done.

According to Tom Peters, "The dumbest mistake is viewing design as something you do at the end of the process to 'tidy up' the mess, as opposed to understanding it's a 'day one' issue and part of everything" (Lawrence 2001). During the course of several usability studies, we learned just how "dumb" this practice really is. We found that soldier–system interactions implemented early in system development were too costly and schedule-impacting to change later in the development process. We also found that the inability to make these changes affected the ability of troops to use these systems when they made their way to operational test. Among the problems encountered and reported through various Army channels were the inability to load mission plans without a standalone computer, confusing startup screen instructions, intensive data requirements that required specialized training, and rain water seeping into radio mounts when a turret was not properly sealed. One industry executive claims that systems are being designed to set up and connect as easily as it is to plug a mouse into a laptop using a USB cable (Hoffman 2011). Although this has yet to be realized for Army communications equipment, it is achievable if human factors expertise and soldier input are obtained early and iteratively throughout the design process.

Myth 10: Usability testing might uncover a showstopper and kill the program.

It is true that usability testing might uncover a showstopper; however, this is only a concern when done late in the design process. Implemented early on in the design process, iterative testing may turn up showstoppers; however, addressing these issues early in the process will have less of a negative impact on cost and schedule.

Myth 11: Measuring ROI will demonstrate the value of user-centered design.

ROI is a technique that has been used to determine the value of user-centered design and usability engineering. The money spent on usability activities is compared to the savings that resulted from the process; the ratio is one of profit to the cost of achieving the profit (Bias and Mayhew 2005). Among the success stories are claims of increased website sales, visitor traffic, new memberships, user success at finding information, and user satisfaction ratings (Foraker Labs 2011, Wizard UI Consulting 2011), as well as decreased registration times, abandoned shopping carts, and help desk costs (Bias and Mayhew 2005, Foraker Labs 2011). In spite of this evidence, there is little consensus that ROI is an appropriate measure for evaluating the worth of user-centered design. Many of those who espoused the virtues of ROI as an appropriate measure have more recently raised concerns about the weakness of ROI evidence and have called for a new approach (Dray et al. 2005, Jones 2005). Rosenberg suggests that this will involve shifting our focus away from short-term arguments that rely on numbers with low credibility, and which look at factors in isolation and without reference to context, and towards modes of analysis more in sync with the thinking of executives who have to conceptualize product value strategically and over the long-term. (2004: 15)

The ISO 9126 Software Quality Model identifies the following six quality characteristics: functionality, reliability, usability, efficiency, maintainability, and portability, all of which are good candidates for inclusion in a total cost ownership analysis (Black 2009). It is important not to neglect usability in the analysis since usability problems tend to increase overhead costs (for example, when users find systems confusing, time-consuming, error-prone, and difficult to learn). Other factors that warrant inclusion are system cost, the six quality characteristics as they relate to hardware and system accessories, and the system's ease of integration with other integral components and systems. Clearly, this is an area ripe for future research.

Myth 12: It is too early in the process—everything is not thought out.

Design can never begin too early. There are simply too many unknowns that require user input. Furthermore, research can be conducted early in the process so answers are readily available when needed to inform the requirements process and jump start the design process; they also identify information voids about the system's intended users that require further investigation—all of which are necessary to create an effective system.

The Myths: Recommendations for Eradication

Any one of these myths on its own has the potential to thwart the ability to conduct user-centered design. Much of the power of these myths comes from decoupling HSI (and user-centered design practices) from the mainstream acquisition life cycle. HSI needs to be involved early in the front-end requirements process during the Capabilities-Based Assessment (CBA). This is when the mission is defined, capabilities and their attributes identified, gaps prioritized and their associated operational risk identified, non-materiel solutions identified and assessed, and recommendations made for addressing gaps. Clearly, these are best identified, understood, and analyzed by an integrated, multidisciplinary team, which includes human factors engineers and representatives from the user community. The CBA User's Guide (JCS J-8 2009) identifies the following expertise as critical for the creation of a successful CBA: adversary expertise, analytical ability, bureaucratic agility, communications ability, doctrinal knowledge, study design, study management, cost estimation, technical knowledge, and policy knowledge. A noticeable omission is the ability to elicit, formulate, and validate effective system functional requirements. This is problematic since Lindquist (2005) determined that "poor requirements management can be attributed to 71 percent of software projects that fail." Tsumaki and Tamai (2006) found that among the factors contributing to poor requirements are an incomplete understanding of user needs, poor collaboration with users, incomplete domain knowledge, and different views held by different users. These problems led to the development of requirements that were incomplete, incorrect, ambiguous, inconsistent, and unstable. Requirements elicitation techniques include observation, structured and unstructured interviews, card sorting, brainstorming, scenario analysis, ethnographic methods, and rapid prototyping (Maiden and Rugg 1996). Human factors engineers are skilled in the execution of these techniques. Clearly, the knowledge elicitation and requirements formulation skills that human factors engineers possess will serve the CBA well.

Soldier-centered Design Techniques

This part of the chapter discusses two soldier-centered design and evaluation techniques that have had a positive impact on soldier performance: usability testing and participatory design.

Usability Testing

Usability is defined as the "extent to which a product can be used by specified users to achieve specified goals with effectiveness, efficiency and satisfaction in a specified context of use" (ISO 1998). The context of use encompasses understanding users' characteristics, motivations, goals, activities, and other factors that impact effective performance (Pew and Mavor 2007). Therefore, it is critical that these performance-affecting factors are captured and understood early and often during the design process. Usability testing yields results that identify not only what is working well but also what is not, and the risk and severity of these performance-impacting problems. Furthermore, it provides a good opportunity for validating user requirements and their context of use. Lessons learned indicate that the following considerations have contributed to the successful implementation of design recommendations from usability testing in the Army.

Enhancing the external validity of the test   "External validity" refers to the extent to which an observed causal relationship can be generalized across different measures, persons, settings, and times (Calder, Phillips, and Tybout 1982). To enhance the external validity of test results, usability test participants must include soldiers who are representative of the target audience and test scenarios that are operationally relevant and realistic. In the Army, this is accomplished by drawing test participants from the same MOS as those who will eventually use the system. It also entails obtaining test participants who have recent combat experience, since these soldiers will have the most accurate understanding of the system's context of use and will be best equipped to validate the extent to which usability tasks represent typical usage of the system in the real world.

Facilitating stakeholder buy-in   Usability test teams should include stakeholders to expedite their buy-in. For the Army, this means representation from TRADOC, the Program Office (system engineers, developers, and logisticians), and the companies under contract to design and develop the system. After system engineers, developers, and logisticians observed an initial usability study during the design of a manpack radio, they requested to be more actively involved and help conduct them. They felt that one-on-one interaction would provide them with greater insight into the usability issues, their root causes, and their potential mitigations. Following this recommendation, we conducted multiple one-on-one usability studies and design activities focused on different design aspects to enable multiple team members to participate more actively. We learned that the more involved personnel were, the greater our common understanding of the problems identified and the easier it was to identify potential mitigation strategies (Savage-Knepshield 2009). Furthermore, it was critical that system engineers and developers were included because only they completely understood the complex trade-offs required to resolve the identified issues.

Validating UI requirements   Savage-Knepshield (2009) describes four research efforts that were conducted to understand tactical communications and the impact that a proposed design would have on mission and soldier performance. The first study identified mission-essential communications functions, validated functional requirements, and identified what was working well and what was not on currently fielded military radios. Each of three usability studies, conducted subsequently during this iterative process, validated the list of mission-essential radio functions that were initially identified. This is significant because we know that mission requirements are dynamic and evolve as soldiers encounter new threats and operational environments. Validating requirements entailed asking participants to describe how they would use a radio during a mission and identify what their most frequent and important reasons were for using radios. This provided valuable design information about which radio functions were most critical and needed to be easily accessible through the UI.

Gaining insight into the issues by creating a fishbone diagram   A root cause analysis of the factors contributing to a usability issue provides insight into the options available for reducing its severity or eliminating it completely. Figure 13.7 contains an excerpt from a fishbone diagram (also called an Ishikawa diagram, herringbone diagram, cause-and-effect diagram, or a Fishikawa) that provides a structured way to identify, sort, display, and analyze possible causes of a specific usability problem. It organizes the problem space, making it easier to see the larger problems more clearly while identifying the smaller ones that contribute to them. It also facilitates design team discussion, keeping it focused on issues and their possible solutions. As shown in Figure 13.7, there were many factors contributing to the difficulty that soldiers experienced when using the system to send and receive text messages. Two root causes were the message's format and content. The format of a message did not contain expected information and soldiers were unable to customize message content to meet their needs. For example, report formats forced soldiers to fill in numerous mandatory fields from drop-down menus some of which were missing accurate content for filling them in (for example, there was no choice for an improvised explosive device to describe an observed threat). A message's header did not contain a subject field, making it difficult to know at a glance what a message was about and later locate and retrieve it when needed. This was further complicated by the inability of the system to allow the user to choose between local and Zulu time (that is, world time based on Greenwich Mean Time). This forced soldiers, who were not using Zulu time, to perform mathematical calculations to convert Zulu to local time prior to locating a message. This is significant because without a subject to refer to when searching for a particular message, time became an important differentiator. Further information about how this particular usability study was conducted and how soldiers overcame obstacles imposed by this new technology is provided in Savage-Knepshield (2006).

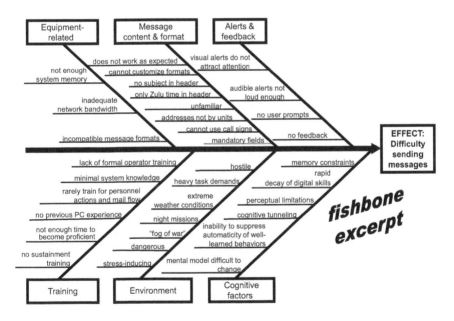

Figure 13.7 Fishbone diagram identifying the root causes of messaging problems with an early version of the FBCB2

Obtaining a comprehensive view of issues through the use of multiple usability measures and post-task queries　　When developing a usability study test plan, it is advisable to incorporate multiple measures to provide a comprehensive view of the system's usability. Consider choosing from among the following measures for each task performed during the test: task completion time, number of assists required to complete the task, whether or not the task was successfully completed, number of errors made, number of button presses, task ease of use (using a five-point rating scale), and cognitive workload (CW) (measured using a customized Modified Cooper–Harper Cognitive Workload Rating Scale, which simplifies the decision tree wording making it more appropriate for the execution of a single task as described in Savage-Knepshield (2009)). These measures provide quantitative data to compare against predetermined usability targets.

When participants rate the ease of use or CW unacceptable for a task, they are asked what change(s) would be required in order for them to rate it acceptable. This provides us with a list of soldier-suggested mitigations for each issue. We refer to the mitigation list when a task does not meet the usability target. This occurs when less than 85 percent of the participants rate a task's ease of use as acceptable (a "4" or "5" on the five-point rating scale) or its CW as acceptable ("1," "2," or "3" on the modified Cooper–Harper Cognitive Workload Rating Scale). We have found it useful to collect both ease-of-use ratings and CW ratings because they are not always positively correlated. Receiving an acceptable ease-of-use rating on a task does not guarantee that the same task will also receive an acceptable CW rating or vice versa, as shown in the following two examples:

- Acceptable Ease of Use and Unacceptable CW: entering a user name and password on a keyboard was considered easy, but the CW was judged unacceptable because it required users to remember a strong user name and password that they had not created.
- Unacceptable Ease of Use and Acceptable CW: one task required pressing a "Next" button 14 times to step through the display screens; the workload of pressing the Next button was judged acceptable but it was not considered easy to have to press the button 14 times.

Focusing on specific aspects of behavior　　Our design questions drive the task selection and other content included in our studies. As discussed earlier, our usability sessions typically have each of four soldiers paired with a data collector. To help ensure that all data collectors are focusing on the same aspects of a participant's behavior and capturing the same data in the same manner, we create a structured usability questionnaire (Figure 13.8) with a corresponding data collector guide (Figure 13.9).

In addition to the content contained on the questionnaire, the guide also contains dedicated space for the data collector to annotate observations, inferences, and utterances (that is, verbatims) as well as to capture actions taken, task time, number of requests for assistance, and questions asked by the participant.

Usability Study #2

Sub ID: _____   Jan 2009

TASK A: MAINTAINER TASKS

1. Power on the radio and perform the Built-In-Self-Test. When you are done with the Built-in-Self-Test the radio will display results of its Power-On-Self-Test (POST).

   Stop when you reach the **Restore Waveforms screen**.

IT WAS EASY TO...	Strongly disagree	Disagree	Not sure	Agree	Strongly agree	CW rating
Power on the radio.	1	2	3	4	5	____
Verify that screen colors and patterns are working properly.	1	2	3	4	5	____
Verify that the audio is working properly.	1	2	3	4	5	____

Figure 13.8   An excerpt from a structured usability questionnaire

Usability Study #2
Data Collector Guide

Sub ID: _____   Jan 2009
Observer: _____

TASK A: MAINTAINER TASKS

1. Power on the radio and perform the Built-In-Self-Test. When you are done with the Built-in-Self-Test the radio will display results of its Power-On-Self-Test (POST).

   Stop when you reach the **Restore Waveforms screen**.

TASK	TIME (min:sec)	# of Assists	Soldier Comments/Quotes	Observer comments (Behavior, attitude & actions taken)
Press the green power button near the top, right corner.	:			

Unacceptable Ease of Use Rating? (1 – 3) _____	Unacceptable CW Rating? (4 – 10) _____

Why did you rate it a < *unacceptable rating* >?
What changes are needed so you would rate it acceptable?

Figure 13.9   An excerpt from a corresponding usability data collector guide

Use of a standardized guide ensures that data collectors, who may not be well versed in the system's functionality, have the information necessary to provide assistance to participants if requested. Tape recorders are used to record participant verbatims and post-task discussions. Contents are transcribed at the end of each day by each data collector for the participants whom they ran through the study. Digital cameras are also used to capture both verbal descriptions and visual images of UI issues for inclusion in usability briefings. Verbatims from soldiers are very powerful when conveying the severity of an issue to those who did not benefit from attending the usability study.

Iterative usability testing Initially, our usability studies use low-fidelity software simulations and early prototypes before transitioning to higher-fidelity representations when technology demonstration units and production-quality systems become available. Iterative testing provides an avenue for assessing enhanced functionality when it becomes available later in the design process, ensures that design changes are truly acceptable, and makes certain that solving one problem does not inadvertently create another. When designing the backlighting functionality of the manpack radio, it took three iterations of testing to know for certain that it would meet the varied needs of multiple soldier populations. For example, dismounted soldiers traveling on foot patrols have very different backlighting needs than do soldiers who travel inside vehicles. Our design challenge entailed meeting the varied needs of our user populations without making the feature difficult to use. Accomplishing this illustrates how important it is to understand the context of use even for a seemingly simple feature such as backlighting. Table 13.1 contains a brief description of the user needs for backlighting and their corresponding system requirements.

Participatory Design

Participatory design is an approach to system evaluation, design, and development that incorporates the active involvement of intended users in the design and decision-making processes. It actively engages end-users in the process of solving design problems. One design problem faced by the manpack radio design team was to determine how best to prioritize the items within its menus. Another was the lack of symbology in Military Standard 2525C, which was needed for labeling keypad buttons. Initially, one system engineer was skeptical and believed that results from the card sorting and icon assessment tasks would not provide a clear design direction; it was believed that strong patterns of agreement among the participants would not emerge. However, results demonstrated fairly strong agreement among the participants, identified a menu tree structure that was well organized, and soldier-selected icons bore a family resemblance. As a result, the team's trust in participatory design techniques was gained. An overview of the methodology and results of each technique follow.

Table 13.1   User needs and corresponding system requirements for backlighting

User needs for backlighting	System requirement
Activation and deactivation must be quick and easy.	Activation and deactivation shall be achieved by a single button press (when the keypad is unlocked).
Deactivation must be controlled by the user.	Once activated, backlighting shall require user intervention for deactivation.
Intensity must be adjustable to accommodate varied mission requirements and environmental light conditions.	Backlighting intensity shall be adjustable and settings shall be savable by the user.
Ten levels of intensity are needed by the user.	Ten levels of backlighting intensity shall be provided including Off.
Inadvertent activation should be prevented when exercising light discipline.	The keypad shall be lockable to prevent inadvertent feature activation. The factory setting for backlighting shall be "Off."
Must be able to wear night vision goggles (NVG) when conducting missions at night.	Low levels of backlighting intensity shall not cause flaring when NVGs are worn.
Reverting to a default factory setting for backlighting will not meet mission needs when light discipline must be maintained.	Upon activation, the last saved intensity level shall be used.
Independent intensity control of the display and keypad is required.	Backlighting intensity levels for the display and keypad shall be individually controlled.
Real-time feedback must be available when adjusting backlighting intensity.	Each change in intensity shall be shown in real time on the display or keypad as its setting is adjusted.
It must be easy to change the display contrast when adjusting backlighting.	Both display backlighting and contrast shall be adjustable from the same dialog box.

Card sorting to prioritize items within menus   The design team defined menu items and their hierarchy in the structural architecture, but an understanding of how best to prioritize the items within each menu was lacking. Given this gap, all menu items had been listed in alphabetical order (Figure 13.10). We suspected that this might not be optimal and that soldiers would know best how the items within each menu should be organized based on their needs, preferences, and previous experience with similar radios. The human factors engineer convinced the team to conduct a participatory design session with soldiers in conjunction with one of the usability studies to determine how best to order the items within their menus.

Menu items in alphabetized order	Mean	Median	Mode	Warfighter recommended hierarchy
1. Main Menu	Mean			1. Main Menu
Chan Setup	3.05	3.00	2	Select Mode
CIK	5.63	5.00	4	Configure
Configure	2.63	3.00	1	Chan Setup
Help	8.37	9.00	9	CIK
Keyfill	4.47	5.00	5	Keyfill
Maintenance	6.42	6.00	6	Maintenance
Select Mode	2.89	3.00	1	Zeroize
User Admin	5.05	6.00	7	User Admin
Zeroize	6.47	7.00	6	Help
2. Chan Setup				2. Chan Setup
Group	1.95	2.00	2	Preset
Preset	1.58	1.00	1	Group
SCAN	3.53	4.00	4	Waveform
Waveform	2.95	3.00	3	SCAN
3. Configure				3. Configure
Clock	3.26	3.00	5	Keys
GPS	3.37	3.00	4	Preferences
Keys	2.42	2.00	1	Network
Network	3.11	4.00	4	GPS
Preferences	2.84	2.00	1	Clock

Figure 13.10 Pre-test (alphabetized) and post-test (warfighter prioritized) menu items

A closed card sorting technique was used since the menu categories had been pre-defined and we wanted to learn how the soldiers would sort content items within each category. Each menu item was typed on a 3 × 5 inch index card creating a total of 100 cards. Menus were shallow in that they did not exceed three levels, and contained no more than nine items each. Menu items were coded on the index cards using font size and color to allow us to quickly reassemble them in the proper order within categories in preparation for the next participant.

Thinking about the importance of each item's function and the frequency with which they would use it, soldiers were asked to hierarchically organize the menu items. They were also asked to suggest better labels for the menu items if any came to mind. Soldiers began with the highest-level menu and organized its contents working their way through the groups of cards until all had been sorted. Soldiers were asked to think aloud while they worked so we would understand their rationale behind their organization schemes. On average, it took about 45 minutes to complete the card sorting. Using a sheet of paper that contained all the menu and submenu items grouped by level and function, the data collector recorded each item's relative position in the hierarchy (that is, 1, 2, 3, 4, 5, 6, 7, 8, or 9 for a nine-item menu). Next, the mean, median, and mode were calculated for each item's ranking. The mode was used to rank order the menu items and when there was a bimodal distribution, the mean was used as the tie breaker as shown in Figure 13.10.

Evaluating the appropriateness of button labels (text and icons)   The design team used its judgment to select roughly five strong candidate icons and one text label for each function accessible through a keypad button. Text labels were included because icons do not necessarily help users find an item better than text labels (Wiedenbeck 1999). A significant constraint in the selection process was the small size of the buttons, which limited icon size to 16 by 16 pixels. Next, a two-page questionnaire was created. At the top of the first page was the function's name (for example, Volume) followed by a brief description of its purpose (for example, "Press this button to adjust the audio output volume of the radio. Use the audio controls on your headset, push-to-talk, or other audio output equipment to fine tune volume levels"). On the left side of the page was an illustration of the radio's UI with an arrow indicating which button would be used for this function and underneath this were the icon choices labeled with a single upper case alphabetic character (for example, J, R, H) (Figure 13.11). To the right of the illustration were boxes in which participants indicated which icon was their first choice, which was their second, and so on. They also indicated how well their first choice represented the function using a five-point rating scale with anchors of "very well" and "not at all." On the second page, participants were asked to rate the appropriateness of each icon using a 10-point rating scale. We found that the representativeness and appropriateness of an icon were highly correlated. Therefore, we discarded the representativeness rating data and used the more finely-grained appropriateness rating combined with the ranking data to isolate the best icon candidates. Figure 13.11 contains a one-page form without the representativeness rating section.

Figure 13.11 contains the final icon and text labels as they appear on the redesigned keypad. Icon "R" was selected as the first choice by 37 percent of participants and received an appropriateness rating of 8.9. The final label set contained three text labels (ESC, MENU, and PRE) and three icons (volume, backlighting, and keypad lock). Thus far, subsequent testing has not revealed any usability issues with any of these selections.

Conclusions

Designing systems and their UIs for military applications is an extremely complex undertaking, which is made even more challenging by the late stage at which MANPRINT and HFE are included in the process, typically after UI and human performance requirements, if any, have been identified, agreed upon, and incorporated into contractual documents. Becoming involved at this late stage ensures that we continue to feed the myths that we seek to debunk.

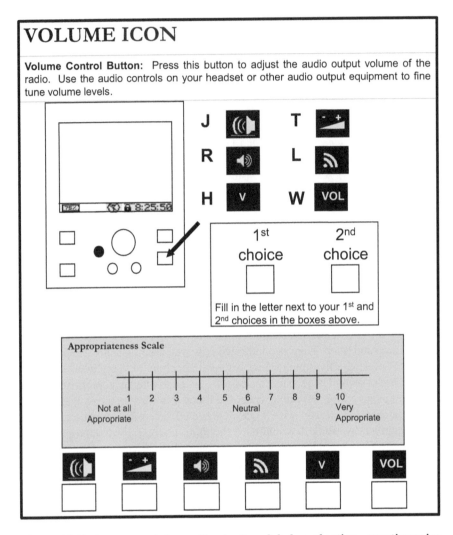

Figure 13.11 An excerpt from the button label evaluation questionnaire (shown here as it was redesigned post-test)

Our task becomes one of redesign. The Department of Defense (DoD) recognizes that the UI is a critical component of design and the key to effective HSI (DAU 2006). Ensuring the design of an effective UI is one of the key contributions that systems engineering in collaboration with HFE should make to the design process. HRED's pilot program, "Moving MANPRINT to the Left," holds great promise to realize the benefits of this partnership by teaming human factors engineers with system engineers early in the acquisition process. This will enable human factors engineers to contribute to the analysis and descriptions of the key boundary conditions and operational environments that impact how systems can be employed to satisfy a mission need. Among these key boundary conditions are manpower, personnel, training, environment, safety, occupational health, human factors, habitability, and survivability—considerations that have a major impact on system performance and life-cycle costs. Unlike commercial design efforts that focus on capturing market share, the designers of military equipment must focus on preserving and protecting the lives of their users, friendly forces, and the civilian populace. The stakes are high for our users; their lives and the lives of others ultimately depend upon their selection of the "right" menu item, the depression of the "right" button, and the selection of the "right" control to override/interrupt an automatic firing sequence. Successful designs are the result of an integrated, iterative design process conducted by a multidisciplinary team that understands not only user needs and requirements but also the mission context in which the equipment will be used. Critical to the design of a successful UI is the inclusion of target end-users—the soldiers who will actually use the system once it is deployed to the field—throughout the design process. This will ensure that the system is designed to support their existing beliefs, attitudes, and behaviors as they relate to the tasks that the system is being designed to support. Soldier-centered design builds a bridge between those creating military products and the soldiers using them. Debunking long-held myths against engaging in any practice is difficult to overcome. However, by engaging design teams in user-centered design practices and sharing lessons learned from successful case studies, we will as a community broaden the awareness that acquisition is not just about the technology, it is about how soldiers work, live, and use the technology to accomplish their mission goals.

References

Abras, C., Maloney-Krichmar, D. and Preece, J. 2004. User-centered design, in *Berkshire Encyclopedia of Human–Computer Interaction*, edited by W.S. Bainbridge. Great Barrington, MA: Berkshire Publishing Group.

Azzara, C.V. 2010. *Questionnaire Design for Business Research*. Mustang, OK: Tate Publishing and Enterprises.

Becker, D. 2011. Usability testing on a shoestring: Test-driving your website. *ONLINE Magazine*, 35(3) (May/June), 38–41. [Online.] Available at: http://www.infotoday.com/online/may11/Becker-Usability-Testing-on-a-Shoestring.shtml (accessed: July 19, 2012).

Bias, R.G. and Mayhew, D.J. 2005. *Cost-Justifying Usability: An Update for the Internet Age*. 2nd Edition. San Francisco, CA: Morgan Kaufmann.

Black, R. 2009. *Management the Testing Process—Practical Tools and Techniques for Managing Software and Hardware Testing*. 3rd Edition. Indianapolis, IN: John Wiley.

Booher, H.R. (ed.) 2003. *Handbook of Human Systems Integration*. Hoboken, NJ: John Wiley.

Booher, H. and Minninger, J. 2003. Human systems integration in Army systems acquisition, in *Handbook of Human Systems Integration*, edited by H.R. Booher. Hoboken, NJ: John Wiley, 663–98.

Bosert, J.L. 1991. *Quality Function Deployment: A Practitioner's Approach.* Milwaukee, WI: ASQC Quality Press.

Bradburn, N.M., Sudman, S. and Wansink, B. 2004. *Asking Questions: The Definitive Guide to Questionnaire Design—for Market Research, Political Polls, and Social and Health Questionnaires.* San Francisco, CA: Jossey-Bass.

Calder, B.J., Phillips, L.W. and Tybout, A.M. 1982. The concept of external validity. *Journal of Consumer Research*, 9, 240–44.

Charlton, J.W. 2003. Digital battle command: Baptism by fire. *Armor*, 112(6) (November–December), 26–29, 50.

Chrusch, M. 2000. The whiteboard: Seven great myths of usability. *Interactions*, 7(5), 13–16.

DAU 2006. *Defense Acquisition Guidebook.* [Online.] Available at: https://dag. dau.mil/Pages/Default.aspx (accessed: July 24, 2012).

Department of the Army 2001. *Army Regulation 602–2: Manpower and Personnel Integration (MANPRINT) in the System Acquisition Process.* [Online.] Available at: http://www.apd.army.mil/pdffiles/r602_2.pdf (accessed: July 19, 2012).

——. 2007. *Army Regulation 10–87: Organization and Functions Army Commands, Army Service Component Commands, and Direct Reporting Units.* [Online.] Available at: http://www.fas.org/irp/doddir/army/ar10-87.pdf (accessed: July 19, 2012).

Donahue, G.M., Weinschenk, S. and Nowicki, J. 1999, July 27. *Usability is Good Business.* Available at: [Online]. Available at: http://half-tide.net/ UsabilityCost-BenefitPaper.pdf (accessed: July 24, 2012).

Dray, S., Karat, C-M., Rosenberg, D., Siegel, D. and Wixon, D. 2005. Is ROI an effective approach for persuading decision-makers of the value of user-centered design? *Proceedings of CHI 2005*, Portland, OR, April 2005, 1168–1169.

First Insights 2011. *Usability testing.* [Online.] Available at: http://www. firstinsights.com/usability_testing.htm (accessed: July 24, 2012).

Foraker Labs 2011. *Usability ROI—Case Studies.* [Online.] Available at: www. usability first.com/about-usability/usability-roi/case-studies (accessed: July 19, 2012).

GAO (Government Accountability Office) 2006. *Defense Acquisitions Restructured JTRS Program Reduces Risk, but Significant Challenges Remain.* GAO-06-955, September, 2006. Washington, DC: GAO.

Gentner, D. and Stevens, A.L. (eds) 1983. *Mental Models.* Hillsdale, NJ: Lawrence Erlbaum.

Gray, D.D. 2009. *Afghanistan, Iraq: Different Wars.* [Online: Breitbart.] Available at: http://www.timesfreepress.com/news/2009/nov/29/afghanistan-iraq-different-wars/?mobile (accessed: July 24, 2012).

Grudin, J. 1989. The case against user interface consistency. *Communications of the ACM*, 32(10), October, 1164–1173.

Gupta, M. 2004. The utopia of intuitive services and products: Interview with Dr. Nico Pals and Joke Korte from TNO, December, 2004. [Online: EURESCOM mess@ge.] Available at: http://archive.eurescom.eu/message/messageDec2004/Interview_with_Dr_Nico_Pals_and_Joke_Korte_from_TNO.asp (accessed: July 19, 2012).

Halasz, F.G. and Moran, T.P. 1983. Mental models and problem solving in using a calculator. *Proceedings of the CHI'83 Human Factors in Computing Systems*, Boston, MA, December 1983. NY: ACM.

Hoffman, M. 2011. US Army seeks easier communications in vehicles. *Defense News*, July 17. [Online.] Available at: http://www.defensenews.com/story.php?i=7120069 (accessed: July 19, 2012).

ISO 1998. *ISO 9241: Ergonomic Requirements for Office Work with Visual Display Terminals (VDTs)—Part 11: Guidance on Usability.*

ISO 1999. *ISO 13407: Human Centered Design Processes for Interactive Systems.*

JCS J-8 2009. *Capabilities-Based Assessment (CBA) User's Guide, Version 3*. March, 2009. Available at: http://www.dtic.mil/futurejointwarfare/strategic/cba_guidev3.pdf (accessed: July 19, 2012).

Jones, C. 2005. When ROI isn't enough: Making persuasive cases for user-centered design. *UX Matters*. [Online.] Available at: http://www.uxmatters.com/mt/archives/2007/05/when-roi-isnt-enough-making-persuasive-cases-for-user-centered-design.php (accessed: July 19, 2012).

Lawrence, P. 2001. Tom Peters on design. *@ issue: The Online Journal of Business and Design*, 6(1), June. [Online.] Available at: http://www.cdf.org/issue_journal/tom_peters_on_design.html (accessed: 19 July, 2012).

Lindquist, C. 2005. Required: Fixing the software requirements mess. *CIO Magazine*, 19(4), November 15. [Online.] Available at: www.cio.com/article/14295/Fixing_the_Software_ Requirements_Mess (accessed: July 19, 2012).

Maiden, N.A.M. and Rugg, G. 1996. ACRE: Selecting methods for requirements acquisition. *Software Engineering Journal*, 11, 183–92.

Maxwell, S. 2011. Future force: Army research lab equips the warfighter. *Soldiers*, August, 2011, 4–5.

MIL-STD-1472F 2001. *Department of Defense Design Criteria Standard: Human Engineering*. [Online.] Available at: http://www.public.navy.mil/navsafecen/Documents/acquisition/MILSTD1472F.pdf (accessed: July 19, 2012).

MIL-STD-2525C 2008. *Department of Defense Interface Standard: Common Warfighting Symbology (Revision C)*, November 17, 2008. [Online.] Available at: http://www.mapsymbs.com/ms2525c.pdf (accessed: July 24, 2012).

Nielsen, J. 2003. Return on investment for usability. *Jakob Nielsen's Alertbox*, January 7, 2008. [Online.] Available at: http://www.useit.com/alertbox/roi-first-study.html (accessed: July 19, 2012).

Nielsen, J. 2008. Usability ROI declining, but still strong. *Jakob Nielsen's Alertbox*, January 22, 2008. [Online.] Available at: http://www.useit.com/alertbox/roi.html (accessed: July 19, 2012).

Nielsen Norman Group n.d. Strategies to enhance the user experience. [Online, available at http://www.nngroup.com/services/testing.html (accessed: June 26, 2012).

Norman, D. 1988. *The Psychology of Everyday Things*. NY: Doubleday.

Norman, D.A. 1983. Some observations on mental models, in *Mental Models*, edited by D. Gentner and A.L. Stevens, Hillsdale, NJ: Lawrence Erlbaum, 7–14.

Norman, D.A. and Draper, S.W. (eds) 1986. *User-Centered System Design: New Perspectives on Human–Computer Interaction*. Hillsdale, NJ: Lawrence Erlbaum.

NRC (National Research Council Committee on Pre-Milestone A Systems Engineering) 2008. *Pre-Milestone A and Earlyphase Systems Engineering: A Retrospective Review and benefits for Future Air Force Systems Acquisition*. Washington, DC: The National Academies Press.

Ou, G. 2006. IP subnetting made easy. *TechRepublic*. [Online.] Available at: http://www.techrepublic.com/article/ip-subnetting-made-easy/6089187 (accessed: July 19, 2012).

Pew, R.W. and Mavor, A.S. (eds) 2007. *Human–System Integration in the System Development Process: A New Look*. Washington, DC: National Academy Press.

Purba, S. (ed.) 2002. *Architectures for E-Business Systems: Building the Foundation for Tomorrow's Success*. Boca Raton, FL: CRC Press.

Rosenberg, D. 2004. The myths of usability ROI. *Interactions*, 11(5), 22–29.

Rubin, J. and Chisnell, D. 2008. *Handbook of Usability Testing: How to Plan, Design, and Conduct Effective Tests*. 2nd Edition. NY: John Wiley and Sons.

Savage-Knepshield, P.A. 2006. Adaptive soldiers: Overcoming obstacles imposed by new technology. *Proceedings of the HFES 50th Annual Meeting*, San Francisco, CA, October 2005, 2512–2516.

———. 2009. Applying a warfighter-centric system design process to a Department of Defense acquisition program. *Journal of Cognitive Engineering and Decision Making*, 3(1), 47–66.

Savage-Knepshield, P.A. and Martin, J. 2005. A human factors field evaluation of a handheld GPS for dismounted soldiers. *Proceedings of the HFES 49th Annual Meeting*, Orlando, FL, September 2005, 1719–1723.

Schwartz, M. 2010. Defense Acquisitions: *How DOD Acquires Weapon Systems and Recent Efforts to Reform the Process*. Washington, DC: Congressional Research Service. Available at: http://www.fas.org/sgp/crs/natsec/RL34026.pdf (accessed: July 23, 2012).

Seffah, A. and Metzker, E. 2004. The obstacles and myths of usability and software engineering: Avoiding the usability pitfalls involved in managing the software development life cycle. *Communications of the ACM*, 47(12), 71–76.

Simple Usability n.d. Website. [Online.] Available at: http://www.simpleusability. com (accessed: June 26, 2012).

Skelton, I. 1997. MANPRINT for the US Army. *Congressional Record—House*, H8269–71.

Sudman, S., Bradburn, N.M. and Schwartz, N. 2010. *Thinking About Answers: The Application of Cognitive Processes to Survey Methodology*. San Francisco, CA: Jossey-Bass.

Tsumaki, T. and Tamai, T. 2006. Framework for matching requirements elicitation techniques to project characteristics. *Software Process: Improvement and Practice*, 11, 505–19.

Vicente, K. 2004. *The Human Factor: Revolutionizing the Way We Live with Technology*. NY: Routledge.

Wiedenbeck, S. 1999. The use of icons and labels in an end user application program: An empirical study of learning and retention. *Behavior and Information Technology*, 18(2), 68–82.

Willis, G.B. 2005. *Cognitive Interviewing: A tool for Improving Questionnaire Design*. Thousand Oaks, CA: Sage Publications.

Wizard UI Consulting 2011. *ROI of Usability*. [Online.] Available at: http:// en.wizardui.com/articles-quotes-2/roi-of-usability (accessed: July 19, 2012).

Wroblewski, L. 2008. *Web Form Design*. Brooklyn, NY: Rosenfeld Media.

Chapter 14

Addressing and Resolving Science and Technology Capability Gaps Identified by Warfighters in Iraq and Afghanistan

Raymond M. Bateman, Charles L. Hernandez, and Frank Morelli

During the early part of Operation Iraqi Freedom (OIF) and Operation Enduring Freedom (OEF) US soldiers engaged an enemy who countered advanced technology with improvised solutions. As the war progressed, this pattern of evolving enemy tactics and Coalition countermeasures led to a cycle of new technologies being introduced to both theaters of operation. Soldier adaptations and improvements to those technologies followed, and in turn, new human factors issues emerged. For example, US vehicles deployed during the beginning stages of OIF were not heavily armored. Soldiers during this period (2003–2004) attached metal plating, sandbags, and crew-served weapons to their vehicles to improve offensive and defensive capability relative to enemy fire, but in doing so increased the maximum vehicle weight from an estimated 8,000 lb to 10,000 lb. These adaptations led the Department of Defense (DoD) to begin acquiring foreign-made armored vehicles and modifying vehicles currently in the Army inventory such as the high-mobility multipurpose wheeled vehicle (HMMWV). The additional weight of the armor required the frame, wheels, and driveline to be strengthened; however, over time these modifications proved to be insufficient. Around 2004–2005, AM General, a HMMWV manufacturer, increased its production of improved, armored HMMWVs for the military. These enclosed vehicles offered an improvement in armor protection, but required air-conditioning to operate successfully in the desert. Some vehicle variants were equipped with air-conditioning units mounted inside the crew compartment, which decreased the amount of space available for its occupants. Human factors engineering (HFE) was balanced with measures to improve soldier capability and well-being, but as time passed the venerable HMMWV was modified to the point at which it had reached maximum functional potential. A new vehicle platform solution—the Mine Resistant Ambush Protected vehicle—emerged to further improve protection for soldiers and offer better mobility. An attempt was made to integrate HFE early during vehicle design. As vehicle requirements evolved in response to ever-changing enemy tactics, constant reevaluation of human factors standards was required to keep pace. Some vehicle adaptations created unanticipated and serious hindrances for vehicle crews such as blocking emergency egress routes.

This chapter principally discusses some of the human system integration (HSI) issues and challenges issues that three civilian scientists and human factors specialists addressed during deployments to OIF from 2003 to 2009. During this time period, the pace of the conflict was fast and placed a heavy demand for rapid human factors assessments in lieu of the more formal assessment process. It was well known that soldiers were continuing to identify warfighting capability shortfalls that needed to be remedied quickly. It was also incumbent on the Army's Research and Development communities to be as responsive as possible in assessing the HSI issues related to these demands when fielding new or improved capabilities. The unceasing demand for professional human factors consultation during material development, production, and deployment to the battlefield is the focus of this chapter. As the experiences that follow will convey, the role of the human factors professional changed fluidly in response to the dynamics and conditions on the battlefield.

Field Assistance in Science and Technology Teams

The Army Materiel Command (AMC), through the Field Assistance in Science and Technology (FAST) Headquarters, deploys FAST Department of the Army civilian (DAC) and uniformed science advisors and Science and Technology Assistance Teams (STATs) to Southwest Asia (SWA) as part of the AMC support to combatant commanders. Civilian science advisors and uniformed science advisors are attached to combatant commanders, major training centers, and senior Army commanders located at major commands worldwide. Additionally, specialized STATs are deployed in Iraq and Afghanistan in direct support to the theater commander. Collectively, the DAC science advisor, the uniformed science advisor, and STATs are called "science advisors."

The AMC FAST activity is a vital communication link between soldiers and the materiel development community and is headed by a senior science advisor. The FAST team provides expert technical advice to commanders and investigates rapid solutions for materiel problems to improve performance, readiness, safety, and training. The science advisors coordinate delivery of urgently needed equipment and assist with field evaluations and product demonstrations. In coordination with the FAST headquarters, the science advisor identifies and conveys new requirements to AMC, the US Army Research, Development, and Engineering Command (RDECOM), and the US Army Research Laboratory (ARL) to enable them to produce solutions to fill capability gaps encountered by soldiers in the field. The FAST team conducts battle damage assessment and evaluates soldier experiences with vehicles, equipment, and current technologies. Ultimately, their mission is to identify capability gaps, potential science and technology (S&T) solutions, and quickly obtain these solutions to provide rapid support to commanders and soldiers.

In OIF, formal human factors evaluations were not practical because the pace of the conflict was fast, and rapid solutions in theater were often necessary. When soldiers have a requirement, they will find a solution, even if it is homegrown (for example, fabricating armor plating from readily available materials). As the war progressed, each subsequent set of science advisors was better able to conduct more thorough human factors assessments and provide better solutions for S&T gaps elucidated by soldiers.

Early in Operation Iraqi Freedom and Operation Enduring Freedom (2003)

The US military has deployed to austere environments many times in the last 20 years for contingency operations such as humanitarian relief and political stabilization. Operating in an austere environment, where entire bases had to be constructed without an infrastructure, proved a challenge for soldiers in OIF and OEF. The challenges included power generation, supply distribution, high temperatures of the harsh Iraqi desert, and long lines of communication between geographically dispersed operating bases. Life-support operations in the desert were initially primitive but quickly improved over time to support critical missions with the addition of DACs and contractor personnel. Deploying civilians in support of combat operations is in itself challenging, but proved essential to reduce the logistical and support burden on soldiers focused on fighting a war. The FAST teams played a major part in overcoming these and other challenges faced by our soldiers.

Science Advisor to Logistics Support Element (OIF-1)

AMC uses the Logistics Support Element (LSE) to provide deployed units logistics assistance to improve materiel readiness and soldier quality of life. The LSE, with personnel assigned from AMC, drew on this capability to provide Army units with the ability to meet unique requirements and resolve technical problems encountered. Expertise from the LSE personnel supported deployed units in areas such as maintenance, electronics, contracting, ammunition surveillance, and equipment retrograde.[1] Soldier-identified capability gaps were documented by the science advisor through observation and participation in the soldiers' operational environment by visiting the different bases and camps in Iraq. Tests and demonstrations were conducted to provide soldiers with the needed S&T solutions to overcome the evolving challenges and threats that they encountered.

The LSE provides a seamless logistics system that spans the strategic, operational, and tactical levels to provide commanders on the battlefield access to robust AMC capabilities that hasten the development of solutions for technical

---

1 Retrograde is a process for the movement of equipment and materiel from a deployed theater to a Reset (that is, replace, recapitalize, or repair) program or to another theater of operations to replenish unit stocks or satisfy stock requirements.

shortcomings. The mission of the LSE is to enhance readiness through the integrated application of AMC's logistics projection of technological capabilities to deployed units, primarily in areas related to technical assistance, supply, and maintenance. The LSE is task-organized for a theater of operations based on specific mission requirements and priorities of the combatant commander. The LSE comprises individuals with varied occupational specialties, thus providing the FAST team access to world-class expertise for a rapid solution to their S&T-identified capability gaps. The FAST team facilitated quantification of problems so issues were communicated correctly to the FAST headquarters, which assigned the issues to Quick Reaction Coordinators for further resolution. The Coordinators work to resolve problems and facilitate communication between the FAST team and the engineers and scientists at the different Army laboratories and research centers.

On arrival in Iraq, the FAST team was briefed by the AMC Chain of Command, so they would understand the current priorities and known problem areas. The initial assessment of the situation by the FAST Science Advisors in the early part of OIF was based on observations and visits to units in Kuwait and Iraq from April to October 2003. Visits were coordinated through the LSE with each of the major units in Baghdad, Balad, Tikrit, Mosul, and many of the smaller Forward Operating Bases (FOBs). Of utmost importance were initial discussions with commanders, which were followed by visits with soldiers in the subordinate units. One of the early challenges encountered was the dramatic increase in usability issues caused by wearing Personal Protective Equipment, which made movement inside vehicles difficult. In general, operations were more difficult due to the harsh environment which affected all aspects of operations due to high temperatures, sand, and dust effects on mechanical and electronic equipment. Peacetime practices for requesting and waiting for parts would not suffice in this environment. Commanders maneuvered quickly into Iraq to conduct continuous full-spectrum operations. A consequence of the rapid and sustained maneuver was accelerated wear on equipment. By 2002, 75 percent of vehicles were beyond their service life. Items such as track for tanks, personnel carriers, and artillery wore out at an accelerated rate because of the harsh desert conditions and high mileage—dust became an abrasive paste and wore out track quickly (Figure 14.1).

Calculations used in the Manpower Requirements Criteria were identified to have serious shortcomings by the Army Audit Agency; and calculations for part reliability, such as mean time to repair and mean time to failure, were overly optimistic. The need for new calculations (predictions) and materiel created a surge in logistics demands during the initial build-up and deployment phase for OIF. More man hours were necessary for repair and maintenance than predicted by the Manpower Requirements Criteria process, and critical force structure decisions based on the initial direct maintenance man hours were incorrect. The Army Forces Command Inspector General, under the auspices of his Commander, initiated a study in 2002 to review the status of materiel readiness. During the study, they inspected 237 vehicles (in the Continental United States classified

Figure 14.1   Track worn by abrasive desert dust and other ambient particulates

as "fully mission capable" (FMC)) and found only 16 percent to be FMC. The Inspector General then had Organization Contract Augmentation Teams inspect 102 FMC vehicles, of which 15 percent were identified as FMC, with 72 percent having three or more "dead line"[2] deficiencies.

Equipment serviceability, as reported in the Unit Status Report, was not a true indicator of actual readiness. The nearly instantaneous need for additional supplies, repairs, and manpower strained the national supply system. The FAST team's contribution included validating and quantifying reported problems and assisting in identifying and rapidly fielding solutions. They provided new or updated procedures, hardware, and software to improve the supply chain and the availability of motor vehicles and manpower. Over time, these initiatives resulted, for example, in improved maintenance manpower, ballistic protection, fuel availability, and countermeasures to improvised explosive devices (IEDs). OIF and OEF demonstrated that tools, resources, and time are interrelated. Increasing maintenance man hours was achieved by having soldiers work 12–14 hours a day. The additional hours worked resulted in quicker gains, freeing soldiers to perform additional tasks. However, it was not without a cost. Working at this pace in a

2   No longer able to operate because of a fault, breakdown, or loss of power.

hostile environment becomes a serious contributor to human error due to physical, cognitive, and emotional fatigue.

Science Advisor to the Commander, III Armored Corps, OIF-2 (2004–2005)

During OIF-1, measures were identified to enhance soldier performance, safety, and combat training. One major project that had an immediate impact for US and Coalition forces during this period included IED mitigation. Through collaboration with the RDECOM Liaison and III Corps Force Management Directorate, the science advisor was able to start the process of making the Operational Needs Statement a more effective reflection of S&T gaps.

During the second Fallujah campaign Al-Fajr (New Dawn) in November–December 2004, the science advisor cooperated with and directly supported the technical exploitation mission for information operations in Fallujah. This mission focused on using and organizing adversarial information to provide an operational advantage to US forces. Specific contributions to the effort included providing S&T solutions to facilitate the identification of suspected chemical munitions and location of weapons and bomb-making caches. Failure analyses of vehicles destroyed by IEDs were also conducted. Results were provided to the S&T community to assist with identifying weaknesses and redesign of the M1114 (level I armor) and the ARL Door Kit. Another contribution was to help describe the mechanism of soldier injuries. These descriptions were provided to the S&T community so they could improve individual protective gear and vehicle design to better safeguard soldiers. This data also assisted with future vehicle production and integration of advanced armor systems.

Spurred by negative publicity about the level of armor protection, Secretary of Defense Rumsfeld declared that no vehicle would enter Iraq from Kuwait without being armored. Coincidentally at about the same time, an officer stationed in Camp Arifjan, Kuwait, shot a metal plate to demonstrate to his soldiers the capability of the armor. Eleven of the 29 rounds fired defeated the metal plate that he had found in a scrap yard. As a result the Commander, AMC LSE SWA, requested a review of the Level III armor that was being manufactured and installed at Camp Arifjan. The science advisor reviewed the manufacturing process of the level III armor kit and conducted field tests that verified that the kits constructed by AMC did have ballistic properties sufficient to stop 7.62 mm × 39 mm and 5.56 mm × 45mm small arms fire (Figure 14.2). Three plates were shot five times with an AK-47 and with an M4 carbine with no penetrations. An in-theater AR 15-6 investigation later determined that the officer had used a piece of metal similar in appearance to the armor plating, but lacking the ballistic characteristics of the metal used by AMC LSE. The science advisor's actions were able to deflect the concern raised by the officer's impromptu, non-scientific assessment which had put the entire AMC LSE level III armor manufacturing and installation process in jeopardy.

Upper left bullet impacts, shot with an AK-47 (commonly used by insurgents)

Lower right bullet impacts, shot with the M4 Carbine (current issue to US Military)

Figure 14.2   Level III door panel used during field ballistic test at Camp Arifjan

Science and Technology Solutions to Warfighter Needs in OIF-3 (2005)

From July to December 2005, the 10th AMC team (STAT-10) deployed to Iraq. The team included three active-duty military personnel from the Natick Soldier Research Development and Engineering Center and AMC, and one DAC human factors specialist from ARL's Human Research and Engineering Directorate (HRED).

Operating from the Victory Base Complex (VBC), in the vicinity of the Baghdad International Airport, STAT-10 and the outgoing STAT-9 conducted week-long side-by-side orientations and onsite visits to familiarize STAT-10 with the lie of the land (Camps Victory and Liberty and FOB Stryker) and the units present at VBC. Most importantly, these orientations enabled face-to-face transitional coordination with unit points of contact, which proved invaluable in sustaining the momentum for S&T programs already in progress. Within the VBC area of operation, STAT-10 had free and protected access to combat and support units ranging in size from platoon to battalion and brigade. STAT-10 networked with two division headquarters to transit the operational area by helicopter and C-130 aircraft, from Mosul in the North to Basra in the South, in order to reach soldiers and units operating throughout theater.

Accompanying the Soldier to Uncover Requirements and Potential Solutions

STAT-10 found itself immersed in an opportunity-rich environment where units were eager to investigate ways to improve their mission capabilities. It was not uncommon for soldiers and crews to arrive unannounced at the building that housed STAT-10, a structure which also housed the Training and Doctrine Command (TRADOC) and Rapid Equipping Force (REF), to present ideas and request help.

On a daily basis, the team visited with units in motor pools, medical aid stations, and orderly rooms, as well as those preparing for and returning from patrols, to get first-hand information on their capabilities, limitations, and requirements. STAT-10 also integrated with combat support patrols and civil–military support convoys in and around Baghdad. At Camp Liberty, the team built upon a relationship that existed between STAT-9 and an Engineer Battalion that was performing daily route-clearance missions on several of the major supply routes in and around Baghdad. The team leader and human factors specialist participated in a reconnaissance patrol and gained valuable first-hand information about the capabilities and limitations of an Armor Battalion's Scout unit that was conducting missions with up-armored HMMWVs in lieu of tanks and Bradley Fighting Vehicles. This is noteworthy because the HMMWV had significant HSI issues, in part because the vehicle was being used in missions other than those for which it was designed and, therefore, was being pushed to its mechanical limits. During OIF-3, HMMWVs were being retrofitted with heavy-duty suspensions and transmissions to meet the operational demands being placed on them.

Documentation and Data Collection

In a six-month span, STAT-10 provided a large amount of information to the Headquarters, AMC System of Systems Integration (SOSI) office, using AMC-FAST information products that were already in place (that is, the Issue Sheet, Feedback Reports and Assessments, and Requests for Information (RFIs)). These products were generated in theater and electronically transmitted to the AMC SOSI office, which then tasked research and development agencies for the requested information or initiated actions that would result in prototype technologies being delivered into theater for evaluation.

The predominant issues STAT-10 addressed related to the ever-growing counter-IED (C-IED) effort, though it also investigated potential S&T solutions ranging from escalation of force capabilities to enhancing soldier survivability to enhancing sustainability. Altogether, STAT-10 generated 14 Issue Sheets, 19 Feedback Reports and Assessments, and 7 RFIs. These were in addition to the 18 active projects that were inherited from STATs 7–9 and still required in-theater assessment. Formality in the HSI investigative process was desired but the urgency of mission requirements dictated more expedient responses. Every prototype technology investigated by STAT-10 was fielded to units along with a data-collection instrument. The data-collection instrument most widely used was a single-page questionnaire (see Figure 14.3) that was designed to solicit quick feedback on low-density items (that is, equipment fielded in small numerical quantities).

System/Item Name _____ Date _____ Location _____ Unit _____

Rate the following from 1 to 10 (Circle your answers)

	Terrible	Poor	Adequate	Good	Excellent
Performance: Does the system/item perform as expected?	1 2	3 4	5 6	7 8	9 10
Effectiveness: Does the system/item meet requirements?	1 2	3 4	5 6	7 8	9 10
Survivability: Is the system/item adequately built for field use?	1 2	3 4	5 6	7 8	9 10
Training: Are adequate training materials/support available?	1 2	3 4	5 6	7 8	9 10
Reliability: Does the system/item hold up under continual use?	1 2	3 4	5 6	7 8	9 10
Maintainability; Is the system/item easily cleaned and fixed?	1 2	3 4	5 6	7 8	9 10
Supportability; Are parts/procedures/support readily available?	1 2	3 4	5 6	7 8	9 10
Interoperation: How well does this system operate with other Systems, if required?	1 2	3 4	5 6	7 8	9 10
Employment: If this system was fielded, would you use it?	Yes	No			
How would you use it? _____					
Pros: List and explain at least 3 things you like about the system/item. _____					
Cons: List and explain at least 3 things you do not like about the system/item. _____					
Spiral progression/development: What are the recommendations for adaptations or enhancements to make this equipment/system more effective for future use? _____					

Figure 14.3 Sample questionnaire

This questionnaire was sufficient to get soldiers to quickly assess the technology and was generally augmented with post-mission interviews. The duration of time the unit used the prototype technology before providing feedback could be as short as a single mission or as long as a month. Ordinarily, there was no requirement for the unit to relinquish the prototype technology after the assessment period if it was serviceable and useful to its mission.

Finding the Right Solutions for System-of-Systems Issues

STAT-10 investigated complicated and broad-breadth issues using a system-of-systems approach (for example, from the individual soldier's weapons and personal protective ensemble, to the vehicle or system from which he was required to fight, to the robotic vehicle he might also be employing). This is exemplified in our efforts to address mission and capability shortfalls in the C-IED effort as described below.

For Explosive Ordnance Disposal (EOD) units, the team investigated night-vision and face-shield protective cover enhancements for the bomb suit worn by EOD

Figure 14.4   Soldier-designed TALON robot enhancements

personnel. It also investigated enhancements soldiers were making to the TALON robotic vehicle to give it additional capabilities (Figure 14.4). STAT-10 responded by codifying soldier ideas into a FAST Issue Sheet. The sheet defined requirements that would provide the robotic vehicle with extended reach and a standardized tool set to improve its digging, cutting, lifting, raking, and grabbing capabilities.

Within two months, prototype enhancements for the EOD suit and the TALON robot were distributed to a number of EOD units across the Iraqi theater for an assessment that extended beyond the STAT-10 six-month deployment.

The C-IED investigative effort extended to the deliberate vehicle-mounted IED detection missions being conducted by engineer units on a daily basis. For this, the team focused on identifying quick-fix solutions to shortfalls in capabilities for the armored HMMWV and the Buffalo Mine Protection Vehicle (MPV). During OIF III armored HMMWVs were being pressed into the C-IED effort although they were never designed for that mission. The Buffalo MPV was also being issued to engineer units to augment their HMMWV-based C-IED mission packages and was inadequate for that task, resulting in mission capability shortfalls. For example, units that employed both vehicles wanted combinations of white and infrared (IR) light capabilities, and IR imaging devices to enhance C-IED operations at night. STAT-10 evaluated the state of the art for IR lights being employed in theater, determined that they were inadequate (Figure 14.5), and recommended acquiring new IR lights and imagers for in-theater evaluation.

In addition to the C-IED capabilities that resided in the EOD and engineer units, all tactical units that travelled in convoys and performed mounted and dismounted patrols were faced with the threat of encountering an IED threat. As such, C-IED tactics, techniques, and procedures (TTPs) within these units varied, compounding the challenge of identifying a standardized set of capabilities that could be fielded to satisfy C-IED needs across the force. An example was the need for vegetation and garbage removal as a C-IED capability. Units conducting patrols and support convoys along less-improved roads found that insurgents were using garbage and heavy vegetation to camouflage IEDs. They requested a solution that would help

Figure 14.5   Inadequate IR and white lights for Buffalo MPV

them remove this camouflage quickly and in large quantities. STAT-10 submitted an RFI for vehicle-mounted directional fans or blowers that could be affixed to the front of tactical vehicles to blow away debris that may hide an IED. As an interim less costly solution (based on a request made by a unit), drip torches were procured commercially and shipped into theater for immediate fielding and evaluation. Receiving units quickly employed the drip torches, dismounting to burn debris and vegetation, but the utility of this process was limited since it left soldiers vulnerable to sniper fire. Ultimately a vehicle-mounted fan was determined more desirable to take advantage of the armor protection of the vehicle and several fans were fielded for evaluation, post STAT-10.

The last aspect of the C-IED mission capability that STAT-10 investigated was viewed by soldiers to be the most promising. During OIF-3 the chance discovery of an IED by any convoy was largely a localized operation wherein the unit detecting such a threat would communicate their finding by radio and await an on-call EOD team to arrive on the scene to eradicate the threat. EOD team reaction times could be as short as a few minutes or much longer leaving the stopped convoy vulnerable to attack. Discussions between STAT-10 and the leadership of several infantry units uncovered a need to combine stand-off identification and interrogation with video and broadcast capabilities. Units wanted the image and data from a suspected IED to be broadcast directly from the sensor to higher headquarters and the EOD Tactical

Operations Center (TOC). With this capability, guidance could be electronically received from the TOC to employ an internal means to dispose of the IED, rather than waiting for an EOD team to perform the task, thereby improving the overall survivability of the detecting unit. STAT-10 transmitted an Issue Sheet that was shared with the REF Headquarters. The AMC SOSI office reacted by presenting the broadcast challenge to the Program Executive Office (PEO) for Command, Control, Communication and Computer Technology and providing the impetus for the US Army Night Vision Laboratory to quickly develop an initial prototype sensor-to-platoon leader transmission capability. Although this capability was not fielded during the STAT-10 time frame, the cooperative involvement by the AMC, PEO, and REF exemplified how the Research and Development community could work together to quickly respond to needs of the warfighter in theater.

Science and Technology Solutions to Warfighter Needs in OIF 09-11 (2009)

Beginning in January 2009, RDECOM deployed its 20th STAT to Iraq. The team consisted of two non-commissioned officers, two field-grade officers, and a DAC scientist from ARL. The basic operational structure and function of the STAT was largely unchanged from previous OIF phases, and many major human factors issues remained as soldiers continued to cope with ever-changing mission challenges. With each new technology or each change to an existing technology, new human factors issues emerged, producing new challenges for the soldiers and human factors engineers to overcome.

FAST STAT-20 Operations

Day-to-day operations in Iraq in 2009 continued to reveal the extreme adaptability and improvisation on the part of soldiers to maintain mission effectiveness. Soldier efforts to remedy equipment shortcomings or adapt to new enemy tactics resulted in new requirements, challenging the Army materiel, research, development, technology, and acquisition communities to accommodate emerging soldier-identified requirements appropriately with quick and effective scientific and technological support.

STAT-20 facilitated technological improvements by methodically visiting units throughout the battlefield. They asked soldiers three basic and essential questions with respect to the status of their equipment and technology, including: (1) What works well? (2) What works poorly? and 3) How have you or would you improve the existing state of your equipment and technology?

As a team, STAT-20 also embedded with units on missions to achieve a deeper understanding of mission demands, technical concerns, requirements, and operational dynamics. While the S&T gaps the team harvested were certainly not limited to human factors and ergonomics concerns, many of the issues encountered were directly related to poor design resulting from failure to consider operational use. The following case studies exemplify some of these HFE shortcomings.

Husky Operator Mission Dynamics Observation

The Husky is a single-occupant vehicle used by combat engineers for route clearance operations in Iraq and Afghanistan. The operator of the Husky vehicle is often the lead vehicle on a route-clearance mission and is frequently the first observer to identify suspicious objects and IEDs along a given route. An experienced Husky operator is a proficient observer, with scanning and perceptual skills for target object discrimination at a level not possible for soldiers without similar, experience-based learned expertise. STAT-20 spoke with numerous Husky operators and identified human factors concerns related to fatigue, detrimental cognitive, perceptual, and attentional resource demands, and conditions that increased susceptibility to hidden IEDs on the battlefield such as heightened operational burdens on this lone-vehicle occupant.

Despite their superior skill, Husky operators reported performance degradations due to cognitive and attentional strain from intense, continuous scanning efforts during prolonged mission durations (for example, 12+ hour missions). This was especially true during long, sequential mission cycles (for example, daily missions for multiple weeks without interruption). In addition, Husky operators described feeling extremely isolated during route-clearance missions and perceiving their environment through a virtual "fishbowl." This appeared to be the result of the Husky's design as a single-occupant vehicle with a small space for the operator within the vehicle. With mission times ranging from 4 hours to in excess of 16 hours, Husky operators felt entrapped—often not afforded the option of exiting on patrol stoppages, given their status as the sole vehicle occupant. Husky operators also reported that the lack of face-to-face communication with fellow soldiers over this time span was emotionally draining. This led to shorter length of effective duty due to "burnout," a characteristic not as prevalent among personnel manning the other route-clearance vehicles in a convoy, all of which are multi-operator vehicles. Self-reported distraction and the resultant risk of task-related inattention were also elevated for the Husky operator. This was caused by the simultaneous tasking to drive, scan the environment for threats, monitor the display output of a sensor package, and communicate new developments to other convoy vehicles when traveling along routes. These conditions unquestionably burden the visuo-perceptual capacity of a single operator, especially over long mission durations. The end result was in many cases a reassignment of skilled and experienced Husky operators to an alternate, non-mission duty (for example, desk assignment), a loss that unnecessarily degraded unit C-IED capabilities. In addition to the aforementioned challenges, the Husky operator was at constant risk of physical injury from undetected explosive devices. If incapacitated during a hostile event, extraction of the Husky operator from the small compartment in which he was confined was extremely difficult. Rescue tools available at this time did not work well in the enclosed, small space and, in fact, introduced a risk of further injury to the soldier. The defensive measures taken by Husky operators for such contingencies caused additional problems. One such tactic was the common practice of loosely wearing tourniquets for self-aid on all appendages (discussed in more detail below).

Husky operators proposed a number of potential solutions to mitigate the challenges they faced. Work schedule solutions included a reduction in Husky mission duration for a given operator, as well as training multiple soldiers in an engineering company to perform Husky duty to provide redundant organic capability. They also provided equipment recommendations based on improved human factors principles, such as vehicular redesign to accommodate an additional mounted operator. This suggestion was founded in the following rationale: a second operator would permit (1) direct face-to-face contact, mitigating the emotional isolation characteristic of Husky duty; (2) shared scanning duty, reducing the visuo-attentional burden on a single operator; (3) shared mission duty allowing for longer mission duration; and (4) less complicated post-IED event extraction of soldiers given a larger compartment. The major shortcoming of the suggested redesign was the additional risk placed on two soldiers rather than one when encountering enemy attack, but the majority of Husky operators that the team spoke to deemed this risk to be acceptable in lieu of maintaining the status quo. STAT-20 played an integral role in relaying the concerns of the Husky operators back to RDECOM scientists and engineers, and was encouraged by the subsequent development of prototype dual-occupant Husky vehicle variants that followed.

Tourniquet Failure Reports and the Request for an Alternative Standard

Another significant HFE problem emerged when STAT-20 spoke with Army medical personnel who voiced concern about the effectiveness of a tourniquet in the Army standard issue first aid kit, citing repeated failures during casualty care. The root cause appeared to be a weak adhesion system, exacerbated by ambient dirt and dust accumulating in the hook and pile used to secure the tourniquet. The result was an inability to hold and therefore, retain arterial blood-flow restriction when applied to a wounded extremity. In addition, medical personnel relayed reports that the plastic windlass rod featured on the tourniquet had cracked under the pressure of tightening when applied to extremity wounds. While the amount of tension required to crack the rod likely exceeds the amount necessary to achieve effective blood flow restriction when the tourniquet is properly applied, the fact that such tension was exerted at all in an attempt to achieve reliable arterial blood flow restriction lent credence to a possible functional design deficiency. Confounding the issue were reports of counterfeit versions of the tourniquet entering the medical supply system through non-traditional logistics practices.

While most medical teams had ceased use of the tourniquet system and replaced it with an alternate system deemed more effective, other medical personnel likely continued to use the approved system, unaware of the documented reliability concerns. Moreover, the issue was not accurately understood by some Army personnel based in the US, who suggested following prescribed usage guidelines, like keeping the tourniquet in its original plastic wrapper prior to use to mitigate the adhesion difficulties due to dirt and dust. As alluded to in discussion of the Husky driver's operational constraints, loosely wearing tourniquets on appendages is a best practice dictated by the reality that no one may be available or able to aid in stemming blood flow after an attack, possibly leaving self-aid as a soldier's only option for care. Beyond the Husky operator, wearing tourniquets on load-bearing vests or appendages is common practice for soldiers who regularly deploy on mission outside the confines of a base complex, so that the devices will be visible and accessible by others for administering aid in the event of an attack. STAT-20 supported and adhered to this common anticipatory care tactic, employing it personally when deployed on mission. STAT-20 relayed this information back to combat developers in the US in an attempt to improve knowledge among non-deployed Army medical materiel developers about how the devices were being employed, prompting a maelstrom of activity to investigate the issue. This culminated in a working group to develop improved tourniquet requirements, testing procedures, and selection processes for future tourniquet devices, leading ultimately to DoD policy revisions on future tourniquet development.

Technology Solutions in Conflict

It was common for STAT-20, at the advanced stage of the war during which it deployed, to witness the solution to one problem entering into operational conflict with solutions to other problems, causing new and unforeseen difficulties. Essentially, each solution introduces the potential for novel problems. By identifying strengths, weaknesses, and potential interactions with other systems before implementing a solution, many human factors issues might certainly have been avoided, but the absence of such proactive considerations due to unanticipated tactical developments was prevalent and ultimately may have had a negative impact on soldier mission effectiveness. These cases typically remain undiscovered until they are reported to FAST STAT teams or other forward-deployed advisory groups. Another complicating factor that warrants note is that the influx of new technology solutions is often so rapid that partially redundant and conflicting solutions are introduced in close temporal proximity, with lack of awareness among developers regarding alternate deployed solutions.

Six years of rapid technology development and insertion after the invasion of Iraq beginning in 2003 yielded overlapping and redundant technologies that at times effectively negated design functionality. For example, sizing of some egress hatches at the inception of vehicle design did not account for the volume occupied by a soldier in full modern battle dress. In addition, "standard" mission wear and gear characteristics changed over time. STAT-20 was therefore presented with problems where even soldiers smaller in stature had extreme difficulty negotiating the breach of an open egress/observation hatch due to their clothing and equipment (Figure 14.6). Another example involved a large piece of equipment hard-mounted directly in front of a vehicle side egress hatch, which rendered the opening useless in the event of an emergency situation. This was without question due to the limited space within the vehicle interior where such a device might otherwise be mounted, but its utility was at the functional expense of a basic escape portal. This was very disconcerting to the soldiers who manned that vehicle on a daily basis (Figure 14.6), given the threat posed to their ability to evacuate.

Figure 14.6   Soldier attempting to egress through a vehicle's overhead hatch

The tendency to incorporate a new solution without properly considering the implications to extant systems may have been the core issue of the STAT-20 mission. Soldiers were repeatedly thwarted by this practice, which is analogous to applying multiple bandages to a wound that realistically needs more comprehensive intervention instead of additional patching. When an insertion is "home grown," or developed and inserted rapidly, it may work well for a brief period, but will often reveal shortcomings in time due to lack of careful consideration for related systems and missions. For example, when such systems arrive in theater without associated reliable training systems, training personnel and maintenance, or unit-organic repair support, there is great difficulty in employing the system as intended. Instead, detrimental impacts on equipment function and associated safety risks to soldiers who use the systems become the norm. STAT-20 worked diligently to identify such issues, canvassing each of Iraq's geographic sectors to gain as representative a sampling as possible of soldiers, military occupational specialties, and mission requirements to validate technological challenges, judge their applicability to other theaters (for example, Afghanistan) and the wider Army community, and facilitate the sharing of information among widely dispersed units with common objectives, missions, and human factors concerns.

Conclusion

Working with US soldiers, civilians and Iraqi and Coalition partners, the science advisors who deployed in support of the FAST mission have addressed many capability gaps and helped to generate a wide variety of S&T solutions to forward theaters of operation. This work continues, and as each individual scientist or engineer returns to their respective duty station from volunteering to deploy on a combat tour, the knowledge, interests, and skills borne out of their assignment enriches the organizations they work for as they share their new experiences and expertise with colleagues. Much of the work that begins in places like Iraq or Afghanistan, such as less-than-lethal technology or predictive modeling and analysis, persists in the work of other organizations such as the Defense Advanced Research Projects Agency and the Army Intelligence and Security Command, as they continue to seek solutions to the problems soldiers encounter and find novel applications for the technologies that are produced. STAT teams continue to serve in theaters of war as capable, small-scale units with large-scale productivity, identifying problems and contributing to solutions in collaboration with superb personnel within the research and development community and most importantly, with the adaptive and inventive soldiers they work with during the course of their combat tours.

Chapter 15

Immersive Simulations for Dismounted Soldier Research

Kathy L. Kehring

Earlier chapters have discussed the complexity of advanced technologies being developed for the military, and the operational and environmental challenges faced when using these technologies. It is critical to consider human performance issues when designing new systems, and to have the most effective impact, it is best to identify issues early in the design phase. However, evaluating prototype systems and concepts for the military and measuring human performance in the operational environment they are being designed for has always been a challenge. Building prototype systems can be costly and time-consuming. The operational environment of field studies can be difficult to control or manipulate and data collection options may be limited. Weather and time of day can affect repeatability. Conditions may be too hazardous, as the safety of the test participants is always of utmost concern.

Immersive simulations of operational scenarios can offer a means for repeatable, event-driven research, where soldiers can be placed in near-realistic and stressful or seemingly dangerous operational situations with no risk to their safety. Simulations for dismounted soldiers have been improving over the years but can vary widely in the level of immersion offered. This chapter highlights the features, data-collection capabilities, limitations, and advantages (over field studies) of some of the immersive simulators being used by the US Army for military research applications. Also discussed are examples of their use in evaluating systems and concepts for the dismounted soldier.

Attributes of Immersion

Immersion, in the cognitive sense, refers to a state of intense mental focus on something that reduces your awareness of other aspects of your immediate surroundings. In simulation, it is this intense focus, usually enhanced by multisensory stimuli and interaction with the simulated environment, that causes you to momentarily place yourself in this other world environment. Some early applications of immersive simulation were primarily for entertainment purposes and therefore did not require the scientific-based artificial reality necessary for training and research. The first simulators developed for military use were typically vehicle or mounted environments, mostly for training applications.

Mounted operations are easier to simulate as a soldier's physical contact with the environment is usually limited to a confined workspace within the vehicle, called the "crewstation" by the military. The world outside of the crewstation is seen through out-the-window displays, which can be easily simulated on computer displays. The dismounted soldier, on the other hand, has direct multisensory interaction with the terrain and environment, which is more difficult to accurately represent in simulation (Campbell, Knerr, and Lampton 2004). Achieving a certain level of immersion in dismounted simulations may be essential to obtaining valid research results; however, realistically simulating all aspects of sensory interaction with the environment may be unnecessary. But what are the essential qualities and capabilities needed in an immersive dismounted soldier simulator?

The fidelity of the sensory immersion or how closely it replicates real-world conditions can vary greatly between systems that are still considered to be *immersive*. The visual stimuli can vary in size, field of view, resolution, quality, and level of detail in the images provided. Sound may be presented monophonically, stereophonically, fully localized (the perceived 3D location of the sound matches the spatial location of the source in the simulated environment) or not provided at all. The method with which the user interacts with or controls objects in the environment can be simplified to joystick or other handheld controller inputs. Realistic props such as weapon systems may also be incorporated for controller input. Dismounted maneuvers through the environment can be accomplished by hand-activated controls or performed on 3D motion interfaces that provide more realistic physical movements and workload. The fidelity required to cause performance effects that can transfer to real-world conditions depends on the type of tasks being performed and should be carefully considered. A simplified interface may be adequate for some tasks; however, more training and cognitive effort may be required in the use of the simple interface than if it was a more realistic replica. Tasks should be executed in the same manner in the simulated environment as they would be in the real environment. This requires test participants with real-world experience to perform as they would in the real environment. Removing the risks of operating in the real environment should not alter their behavior. To assure the reliability of the results from simulation-based research, the simulators should be validated against real-world conditions whenever possible to obtain an objective measure of their equivalence. Ideally, the simulator should achieve full performance fidelity, where a task performed by a soldier in the simulator has the same result as when performed in the real world according to any objective measure of performance (Levison and Pew 1993).

Immersive Simulation of Dismounted Soldier Tasks

Shoot, move, and communicate is a soldier's most basic directive (Deweese 2010). These fundamental functions are vital to the success of a mission. Although the complexity of these functions for the dismounted soldier seems to increase from move to shoot to communicate, ironically the complexity of realistically simulating these tasks seems to follow the opposite trend. Communication, or the transfer of information, can occur in various modalities. The level of interpretation needed to understand the information and respond or react appropriately may require some training and expertise. The method of obtaining this information, from the sensors designed to capture it to the systems developed to transmit and display it, can be a very complex and involved process. However, emulating this information capture, transfer, and display is generally relatively easy in the digital world. Shooting and successfully engaging targets with a weapon require skill and training. Although the underlying mechanics of a small arms weapon may not be as complicated as some information systems, accurately simulating the characteristics of weapon firing so it feels and operates like that in the real world presents some challenges. Moving, for a dismounted soldier, involves performing maneuvers such as walking, running, crawling, or climbing, which are innate to most people yet present the greatest challenges to realistically recreate this type of natural physical control and effect through simulation.

Dismounted Soldier Communications Research in Semi-immersive Simulations

Perhaps the simplest simulators used for research that may be considered immersive comprise realistic, interactive graphical environments displayed on a desktop computer monitor. Many such simulators are derived from commercial off-the-shelf militaristic games. These games can be customized to provide a controllable, cost-effective simulation to evaluate information system concepts (Wiley, Scribner, and Harper 2009) and multiplayer team interaction (Bowyer, Poltrock, and Waggett 2010). In many cases, when the research is mainly focused on communication issues, for example, how information is presented, how effectively it is received and interpreted, or how teams interact, these lower-fidelity systems are used. This type of simulator may be adequate for early exploration of the cognitive aspects of communication when the fidelity of the secondary or physical tasks may have less effect on the primary research interest. For training simulations, Knerr (2007) found that even though a simulator may not allow soldiers to perform many of the physical actions in the way they need to perform them in the real world, it does allow them to practice and learn the cognitive skills they need such as mission-planning, maintaining situational awareness, and decision-making.

The first-person-shooter game, Tom Clancy's *Rainbow Six 3: Raven Shield®*, produced by UbiSoft Entertainment, Montreuil-sous-Bois, France, was adapted for use by Wiley, Scribner, and Harper (2008) to examine team-leader mission performance under different visual display concepts. The operational scenarios included a reconnaissance and an assault mission. The goal of the reconnaissance mission was to avoid all enemy while navigating to the objective area. The goal of the assault mission was to find and eliminate all enemy in and around the objective area. Eight different yet characteristically similar terrain environments (same size area with similar amounts of tree cover, buildings, obstacles, and hazards) were required. The embedded characteristics of the game readily enabled the creation of the scenarios, terrain, and advanced display concepts. These characteristics included an environmental editor for creating terrain object models, computer-driven militaristic behaviors of friendly and enemy characters, multiple weapon models, a built-in digital map display, and real-time aerial views of the terrain. Creating and controlling these same features for a field study could be difficult and cost-prohibitive. A standard keyboard and mouse served as the interface to the simulation. Training was required, as well as a reference template of the keyboard and mouse functions, to ensure the test participants understood how to operate the interface to carry out the missions. Through the simulation the researchers were able to show how providing the team leader with advanced visual displays of battlefield information enhances SA and some performance measures, but at an increase in mission completion time. Adding a team member to serve as an information manager, who examined and filtered the information provided to the team leader, enhanced the team leader's SA, survivability, and lethality with a reduction in workload and no increase in completion time. However, this concept would require an increase or reallocation of manpower.

The same simulator was used in a follow-on study to examine the effect that certain wearable display characteristics, including display modality (visual, auditory, or both), display location (forearm-mounted or helmet-mounted), and display configuration (digital map only or digital map plus unmanned aerial vehicle (UAV) feed), had on team leader mission performance (Wiley, Scribner, and Harper 2009). The same terrain models created for the previous study could be reused with slight modifications. In this study, visual and auditory displays did have a significant effect on mission performance by enhancing SA and increasing survivability; however, no significant differences were found for display location or display configuration. A keyboard and mouse were used as the interface; therefore, firing a shot did not involve aiming an actual weapon. Any potential interference between the display location and firing the actual weapon would not have been detected.

PC gaming technology has also been employed to study team collaboration under an International Technology Alliance in Network and Information Sciences sponsored by the US Army Research Laboratory (ARL) and the UK Ministry of Defence (Bowyer, Poltrock, and Waggett 2010, Masato et al. 2008). The commercially available first-person shooter game, *Battlefield 2*™ (Electronic Arts Inc., Redwood City, CA) was a simulation platform used in this research. *Battlefield 2*™ was chosen because it allows for collaboration among players, establishing a chain of command, simulating tactical and strategic military situations, easy modification for data capture, and a level of immersion beyond a simple game (Bowyer, Poltrock, and Waggett 2010). As with many customizable commercial games, modifications to *Battlefield 2*™ have been created by user groups to enhance the simulation. One notable modification selected for use in this research is called "Project Reality." The "Project Reality" modification was created by a group of game developers, some with military experience, who aimed to provide a more realistic combat environment and improve the military features of this game. The modifications included added operational features, improved accuracy of the weapon models, and the removal of many of the onscreen aids that are not available in a real battlefield.[1]

Other changes were made to the standard rules of the game. These included not allowing the use of vehicles and not allowing players that were killed to reenter the scenario. This feature, generally allowed in game play, can encourage risky behavior that would not occur on the battlefield. Additional external software components and scripts were developed and added to create a simulation environment for observing and analyzing military collaboration and evaluating the effects of adaptive mission-planning technologies on performance (Masato et al. 2008).

The Technology Alliance is proposing the development and use of adaptive technologies to facilitate rapid adjustments to plans in response to changing operational conditions. They aim to develop methods for automatically constructing and maintaining dynamic models of team behavior such as communications, weapon use, and movement (Bowyer, Poltrock, and Waggett 2010). The simulation provides a vehicle to explore these concepts by monitoring all communications, locations, and movements of the participants and comparing them against the mission plan and objectives to provide suggestions to improve their progress. Since the actual technology is yet to be developed, collecting and analyzing this data during a live field study and providing real-time feedback would be very difficult.

These simple desktop simulations could be used to study communication and information system characteristics and their effect on SA and other measures of mission performance. However, conclusions on issues such as the recommended physical size and location of equipment should not be drawn from this type of simulation, in which the participant is comfortably seated while navigating through the terrain. Likewise, hits and kills may be used to indicate that targets were detected and identified, but other shooting performance measures such as aiming accuracy or workload should not be used because firing a weapon with these interfaces bears little resemblance to the actual task. As the information technology concepts, modalities, and tactics are further developed, the interfaces controlling locomotion and performance of the secondary tasks should be more realistically simulated to evaluate the effects on performance more comprehensively.

Dismounted Soldier Shooting Performance Research in Higher-Level Immersive Simulations

Immersive simulators used in research concerned with issues affecting the shooting performance of dismounted soldiers should incorporate a shooting interface that physically resembles the weapon being evaluated in both form and weight so that

---

1   These modifications are available for free download at http://www.realitymod.com (accessed: July 25, 2012).

the same techniques for aiming and firing are used and the effort and fatigue of holding the weapon are realized. Aiming at a target should require shouldering the weapon and lining up the target in the weapon sights. The same force should be required to pull the trigger and the recoil resulting from the shot should feel the same. The field of view should allow adequate range to scan for targets in the environment.

The military uses several small arms shooting simulators, primarily for marksmanship training, that have large visual displays and either demilitarized weapons or accurate prototypes for the user interface. For training purposes, the realism and feedback provided by many of these simulators may be adequate. In this case, the simulated tasks do not have to be modeled with exact precision; they merely have to instill skills that can transfer to the real-world performance of that task. For research concerned with measures of shooting performance, the simulator should exactly replicate the weapon interface and response. More precise measurements than the training feedback may be necessary to analyze the event such as the exact location of the round, time of target exposure, time of trigger pull, and aiming location throughout the target acquisition and engagement.

One device that has been proven useful for ARL's Human Research and Engineering Directorate's (HRED) research efforts is the Dismounted Infantryman Survivability and Lethality Test Bed (DISALT). The DISALT (Figure 15.1), developed by the Naval Air Warfare Center Training Systems Division, Orlando, FL, was originally designed to serve as a marksmanship trainer for shipboard operations (Scribner, Wiley, and Harper 2007). DISALT was judged to be suitable as a research tool based on the weapon representation, tracking accuracy, and high-fidelity data capture. It employs the DI-GUY™ Scenario (Boston Dynamics, Waltham, MA) software application to create and display custom target scenarios ranging from standard pop-up silhouettes to realistic moving characters that can represent friend, foe, and non-combatants. The visual environment is forward-projected onto a large-screen display. Instrumented, demilitarized small arms weapons are used for the shooting interface so the form and controls are identical to what would be used in live fire. A cable mechanism is employed to raise the muzzle of the weapon when fired, producing a realistic recoil effect. The level of recoil can be incrementally set to match the recoil of the small arms weapon being used. Weapon aim point is detected using an infrared emitting diode and collimator lens. Shot location is derived from ballistic calculations of the trajectories of the rounds (Kerrick, Hatfield, and Allender 2007).

To verify that the DISALT would yield effects similar to live-fire shooting, Scribner, Wiley, and Harper (2007) conducted a validation study to compare a live-fire versus DISALT simulated pop-up target-shooting scenario. The outdoor small arms experimental range used for the live fire was modeled for use as the DISALT environment. The apparent size and perceptual angle of incidence of each target from the firing point in the simulated environment were matched to those in the firing point of the real environment to assure the visual and psychomotor workload were the same for aligning the weapon to the targets in both environments.

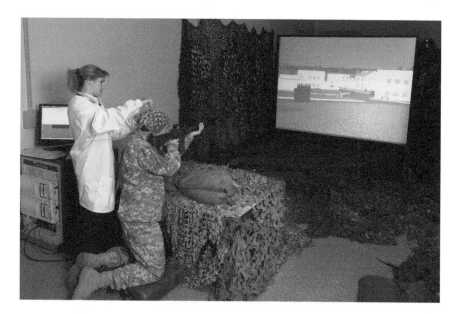

Figure 15.1   The DISALT

The shooting performance data collected included hits, time to first shot, and radial aiming error from the center of mass of the target. Subjective data on workload and stress were also collected. No significant differences were found for hit percentage, workload, or stress. However, the reaction time, or time to first shot, was significantly longer and the radial error of the shots was significantly less in the simulator than at the live range. Nonetheless, the study shows a strong relationship between live fire and shooting in the DISALT and supports the validity of the DISALT for some important aspects of shooting performance, namely, target hits, workload, and stress.

The DISALT has been used to study a number of issues involving shooting performance. Scribner, Wiley, and Harper (2005) conducted a study to determine if there were any shooting and secondary task performance differences caused by secondary task display modality and auditory alert cues for soldiers shooting under additional cognitive workload. Visual (helmet-mounted and forearm-mounted) and auditory displays were used to provide secondary task workload while the soldier engaged in a friend or foe target identification and engagement scenario. This study was a follow-on to related studies conducted on a small arms, live-fire experimental range, which indicated that secondary task performance decreases (Scribner and Harper 2001) and fratricide rates increase (Scribner 2002) with higher levels of workload induced by secondary tasks. The goal of this line of research is to identify display characteristics that could minimize these effects. The results from the simulator study show that shooting performance was hindered by the use of a helmet-mounted display (HMD). This would indicate that the HMD interfered with the aiming and firing of the weapon. This differs from the results of the study by Wiley, Scribner, and Harper (2009), discussed in the previous part of this chapter, where no difference in performance was detected between an HMD and a forearm-mounted display; however, a keyboard and mouse were used as the interface, so aiming and firing did not involve holding a weapon up to the eye. Shooting performance was best with the auditory display and secondary task performance was best with the HMD. The authors suggested that the best possible trade-off between shooting and secondary task performance was the forearm-mounted display with an auditory alert cue.

To further explore how cognitive workload affects shooting performance, the DISALT was used in conjunction with electroencephalographic (EEG) measures. The ability to assess the neurocognitive demands of soldiers in an operational scenario may provide significant insights to understanding soldier performance. However, the ability to reliably measure brain dynamics of soldiers in operational environments can be a major challenge due to inherent artifacts in these environments (Kerrick, Oie, and McDowell 2009). The DISALT provided a controlled and validated environment to record the dynamic cortical processes of soldiers during reactive shooting in simulated scenarios (Kerrick, Hatfield, and Allender 2007). The scenarios varied in task demand by load (single or dual task), decision (no decision or friend/foe decision), and target exposure time (short or long). The EEG was synchronized to the onset of the target exposure to allow for analysis of the cognitive evaluation processes and reactive motor preparation during the shooting event. The shooting performance measures followed the expected trends, decreasing with increased task demand. Detailed discussions of the analysis of the EEG data are beyond the scope of this chapter; however, data collected from this study confirm that attention is: of limited capacity; is temporally and spatially distributed among cortical networks, which oscillate at different frequencies to enable both parallel processing of sensorimotor information and sequential processing of cognitive demands; and is influenced by task demands and perceived effort. This provides evidence that mental processes can be quantified and that soldier performance can be evaluated objectively in terms of the impact on cognitive load.

The effects of the physical characteristics of the weapon on shooting performance have been explored using the DISALT. In one study, the weight and vertical center of mass of an M16A2 rifle were modified to represent current and estimated future assault rifle concepts (Harper, LaFiandra, and Wiley 2008). The weight was modified by adding 1.5 kg, 3.0 kg, and 4.5 kg to the center of mass. The vertical center of mass was moved from the weapon's natural vertical center of mass to 4 cm higher than the natural center of mass and 4 cm lower than the natural center of mass. To examine the full target engagement event, capturing the aiming location prior to firing the weapon was necessary. This can be difficult to capture on a live-fire range but is attainable in the DISALT. Using a simulated weapon also eliminated the safety concerns associated with modifying a real weapon to use for live fire. The results from this study indicated that the change in vertical center of mass did not affect aiming error or engagement time. The change in weight did not affect engagement time, but adding the 4.5 kg weight did significantly increase the radial aiming error over the two lighter weights. Along with the shooting performance measures collected by the simulator, collecting motion data of the shooter and the weapon using a camera-based motion analysis system (Vicon Motion Systems, Inc., Lake Forest, CA) has been explored. Passive reflective markers placed on the shooter and the weapon enable the capture of the movement of the weapon with respect to the shooter to further analyze the aiming, firing, and recoil events. This type of optical motion analysis is difficult to perform in an outdoor setting such as a firing range, because light level changes, restrictions on camera placement, and other objects in the camera view can affect data collection.

ARL-HRED has also acquired another small arms training simulator to investigate issues involving judgmental use of force, SA, and shooting performance. The Immersive Combat Readiness Simulator, based on the commercially available VirTra System (VirTra Systems, Inc., Tempe, AZ), provides a 300° field of view using five rear-projection displays with true high-definition resolution video scenes. The immersive simulator uses laser tracking to track up to six wireless weapons in the simulator at the same time. Weapon recoil is replicated using pressurized cartridges embedded in the butt stock of the demilitarized weapons. Movement within the environment is limited to the confines of the display screens—approximately 100 sq ft. The participants can wear a Threat-Fire™ belt that can apply an electric shock of up to 80K volts to the wearer to safely simulate being shot. This type of feedback can add to the level of stress experienced by the participants and may increase their immersion in the scenarios.

The Squad Synthetic Environment (SSE) at the US Army's Maneuver Battle Lab (formerly called the "Soldier Battle Lab") was designed for squad-sized, force-on-force, multiplayer interactive simulations. Although not designed primarily as a shooting simulator, the SSE does provide a realistic weapon interface. Weapon-tracking and user posture are acquired by an inertial/acoustic motion-tracking system (Intersense, Inc, Billerica, MA) using a single sensor on the weapon and another on the head. Each soldier is confined to a 10 ft square cubicle with a rear projection display on one wall and opaque sound attenuating fabric on the other walls to block out the outside environment (Campbell, Knerr, and Lampton 2004). Movement through the simulated environment beyond the 10 ft square is controlled via a thumb-controlled joystick on the weapon. Spatialized battlefield audio is presented over a speaker system and communication channels can be simulated between team members. The SSE uses commercially available distributed interactive simulation software called the Soldier Visualization Station (SVS®, Advanced Interactive Systems, Seattle, WA). SVS®, initially developed by Reality By Design (Boston, MA) for the US Army in 1997 (Lockheed Martin Corporation 1998), has continually evolved over the years, responding to the needs of the military user community to provide increasingly more realistic battlefield capabilities and effects. SVS® currently includes tools for custom scenario generation, computer generated forces, dynamic terrain, stealth-viewing of the battlefield, and an after-action review (AAR) capability. Along with the screen display of the environment, SVS® can present a simultaneous HMD display to simulate an integrated helmet assembly. This can be used to simulate views from weapon-mounted camera sights or other similar displays.

The SSE has been used in various research capacities. Redden (2002) used the SSE to conduct a study to investigate the critical information requirements for members of an infantry platoon operating in an urban environment. The goal was to determine the information most needed for different echelons (squad member, team leader, squad leader, and platoon leader) to develop information system design requirements for the different platoon members. A validation study was conducted prior to this study to ensure the data collected from the SSE would be comparable to data collected in a live Military Operations in Urban Terrain (MOUT) environment. A model of the MOUT environment was used as the terrain in the SSE. The same missions were executed in both environments with questionnaires administered during pre-planned pauses to obtain ratings of the importance of specific types of information to each soldier. The rank order of the information criticality ratings was found to be consistent between the simulated and live environment. All of the information was deemed more critical in the SSE than in the live environment but by a fairly consistent amount. When the mean difference was subtracted from the ratings obtained in the SSE, there was no significant difference. Based on these findings, the SSE was considered a valid research environment to place soldiers in a realistic scenario and determine the information that is most important to their position in the platoon. The results from the study indicated that within each echelon only the information that has a direct effect on their actions is critical. Thus, the information immediately provided should only include items the platoon member can affect or that would influence decisions within their scope. Squad members were mainly concerned with localized information on their current objective. The criticality of information beyond the proximity of the objective increased as the level of leadership increased. The SSE has also been used by Pleban et al. (2001) and Strater et al. (2001) as a test bed for developing situational awareness measurement tools. It is difficult to assess the situational awareness of dismounted soldiers in field exercises, as it typically involves halting the action at multiple times to ask probing questions. This can be distracting to other elements of the exercise and may be difficult to precisely time and control. The simulation environment allows for greater control and timing of events than live exercises.

The simulators mentioned previously may be considered more immersive than the desktop variety discussed earlier because they provide a larger field of view and they more realistically and accurately simulate the firing of a weapon, albeit mostly from a stationary position. This may be adequate for simulating range-firing, defensive firing, or operations in close quarters where the participant's location in the environment does not change much. For operations that require locomotion through the environment such as navigation, movement to contact, or clearing a building, a mobility platform that promotes realistic physical movement should be used.

Dismounted Soldier Research with Physical Movement in Fully Immersive Simulations

Incorporating realistic effects of physical workload due to traversing terrain (such as fatigue, psychomotor control, and spatial relationships) into dismounted soldier simulators and enhancing their psychosomatic immersion into the environment may increase the validity of research conducted using these devices. Having an interface where the user naturally interacts with the simulated environment in the same manner that they do in the real world would eliminate the cognitive processes needed to translate these interactions into other artificial means. Results from a study by Zanbaka et al. (2004) suggest that performance in problem-solving and interpretation of material is improved when naturally walking through a simulated environment as opposed to using other techniques based on hand controls and head orientation. Immersive simulations with full body controlled movement (rotation and translation) have shown improved performance of complex spatially oriented tasks and acquisition of spatial knowledge (Ruddle and Lessels 2006, Chance et al. 1998). But creating such an interface, one that allows soldiers to naturally move (for example, walk, jog, run, or crawl) and turn in the simulated environment, and that precisely responds to their dynamic motions with no perceptible delay, is not a simple matter.

Over the years there have been numerous attempts to create a more realistic mobility interface for immersive simulations. These designs involved concepts ranging from walking in place to unicycle-type devices to standing bipedal devices to treadmill-based devices (Crowell et al. 2006). Many of these were federally funded initiatives with the primary focus to develop training platforms for dismounted soldiers; however, these efforts have also contributed to advancing the state of the art in mobility platforms for research use (Knerr 2007, Campbell, Knerr, and Lampton 2004, Crowell et al. 2006, Lockheed Martin Corporation 1998). Through these initiatives, a series of engineering and user experiments were conducted to evaluate the evolving simulation technologies and assess their usefulness for dismounted soldier training and research.

One of the technologies that did show promise as a mobility platform for dismounted soldier simulations research was the Omni-Directional Treadmill (ODT), because it allowed for a more natural gait and realistic physical workload to move through the simulated environment. The users could walk in any direction without leaving a confined (4 ft × 4 ft) workspace. This was accomplished using two perpendicular belts, one inside the other. The outer belt consisted of rollers that rotated perpendicular to the direction of travel of the belt. The rotation of the rollers was caused by contact with the inner belt as it moved. A mechanical tracking arm tethered to the user was used to determine their position. As the user moved from the center, they were drawn back through movement in one axis due to the movement of the outer belt and in the other axis from the rotation of the rollers, both occurring simultaneously. There were no serious safety concerns with the original ODT; however, it did have other operational issues. Users sometimes experienced instability and had difficulty maintaining balance during certain maneuvers. The control system was prone to false starts or attempting to re-center a user who was merely leaning or trying to turn in place. It also tended to overshoot the center when the user stopped walking, causing the belts to then reverse direction and overshoot again, oscillating about the center. Changing the direction of travel while away from the center of the ODT also caused users to feel unbalanced. The ODT's control system was designed to return the user to the center of the ODT along a vector from the user's center of mass to the center of the active surface. If the direction of travel of the user was not in line with this vector, the user would experience some lateral motion while being pulled back to the center, causing some instability. The tethered tracking mechanism used to determine the user's position restricted movement and placed noticeable forces on the user. At the time the ODT was first developed, the accuracy and speed of available tracking technologies was not really sufficient for this type of real-time control. The ODT was also loud enough to require hearing protection when moving at higher speeds, had a small workspace that could limit stride length, and did not permit kneeling or crawling.

The ODT has since undergone several design upgrades that have improved upon these earlier issues. The most recent upgrade of the ODT (MTS Systems Corporation, Eden Prairie, MN, and Virtual Space Devices Inc., Bloomington, MI) now has an 8 ft × 8 ft working surface that allows the user to walk, jog, or crawl in any direction while the treadmill belts return the user back to the center. Unlike a conventional treadmill, the speeds at which the ODT's belts move are in response to the movement of the user rather than being set to a constant value. To achieve omnidirectional movement, the upgraded ODT incorporates a series of 80, 4 in-wide mini-belt segments attached perpendicularly to the underlying main belt. The main belt travels in the x-axis of the working surface. The mini-belts travel with the main belt but also operate like mini-treadmills to generate movement in the y-axis. The operation of the mini-belts (y-axis) and the main belt (x-axis) are controlled by two separate 25 hp motors so they can move independently and simultaneously to re-center the user. The control system driving the belts uses a camera-based motion-tracking system (Vicon Motion Systems, Inc., Lake Forest, CA) to capture the movement of the user. Eight digital video cameras mounted to a frame above the screens track the position and orientation of clusters of passive, reflective markers attached to the user's lower back, head, and weapon. The markers on the back are attached to a resin plate held in place by a neoprene belt worn around the waist. These markers are used to capture the user's speed and heading to determine the belt movement needed to re-center the user. The markers on the head are attached to a helmet worn by the user and are used to capture the eyepoint to appropriately adjust the viewing perspective of the visual display. Markers placed on the weapon are used to determine the aim point for target engagements. Additional markers can be added to track other objects or body segments of interest. The number of markers is limited only by the number that can be positioned such that they are distinguishable from one another in the camera views.

The control system incorporates a washout algorithm and fuzzy logic systems to determine the optimal re-centering motion of the belts (Hessburg and Clark 2005). The washout algorithm attempts to allow the users to feel the expected motion cues of their various gaits while minimizing their perception of the movement needed to maintain their position in the workspace. Fuzzy logic is used to determine the gait of the user (walking forward, walking backward, side-stepping, or crawling) and adjust the parameters of the washout algorithm accordingly. Some initial practice is usually required before users feel comfortable walking on the ODT, although achieving a natural gait is easily attainable for most who have tried. The user can operate the ODT completely untethered; however, as a safety precaution, a harness connected to an inertia reel is worn to keep the user from hitting the surface should they fall.

ARL-HRED has integrated the upgraded ODT with visual and auditory displays and a realistic weapon interface to create a fully immersive dismounted soldier simulator called the Immersive Environment Simulator (IES) (Figure 15.2). The visual environment of the IES is presented on a four-sided RAVE II display (Fakespace Systems, Inc., Kitchner, Ontario, Canada) providing a full 360° field of view. Each of the four screens is 12.5 ft × 10 ft with rear projected images from very bright (6,000 lumen) projectors (Christie Digital Systems, Inc., Cypress, CA). Acoustical foam panels cover the walls and ceiling around the simulator to reduce sound reflections, creating a simulated free-field environment. Forty-four speakers are mounted around the walls, ceiling, and top of the screens to provide directional sound. The visual and auditory displays are driven by the SVS® simulation software that is also used in the SSE described earlier. The software provides the same environmental features and effects but was modified to integrate with the IES systems, allowing for camera-based tracking, obtaining speed and heading information from the ODT, and providing a multiscreen display. A demilitarized small arms weapon is used for the shooting interface.

To validate the IES as a suitable simulator for research use, Boynton, Kehring, and White (2011) conducted a study to compare walking on the ODT in the IES to walking overground for selected biomechanical and physiological variables and quantify any significant differences. Objective measures of the test participants' gait and physical exertion were collected while they walked on a circular course laid out in a large room and while they walked on the ODT through a simulated version of that same room in the IES. Subjective ratings of perceived level of exertion were also obtained. Twenty-three reflective markers were placed on the lower torso and lower extremities of the test participants at various anatomical landmarks. The camera-based motion analysis system captured 3D position data of these markers for biomechanical analysis of their gait. Energy expenditure was

Figure 15.2   The IES

measured using a portable cardiopulmonary exercise-testing device, the Cosmed K4b^2 (Cosmed USA, Chicago, IL). The Cosmed K4b^2 performed gas exchange analysis on a breath by breath basis through a face mask worn by the test participant and also collected the heart rate from a Polar® heart-rate monitor worn around the participant's chest. The trials were conducted at two different speeds, 2.5 mph and 3.0 mph, and the test participants traveled at those speeds for 4 min before data was collected to achieve steady-state energy expenditure.

The results indicate that the gait mechanics associated with walking on the ODT were generally comparable to those from walking overground; however, at the higher speed (3 mph) a decrease in swing time of the left and right leg, an increase in left cadence, and a decrease in left stride length were found with the ODT trials. Although these differences were found to be significant, the magnitudes of the differences were relatively small and consistent with previous studies comparing walking on conventional treadmills to walking overground. The sagittal plane range of motion of the ankle was significantly less in the ODT trials. Although the overall range of motion at the hip and knee were similar between the over-ground and ODT trials, there was a tendency for the joints to be more flexed throughout the gait cycle on the ODT. This was likely due to an attempt by the central nervous system to increase stability on the moving surface of the ODT by lowering the center of gravity. The energy expenditure was more than 20 percent greater walking on the ODT than walking at the same speed overground. In fact, the energy expended while walking on the ODT at 2.5 mph appears to be roughly equivalent to walking 3.0 mph overground. Subjective ratings of perceived exertion were also significantly higher for the ODT trials. Although the differences in the kinematics of walking on the ODT and walking overground were minimal, each could be contributing to this greater metabolic cost. Another factor that may also be contributing to this increase is an unnatural arm swing. Although not measured in this study, it was noted by the researchers that some test participants did not appear to swing their arms as naturally while walking on the ODT as they did when walking overground. This could also have contributed to increased energy expenditure and reduced stride length. The researchers concluded that the ODT does permit the user to walk through an immersive environment in a natural way, but with a significant increase in metabolic cost and perception of physical exertion. These differences must be taken into account when comparing the results of research or training efforts using the ODT to those of real-world activities.

Follow-on studies are planned to try to identify the underlying causes of the increased physiological demand associated with walking on the ODT. This information could aid in developing further improvements to the control system and more optimal design characteristics for mobility platforms that permit closer replication of real-world physiological demands. A pilot study has been initiated to capture electromyography (EMG) measures and arm-swing motion data to explore potential reasons for the differences. Other planned areas of research for the IES include investigations of the effects of speech communications, information systems, robotic control, and load carriage on dismounted soldier performance while on the move.

Future Enhancements for Increasing Immersion

Immersive simulation technology for the dismounted soldier has progressed enough to prove useful for research purposes. The characteristics of the immersive simulators discussed in this chapter are summarized in Table 15.1. This list is by no means all-inclusive; there are numerous immersive simulators in existence and new ones are continually evolving and improving. Since the first immersive simulators were introduced, there have been significant advances in the ability to provide greater immersion through the visual and auditory displays for dismounted soldier simulations, but other areas of sensory displays and interaction are still progressing. Motion-tracking technology has improved, allowing for more accurate response from simulator interfaces (such as weapon-aiming) and enhanced representation of the participant's avatar in the simulated environment. Mobility interfaces such as the ODT show the potential for future dismounted mobility platforms to approach completely natural physical motion. For the time being, there is still work to be done in replicating a more accurate physical workload and incorporating the effects of grade and obstacles in the terrain. The cost, facility requirements, and support for this type of mobility interface are currently not practical for many organizations. Further development is needed to

Table 15.1   Characteristics of immersive simulators discussed in this chapter

Simulator	Sensory Cues				Interface		Research Areas
	Visual	Auditory	Tactile	Mobility	Weapon	
Rainbow Six 3: Raven Shield®	Desktop monitor	Stereo	None	Keyboard and mouse	Keyboard and mouse	Communication and information systems
Battlefield 2™	Desktop monitor	Stereo	None	Keyboard and mouse	Keyboard and mouse	Multiplayer communication and interaction
DISALT	Forward projection single screen	Stereo	None	None	De-militarized weapon with recoil	Precise measures of shooting performance
ICRS	High definition rear projection 300° surround	Stereo	Threat fire feedback via electric shock	None	De-militarized weapon with recoil	Judgmental use of force, situational awareness
SSE	Rear projection single screen	Stereo	None	Weapon Joystick	De-militarized weapon	Information requirements, situational awareness
IES	Rear projection full 360° surround	Localized	None	Omni-directional treadmill	De-militarized weapon	Effects of physical workload on performance

optimize the design and bring the cost down. Tactile interaction with the simulated environment has proven to be a challenge but is still desirable for a truly immersive experience. Speech recognition remains an obstacle to more natural control and communications with computer-generated entities. The automated behavior of these entities is an area in which much effort is currently being placed to reduce the number of live participants needed to control the entities without reducing their realism. Improvements in these areas will create new, innovative, and ever more realistic platforms for future research efforts.

References

Bowyer, H., Poltrock, S.E. and Waggett, P. 2010. *Collaboratively Developed Extensions to Commercial Game Environments for Research and Analysis of Coordinated Team Activities*. (ITA Technical paper). Available at: https://www.usukitacs.com/papers/5362/details.html (accessed: July 26, 2012).

Boynton, A.C., Kehring, K.L. and White, T.L. 2011. *Biomechanical and Physiological Validation of the Omni-Directional Treadmill (ODT) Upgrade as a Mobility Platform for Virtual Environments*. (Technical Report ARL-TR-5510). Aberdeen Proving Ground, MD: US Army Research Laboratory.

Campbell, C.H., Knerr, B.W. and Lampton, D.R. 2004. *Virtual Environments for Infantry Soldiers*. (ARI Special Report 59). Alexandria, VA: US Army Research Institute for the Behavioral and Social Sciences.

Chance, S.S., Gaunet, F., Beall, A.C. and Loomis, J.M. 1998. Locomotion mode affects the updating of objects encountered during travel: The contribution of vestibular and proprioceptive inputs to path integration. *Presence: Teleoperators and Virtual Environments*, 7(2), 168–78.

Crowell, H.P., Faughn, J.A., Tran, P.K. and Wiley, P.W. 2006. *Improvements in the Omni-Directional Treadmill: Summary Report and Recommendations for Future Development*. (Technical Report ARL-TR-3958). Aberdeen Proving Ground, MD: US Army Research Laboratory.

Deweese, N. 2010. *Alpha Soldiers Train to Shoot, Communicate*. Available at: http://www.army.mil/-news/2010/01/07/32608-alpha-soldiers-train-to-shoot-communicate/ (accessed: July 25, 2012).

Harper, W., LaFiandra, M. and Wiley, P. 2008. The effect of weapon mass and center of mass on shooting performance with an assault rifle. *Proceedings of the 13th Annual International Conference on Industrial Engineering Theory, Applications, and Practice*, Las Vega, NV, September 2008.

Hessburg, T.M. and Clark, A.J. 2005. Intelligent control of a planar treadmill ambulatory simulator. *Proceedings of the ANNIE 2005 Conference*, St. Louis, Missouri, November 6–9, 2005.

Kerrick, S.E., Hatfield, B.D. and Allender, L.E. 2007. Event-related cortical dynamics of soldiers during shooting as a function of varied task demand. *Aviation, Space, and Environmental Medicine*, 78(5), Section II, B153–64.

Kerrick, S.E., Oie, K.S. and McDowell, K. 2009. *Assessment of EEG Signal Quality in Motion Environments.* (Technical Report ARL-TR-4866). Aberdeen Proving Ground, MD: US Army Research Laboratory.

Knerr, B.W. (2007). *Immersive Simulation Training for the Dismounted Soldier.* (Study Report 2007-01). Alexandria, VA: US Army Research Institute for the Behavioral and Social Sciences.

Levison, W.H. and Pew, R.W. 1993. *Use of Virtual Environment Training Technology for Individual Combat Simulation.* (Technical Report 971). Alexandria, VA: US Army Research Institute for the Behavioral and Social Sciences.

Lockheed Martin Corporation 1998. *Dismounted Warrior Network Enhancements for Restricted Terrain.* (Advanced Distributed Simulation Technology II, Dismounted Warrior Network DO #0055, CDRL AB01, ADST-II-CDRL-DWN-9800258B). Orlando, FL: US Army Simulation, Training, and Instrumentation Command.

Masato, D., Norman, T.J., Poltrock, S., Bowyer, H. and Waggett, P. 2008. Adaptive military behaviour in a collaborative simulation. *Proceedings of the SISO 2008 European Simulation Interoperability Workshop*, Orlando, FL, September 2008.

Pleban, R.J., Eakin, D.E., Salter, M.S. and Matthews, M.D. 2001. *Training and Assessment of Decision-Making Skills in Virtual Environments.* (Research Report 1767). Alexandria, VA: US Army Research Institute for the Behavioral and Social Sciences.

Redden, E.S. 2002. *Virtual Environment Study of Mission-Based Critical Information Requirements.* (Technical Report ARL-TR-2636). Aberdeen Proving Ground, MD: US Army Research Laboratory.

Ruddle, R.A. and Lessels, S. 2006. For efficient navigational search, humans require full physical movement, but not a rich visual scene. *Psychological Science*, 17(6), 460–65.

Scribner, D.R. 2002. *The Effect of Cognitive Load and Target Characteristics on Soldier Shooting Performance and Friendly Targets Engaged.* (Technical Report ARL-TR-2838). Aberdeen Proving Ground, MD: US Army Research Laboratory.

Scribner, D.R. and Harper, W.H. 2001. *The Effects of Mental Workload: Soldier Shooting Accuracy and Secondary Cognitive Task Performance.* (Technical Report ARL-TR-2525). Aberdeen Proving Ground, MD: US Army Research Laboratory.

Scribner, D.R., Wiley, P.H. and Harper, W.H. 2005. The effect of various display modalities on soldier shooting and secondary task performance: Strategies to maintain combat readiness during extended deployments—a human systems approach. *Proceedings RTO-MP-HFM-124 Meeting.* Neuilly-sur-Seine, France: RTO, 4-1-4-10.

Scribner, D.R., Wiley, P.H. and Harper, W.H. 2007. *A Comparison of Live and Simulated Fire Soldier Shooting Performance.* (Technical Report ARL-TR-4234). Aberdeen Proving Ground, MD: US Army Research Laboratory.

Strater, L.D., Endsley, M.R., Pleban, R.J. and Matthews, M.D. 2001. *Measures of Platoon Leader Situation Awareness in Virtual Decision-Making Exercises*. (Research Report 1770). Alexandria, VA: US Army Research Institute for the Behavioral and Social Sciences.

Wiley, P.W., Scribner, D.R. and Harper, W.H. 2008. Manipulation of a commercial game to provide a cost effective simulation to evaluate wearable information systems. *Proceedings of the 13th Annual International Conference on Industrial Engineering Theory, Applications, and Practice*, Las Vegas, NV, September 2008.

Wiley, P.W., Scribner, D.R. and Harper, W.H. 2009. Evaluation of wearable information systems during route reconnaissance and assault missions using commercial games. *Proceedings of the 14th Annual International Conference on Industrial Engineering Theory, Applications, and Practice*, Anaheim, CA, October 2009.

Zanbaka, C., Lok, B., Babu, S., Xiao, D., Ulinski, A. and Hodges, L.F. 2004. Effects of travel technique on cognition in virtual environments, *IEEE Virtual Reality Conference* (VR 2004), Chicago, IL, March 2004.

Chapter 16

Tactical Ergonomics: Applying Manual Material Handling Analysis for Current Operations

Andrew Bodenhamer and Bradley Davis

Introduction

In the practice of human systems integration (HSI) within the US Army's acquisition community, the need to analyze manual material handling practices is frequently encountered. Proper ergonomic design is critical to minimize musculoskeletal injury risk, reduce fatigue, reduce required manpower, remove task specific personnel restrictions, and ensure maximum user–system compatibility. Although this field of physical ergonomics has been well established in the industrial environment for some time, the application of existing heuristics and best practices to the soldier in a field environment is not a "cut-and-paste" affair. Specifically, the broad-ranging mission of the soldier frequently makes analyses of soldier material handling tasks unique and solutions difficult to enact. Most military equipment must be used within extreme variations in terrain and environment, with the requirement for minimal additional equipment and its associated costs. Furthermore, military equipment tends to be heavy by necessity and constructed using rugged and armored materials.

The Army employs vast numbers of tactical vehicles and other cargo carrying machines, but from the commander's perspective, the soldier is an optimal material handling system—fit and adaptable to rapid changes on the battlefield. Manual material handling will likely be a part of many soldiers' daily lives for the foreseeable future. Even new Army systems, such as the Line of Communication Bridge, that are designed for construction with material handling equipment still require a fallback method of primarily manual construction.

This chapter outlines selected manual material handling activities and suggests a course of analysis to rapidly identify ergonomic issues and direct possible solutions. Presented are a selection of analysis tools and related case studies, focused on two broad categories of activities that nearly all soldiers experience: "lift-and-carry" and "push/pull."

Manual Material Handling

Manual material handling (MMH) is the use of human energy to perform physical tasks that may consist of lifting, lowering, pushing, pulling, carrying, and holding activities. The design of MMH tasks must be limited by the maximum exertion or work rate available from the soldier population and the acceptable risk of musculoskeletal injury posed by the activity. Unfortunately, the forces generated within the body at levels less than maximum exertion can pose a risk for musculoskeletal disorders (Chaffin, Andersson, and Martin 1999). Sanders and McCormick (1993) identify six categories of MMH risk factors: forceful exertions, awkward work posture, repetitive motions, sustained exertions, localized contact stresses, and whole body or segmental vibrations.

While all of these risk factors may appear in any given military system, this chapter primarily addresses the problems faced by forceful exertions and awkward work posture, as these may be the two most prevalent MMH issues with tactical equipment.

Analysis Process

Developing and following a consistent and simple framework for performing analysis of suspected MMH issues is key to rapidly identifying and resolving these issues. The following five steps may serve as a baseline for most routine ergonomic analyses:

1. Identify the problem. How a MMH analysis is initiated can take many forms. The HSI professional may identify a task that nominally exceeds a known standard. Sometimes, other acquisition team members may identify an area of concern to be examined. At times, simply observing user-testing may reveal tasks that operators have difficulty performing. Regardless of the source, specifying and documenting the boundaries of the MMH task that need to be analyzed should be performed. This includes the "who, what, when, where, why, and how" aspects of the suspected problem.
2. Analyze the severity. HSI professionals begin by identifying and creating any required mathematical models of the static and dynamic forces exerted during the task. This includes the interaction of key dimensions and weights with the forces required to perform the task through the entire range of motion. Analyzing the entire range of every variable is time-consuming and inefficient; it is usually only necessary to find the probable worst-case conditions based upon maximum forces, exertion start and end points, and user anthropometry. It includes identifying if the problem may be with required strength (generally, a problem for 5th percentile females) or poor work posture, such as low starting points for lifts (generally, a problem for 95th percentile males) or both. Except in cases of extreme

tradeoffs in forces required and range of motion—or between other related factors—solving the ergonomic problem for the highest-risk soldier group will usually be acceptable for the entire soldier population. It is during this step that the conditions of the particular MMH task are compared against relevant standards (discussed in the following part of this chapter) to determine the need for intervention.

3. Determine suitable solutions. For most MMH tasks, the obvious solution is to reduce the weight of the object being manipulated. Unfortunately, this is often the least feasible solution given the other requirements of a system's design. A good starting point is to identify the constraints placed by the mission or other elements of the system's design. This is followed by identifying the remaining ways the problem may be solved, such as improving mechanical advantage or improving required postures. Methods of identifying these candidate solutions include following heuristic guidelines (such as ones found on the US National Institute for Occupational Safety and Health (NIOSH) website), performing literature reviews, adapting from similar situations, or brainstorming with a group of subject matter experts (SMEs).

4. Evaluate the proposed solutions. Now that one or more possible solutions have been identified, it is necessary to determine the design parameters for these solutions that will satisfy the relevant standards of concern from step 2 above. It may be necessary to compare two or more competing solutions to identify benefit and cost tradeoffs between the different ergonomic implementations. Sometimes the analyst is faced with an overly confined problem, as many MMH analyses tend to be. Thus, the goal may not be to meet guideline thresholds, but rather to provide the most ergonomically correct outcome given the restrictions of the mission.

5. Define benefits and limitations for the customer. Once the most suitable solution has been identified, additional factors that may justify or influence the implementation of the solution must be identified. If the solution has an acceptable range in one or more variables of its configuration, these guidelines can be provided to the customer in terms of "threshold" and "objective" values. The cost/benefit tradeoff of the solution must also be considered. Costs may be financial or may be a consequence of limitations imposed on the mission by the solution. Benefits may be meeting known standards, reducing injury rates and associated costs, or improving the mission performance in some way. With major MMH interventions, a summary of how the MMH solution influences the doctrine, organization, training, materiel, leadership and education, personnel, and facilities of the system may be needed.

Tools

The following is a basic set of references and tools that will satisfy the analytical needs for many common MMH activities within the military environment.

MIL-STD-1472

MIL-STD-1472F (2001) is the primary military-centric reference for HSI design guidelines. The document contains guidance for maximum weights and forces for many common MMH activities.

For lifting and carrying tasks, section 5.9.11.3 of the MIL-STD provides for three categories with separate values for male and mixed gender populations (Figure 16.1). While the standard takes into account several special cases—such as lifting frequency, oversized objects, and lifts performed by more than one person—it does not take into account posture when determining suitable ergonomic design. Based upon the weight limits alone, an improved ergonomic design for MMH cannot often be shown to be better than the original configuration if the weight of the object being lifted cannot be changed.

HANDLING FUNCTION	POPULATION	
	Male and Female	Male Only
A. Lift an object from the floor and place it on a surface not greater than 152 cm (5 ft) above the floor.	16.8 kg (37 lb)	25.4 kg (56 lb)
B. Lift an object from the floor and place it on a surface not greater than 91 cm (3 ft) above the floor.	20.0 kg (44 lb)	39.5 kg (87 lb)
C. Carry an object 10 m (33 ft) or less.	19.0 kg (42 lb)	37.2 kg (82 lb)

Figure 16.1   MIL-STD-1472F, maximum design weight limit

Section 5.9.11.4 of the MIL-STD addresses push and pull forces for one-handed and two-handed maximum required forces for a selection of exertion- and situation-based categories. The values provided are for 5th percentile males, but allow for a 2/3 multiplier for a mixed gender population. Figures 16.2 and 16.3 address these guidelines.

For the military HSI practitioner, the MIL-STD values have the advantage of being based on a military population. Their primary disadvantage is that they are mostly rigid guidelines and often do not contribute well to modeling, comparing specific tasks, or designing alternative MMH solutions.

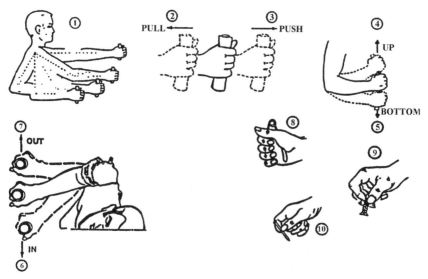

ARM STRENGTH (N)													
(1)	(2)		(3)		(4)		(5)		(6)		(7)		
DEGREE OF ELBOW FLEXION (deg)	PULL		PUSH		UP		DOWN		IN		OUT		
	L*	R*	L	R	L	R	L	R	L	R	L	R	
180	222	231	187	222	40	62	58	76	58	89	36	62	
150	187	249	133	187	67	80	80	89	67	89	36	67	
120	151	187	116	160	76	107	93	116	89	98	45	67	
90	142	165	98	160	76	89	93	116	71	80	45	71	
60	116	107	98	151	67	89	80	89	76	89	53	76	
HAND, AND THUMB-FINGER STRENGTH (N)													
	(8)				(9)			(10)					
	HAND GRIP				THUMB-FINGER GRIP (PALMER)			THUMB-FINGER GRIP (TIPS)					
	L		R										
MOMENTARY HOLD	250		260		60			60					
SUSTAINED HOLD	145		155		35			35					

*L = Left; R = Right

Figure 16.2 **MIL-STD-1472F, arm, hand, and thumb–finger strength (5th percentile male data)**

HORIZONTAL FORCE[1]	APPLIED WITH	CONDITION (m = Coefficient of Friction)
100N (25 lb) push or pull	both hands or one shoulder or the back	Low traction: $0.2 < \mu < 0.3$
200N (45 lb) push or pull	both hands or one shoulder or the back	Medium traction: $\mu \sim 0.6$
250N (55 lb) push	one hand	if braced against a vertical wall 51–152 cm (20–60 in) from and parallel to the push panel
300N (70 lb) push or pull	both hands or one shoulder or the back	High traction: $\mu > 0.9$
500N (110 lb) push or pull	both hands or one shoulder or the back	if braced against a vertical wall 51–178 cm (20–70 in) from and parallel to the panel or if anchoring the feet on a perfectly nonslip ground (like a footrest)
750N (165 lb) push	the back	if braced against a vertical wall 51–178 cm (20–70 in.) from and parallel to the panel or if anchoring the feet on a perfectly nonslip ground (like a footrest)

[1] May be doubled for two and tripled for three operators pushing simultaneously. For the fourth and each additional operator, not more than 75% of their push capability should be added.

Figure 16.3   MIL-STD-1472F, horizontal push and pull forces exertable intermittently or for short periods of time (male personnel)

NIOSH Lifting Equation

NIOSH developed an equation to calculate the recommended weight limit for a person performing lifting and lowering tasks in order to prevent or reduce work-related lower back pain and disability (Waters, Putz-Anderson, and Garg 1994). The recommended weight limit is then used to calculate a lifting index, which is a normalized value describing the severity of the lifting task. The Revised NIOSH Lifting Equation (RNLE) was designed to protect 90 percent of the mixed (male/female) population based upon factors such as lower back stresses and energy expenditure rates (Waters Putz-Anderson, and Garg 1994). The RNLE takes into account six different variables, called "multipliers," in defining a recommended weight limit (RWL) for lifting and lowering loads: the horizontal location (HM), vertical location (VM), vertical travel distance (DM), asymmetry angle (AM), frequency of lift (FM), and coupling (CM). Each multiplier is found using standard tables. The equation works by starting with a Load Constant (LC), which is 51 lb. Each multiplier provides a fractional reduction (all values are ≤ 1.0) of the load constant such that the recommended weight limit is found by: $RWL = LC \times HM \times VM \times DM \times AM \times FM \times CM$. The lifting index (LI) is then calculated as a ratio of the weight of the subject load (L) divided by the recommended weight limit derived from the lifting equation: $LI = L/RWL$

Since the RNLE is targeted toward a civilian industrial environment and the RWL derived from the RNLE will normally be more restrictive than the MIL-STD, the lifting index, as an absolute metric, is not very useful in military applications. However, the utility of the equation allows the practitioner to set known parameters and identify where each variable is maximized, providing heuristic type information on how and where to optimize the design of a lifting task. These design characteristics include the position of load, body posture while lifting, height to be lifted, frequency of lift, and object characteristics Thus, a solution with a lower LI is preferable to one with a higher LI, regardless of the fact that both LIs may be greater than 1.0.

NIOSH L5/S1 standard and 3DSSPP

The L5/S1 disc in the lower back experiences the greatest moment arm in relation to loads carried in the hands (Chaffin, Andersson, and Martin 1999), thus analyzing and minimizing the load experienced at this point is key to reducing lower back injury risks. NIOSH has recommended that 764 lb of compression of the L5/S1 disk serve as an action (design) limit, and that 1,438 lb of compression serve as a maximum permissible limit (Waters, Putz-Anderson, and Garg 1994). Since this metric is independent of strength capabilities, it is reasonable to assume that these limits are directly applicable to a military population. To easily analyze spinal disc compression, biomechanical modeling software such as the University of Michigan's Three-Dimensional Static Strength Prediction Program (3DSSPP) can be used (Chaffin, Andersson, and Martin 1999). This program requires the user to build a pose for a human figure and input the required hand loads experienced during a MMH task. The software will calculate the forces experienced at each joint (including the L5/S1). It also estimates the civilian population's strength capabilities and whether acceptable balance conditions exist for the modeled lift.

MCAM

The Logistics Management Institute and the US Army Public Health Command have developed a medical cost-avoidance model (MCAM) to estimate the costs associated with the failure to eliminate or control health hazards of Army materiel systems (Bratt et al. 2010). The MCAM uses cost factors for individual health hazard categories to calculate Army materiel acquisition life cycle medical costs. The Army's Health Hazard Assessment Program employs the MCAM to assess many types of health hazards, including heavy lifting, commonly found for materiel involved in the defense acquisition process. Given the parameters of a redesigned MMH task, the MCAM can estimate the per exposure cost savings compared to the original task. This medical cost description can be used to justify any increased materiel cost or training cost incurred by the proposed ergonomic redesign of the MMH task.

Energy Expenditure Prediction

Studying the energy expenditure of an individual completing a MMH task can provide insights into the physical demands of the work. Both the total energy required and the rate at which it is expended are key concerns when designing a MMH task. The worker endurance is primarily limited by the capacity of the oxygen transporting and utilization systems, in other words, a worker's maximum aerobic power (Waters, Putz-Anderson, and Garg 1994). By relating the energy expended in a task to the aerobic power of the individual for endurance effort, an objective assessment can be made of the work capacity of the worker carrying out a task.

It is recommended that 33 percent of the maximum aerobic power of a normal healthy person is the maximum energy expenditure rate that should be expended for an eight hour work day. A value of 16 kcal per minute is taken as the maximum aerobic power of a normal healthy young male for a highly dynamic job (Garg, Chaffin, and Herrin 1978). Thereby, for an eight hour work period, a physical work capacity limit of 5.3 kcal per minute is recommended to avoid undue fatigue.

Three commonly used methods for determining a worker's metabolic rate include spirometry, physiological monitoring, and work modeling. Spirometry involves the direct measurement of oxygen consumption while performing the task. A worker's oxygen consumption is difficult to measure on the job because of the interference between the measuring equipment and the work methods. Physiological monitors such as the BodyMedia® SenseWear® Pro Armband measures skin surface temperature, limb acceleration, and galvanic skin response to estimate the metabolic work performed by the wearer. Work modeling, such as the approach posed by Garg, Chaffin and Herrin (1978) is based on the assumption that a job can be divided into simple elements and the job's total energy expenditure rates can be predicted by knowing the energy expenditure rates of the simple subtasks. Simple factors such as body weight, gender, and time to perform a task element are used to calculate the energy expenditure for a task element. The summation of all the energy expenditure rates of the task elements and the energy required to maintain the posture gives the average metabolic energy expenditure for the job. Equations for predicting the energy expenditure for the task elements are obtained from least square regression analysis. For a complete detailed explanation please refer to Garg, Chaffin and Herrin 1978.

Garg's modeling approach is the theory behind the software Energy Expenditure Prediction Program™ (EEPP). EEPP is a very useful software program developed by the University of Michigan's Center for Ergonomics to predict the energy expenditure rates for manual material handling tasks. The EEPP method subdivides a job is into simple tasks or elements. EEPP then determines the average metabolic energy expenditure rate of the job by calculating the energy expenditures of the simple tasks and the time duration of the job. The main parameters required for estimating the task element energy expenditure are gender of the worker, body weight, weight of the load or force applied, frequency of handling the load, vertical and horizontal range of hand movement, speed of walking and carrying loads, body posture, and time duration of the task. Nine different task elements are considered to subdivide a task using the EEPP method. The nine task elements are lifting, lowering, push/pull, holding, walking, carrying, arm work, hand work, and climbing steps.

Case Studies

The following case studies illustrate a set of rapid ergonomic analyses that apply the tools and process previously described in this chapter.

Case Study 1: Drum upending

Chemical Corps soldiers are required to frequently position and maneuver drums filled with oil for smoke generation. These drums are the standard 55 gallon type with full weights of approximately 460 lb. The dimensions of the drum are 23 in in diameter and 34.5 in, in height. Figure 16.4 illustrates correct drum-lifting posture.

Figure 16.4   Drum-lifting posture

Step 1—Identify the problem   The primary task of concern is lifting the drum from horizontal to vertical orientation. This lift is performed by one person with no mechanical assists. The soldier squats at the end of the drum and grips at the lowest accessible point. The soldier then lifts the drum using leg forces primarily and stabilizes the drum with torso forces. At some point, the center of gravity of the drum will cross over the pivoting edge and the drum will right itself. Prior to the analysis, soldiers reported that the lift is very heavy and difficult to perform by 5–10 percent of the mixed gender military occupational specialty (MOS).

Step 2—Analyze the severity   Before any application of standards or tools, the dynamics of the lift must be determined. For the slow lift of a rigid object with uniform mass distribution that pivots on one end, a simple two-dimensional free body diagram will yield: $F = (W/L) \times (CG_h \cos \theta - CG_v \sin \theta)$, where W is the weight of the object (460 lb), L is the distance from the pivot point to the hand grips (34.5 in), CG_h is the horizontal distance between the center of mass of the object and the pivot point (half of the height, 17.25 in), CG_v is the vertical distance between the center of mass of the object and the pivot point (half of the diameter, 16.5 in), and θ is the angle of lift of the object. Table 16.1 outlines this dynamic load as a function of lift angle.

Table 16.1   Lifting forces required at varying drum orientations

Degrees (from horizontal)	Height (in)	Lifting Force (lb)
0	0.00	230.0
5	3.01	215.0
10	5.99	199.9
15	8.93	182.5
20	11.80	163.7
25	14.58	143.6
30	17.25	122.5
35	19.79	100.5
40	22.18	77.6
45	24.40	54.2
50	26.43	30.4
55	28.26	6.3
60	29.88	−17.8
65	31.27	−41.8
70	32.42	−65.4
75	33.32	−88.6
80	33.98	−111.1
85	34.37	−132.7
90	34.50	−153.3

In Step 1, we identified that at some point, $CG_h \cos \theta = CG_v \sin \theta$ and the object is balanced and will begin to self-right. This point is easily found by simplifying the equation to: $\theta = \tan^{-1} (CG_h/CG_v)$. Solving for this θ, the weight crossover point is found to occur at 56.31° (28.7 in of lift). MIL-STD-1472F states that the maximum load lifted should be 87 lb for males and 44 lb for females. This standard is defined for lifting an object from the floor to a height of less than 3 ft. The drum lift exceeds those limits by a factor of 2.64 for males and 5.23 for females.

Posture modeling with the University of Michigan's 3DSSPP program was performed for 5th and 95th percentile males and females. The maximum lower-back (L5/S1) compression occurs at the beginning of the lift due to the extremely crouched posture and heavy initial load. The max L5/S1 compression is estimated to be 1,736 lb for 95th percentile males, 1,381 lb for 5th percentile males, 1,336 lb for 95th percentile females, and 1,076 lb for 5th percentile females. 3DSSPP predicts the drum lift to have higher disc compression than the NIOSH recommended limit of 768 lb of compression. For the largest males, the lift is predicted to exceed the upper limit of 1,438 lb of compression.

Based upon the analysis, it is proposed that frequent performance of this lifting task most likely would lead to musculoskeletal injury. The lift may be especially hazardous for taller stature individuals. Also, the large lifting load may pose problems for fatigued or smaller stature soldiers. Furthermore, the training non-commissioned officers and other personnel who frequently lift these drums should seriously consider employing means to mitigate the ergonomic risks associated with this lift, especially while performing repetitively in a controlled environment.

Step 3—Determine suitable solutions   The constraints in this case study were that the solution cannot change drum weight or size, the task must not take significantly more time to perform, and the solution cannot be powered, must still allow for individuals of shorter stature to perform the task, and must work in warehouse or field conditions.

The design goals are to apply one or both of the following ergonomic principles to reduce the severity of this lift. First, increase the height at which the lift begins. Second, decrease the force to lift the load.

Fortunately, there are many well known mechanical and procedural solutions to meet these goals. Some candidate examples follow. By rolling one end of the drum onto a small ramp, the initial lift height of the drum will increase and decrease the force required. The same result would also come from rolling one end of the drum into a specifically dug shallow trench. Another simple solution would be to employ a commercial lever bar ("drum upender"). The use of a lever device such as this could decrease the force required as well as provide for a better lifting posture.

Step 4—Evaluate proposed solution   The parameters of the proposed designs must be then identified. For the ramp/trench, what minimum height is needed to get 95th percentile male L5/S1 compression to less than the upper limit

(1,438 lb)? Incrementing the hand position and load in 3DSSPP using Table 16.1 will reveal that at 20° or with an 11.8 in elevation of the lifted end of the drum, a normal posture lift will result in less than 1,400 lb of L5/S1 compression for a 95th percentile male. Using a 1 ft high ramp with a drum this heavy is not likely to be easy. Alternatively, the use of a trench will decrease the lifting start height and require an even greater angle advantage be used. However, the use of a lever bar may still be viable. A commercial drum upender is available that provides a 20 in advantage in horizontal hand position and a 24 in increase in vertical hand position. This makes the lift start require only 145.6 lb. In 3DSSPP, this allows for 95th percentile males to have 690 lb L5/S1 compression, which is less than the NIOSH design limit. This solution is obviously superior to the ramp/trench-based approach in terms of ergonomic suitability.

Step 5—Define benefits and limitations for customer   In this case, the intervention cost is minor and there are no known limitations to the mission through implementing the drum upending bar. Although the solution still has not met the MIL-STD values for design limits, it has at least been able to mitigate the excessive lower back compression that is one of the primary MSI risks and problems with this task. This work can then continue with assisting the customer in developing, acquiring, and fielding the solution.

Case Study 2: Manual operation of up-armored doors

Many tactical vehicles have been constructed or upgraded with heavy armor plating for the crew compartments. The added weight of the armor can drastically increase the forces required to open and close the vehicle doors. While powered door operation assistance systems have been designed and used for certain vehicles, many still require the manual manipulation of the door. The High Mobility Engineer Excavator Type III (HMEE III) is a backhoe-loader used by Combat Engineers. The manually operated armored door on this vehicle weighs 350 lb (Figure 16.5).

Step 1—Identify the problem   While the vehicle is on level ground, the weight of the door is not a problem. Since the force is distributed to the hinges, not the operator, the force required to open and close the door is only 10 lb. However, once the vehicle is parked on a surface that is sloped, as most outside surfaces are, the direction of the force due to gravity will no longer be parallel to the hinges, and the operator will be required to manage the weight of the door under the force of gravity. The issue is then identified as the following: Given the acceptable maximum push/pull forces required of the soldier, what terrain conditions may cause soldiers to not be able to operate the vehicle doors?

It is necessary to analyze the full factorial combination of operational tasks and the environmental conditions present. Four tasks and four slope conditions were considered. The tasks were opening or closing the door from inside the cab, and

Figure 16.5 HMEE III door

opening or closing the door from the exterior of the vehicle. Each of these tasks was also considered for positive and negative side slopes (roll) and positive and negative front/back slopes (pitch).

From the 16 initial conditions, eight critical scenarios were determined. For the other eight, gravity was assisting in the task. The eight critical scenarios are as follows:

- *Opening* the door from the *outside* where the operator must overcome the force of gravity on the door.
  - If opening on a side slope, the issue is the initial force needed.
  - If opening on a front/back slope, the issue is holding the door open at its maximum.
- *Closing* the door from the *outside* where the operator must overcome the force of gravity on the door.
  - If closing on a side slope, the issue is the final force needed to close the door.
  - If closing on a front/back slope, the issue is the initial force needed to close the door.
- *Opening* the door from the *inside* where the operator must overcome the force of gravity on the door.
  - If opening on a side slope, the issue is the initial force needed.

- If opening on a front/back slope, the issue is holding the door open at its maximum.
- *Closing* the door from the *inside* where the operator must overcome the force of gravity on the door.
  - If closing on a side slope, the issue is the final force needed to close the door.
  - If closing on a front/back slope, the issue is the initial force needed to close the door.

Step 2—Analyze the severity   It is prudent to begin the analysis with a list of assumptions. In this case, the assumptions included that the strength capabilities of the soldier remain constant, independent of the slope on which the vehicle is sitting; that the initial static friction force in the door is the same when the door is closed or open; that the soldier will be able to pull/push in manner that allows maximum mechanical advantage, using either or both hands, and with a force that is perpendicular to the desired motion (tangent to the path of rotation); that the mass distribution of the door is constant over its width (that is, the center of gravity of the door is halfway between the hinge and the door edge); and that unless the soldier has a known foot or hand bracing point, they will have a medium level of traction on the standing surface. The critical strength analysis will be determined with a 5th percentile female soldier.

The next task is to create a model of the dynamic forces required to operate the door. This force (F), as determined by a free body diagram is: $F = F_r + (W/H) \times CG \sin \theta$, where F_r is the resistance caused by friction in the hinges (the force needed to move the door on a level surface, 10 lb), W is the weight of the door (350 lb), θ is the ground slope.

CG is the distance perpendicular from the hinge-line to the center of gravity of the door (assumed to be 12.75 in), and H is the distance perpendicular from the hinge-line to the center of where the hand (or hands) grip the door. H can be found to vary based upon the task. In this case, there is an outside handle (H = 19 in), an inside handle (H = 10 in), and the door frame (H = 25.5 in) that can be used.

Since the maximum ground slope is the variable of interest, the above equation can be rearranged to find $\theta = \sin^{-1} (((F\text{-}Fr) \times H)/(W \times CG))$. Before calculating the slopes, the maximum allowable forces must be determined. The most likely configuration from the push/pull tables from MIL-STD-1472F can be used, remembering to use the two-thirds modifier for a 5th percentile female. Performing the analysis for the eight conditions identified yields the following information:

- *Opening, Outside, Side Slope:* Pull using handle, two hands, with foot bracing. Acceptable force = 73.3 lb, H = 19 in. Estimate of maximum slope = 15.3°.
- *Opening, Outside, Front/Back Slope:* Push or pull using door frame, two hands, hold at max, medium traction. Acceptable force = 30 lb, H = 25.5 in. Estimate of maximum slope = 6.5°.

- *Closing, Outside, Side Slope:* Push using door frame, two hands, and medium traction. Acceptable force = 30 lb, H = 25.5 in. Estimate of maximum slope = 6.5° (more if using inertia).
- *Closing, Outside, Front/Back Slope:* Push or pull using door frame, two hands, medium traction. Acceptable force = 30 lb, H = 25.5 in. Estimate of maximum slope = 6.5°.
- *Opening, Inside, Side Slope:* Push using door frame, two hands, with foot bracing. Acceptable force = 73.3 lb, H = 25.5 in. Estimate of maximum slope = 20.3°.
- *Opening, Inside, Front/Back Slope*: Push using door frame, one hand, hold at max, with bracing. Acceptable force = 36.7 lb, H = 25.5 in. Estimate of maximum slope = 8.7°.
- *Closing, Inside, Side Slope:* Pull using handle, one hand. Acceptable force = 37.3 lb, H = 10 in. Estimate of maximum slope = 3.5° (more if using inertia).
- *Closing, Inside, Front/Back Slope:* Pull using handle (unlikely able to reach door frame), one hand. Acceptable force = 37.3 lb, H = 10 in. Estimate of maximum slope = 3.5°.

A critical limitation is found—a very minor slope may hinder closing the door while inside the vehicle.

Step 3—Determine suitable solutions The constraints placed on the solution by the customer are that it will not use power or decrease the armor/door weight. The design goals are subsequently limited to improving the mechanical advantage by providing better handle placements.

Step 4—Evaluate proposed solution The furthest it is possible to place a second inside door handle in horizontal distance from the hinge is approximately 20 in. This increases the acceptable slope to 6.99°, which is very close to the next limitations of three of the four outside door tasks.

Step 5—Define benefits and limitations for customer Even though adding a secondary handle in the near term and providing training regarding the limitation is not expensive, the potential 6.5° slope limitation that remains is still very restrictive. The customer now has justification for pursuing additional routine-operation and emergency-operation door assist devices for this vehicle.

Case Study 3: Assembly of the Medium Girder Bridge

US Army bridge crewmember soldiers face strenuous MMH tasks during the construction of bridges, especially during construction of the Medium Girder Bridge (MGB). The MGB is an aluminum alloy modular bridge that is built in an "erector set" fashion, assembling the bridge by hand, with only hand tools. A typical

MGB configuration spans a 20 m gap, consists of approximately 200 components (20 unique components), weighs approximately 10 t, and is constructed ideally by 24 bridge crewmembers. The MGB construction process involves the unloading of components from preconfigured pallets, carrying these components over varied terrain, and assembling the components at the bridge site. The components can be stacked up to 2.5 m high on the pallets, carrying distances can be greater than 20 m, and some components may need to be lifted up to 2 m off the ground to be added to the bridge.

The components of the MGB range from 60 lb to 595 lb; some lifted individually, others requiring coordination of two to six crewmembers. Construction of the MGB requires lifting, carrying, and positioning of components. Other than two simple carrying tools, there are no mechanical or powered assists used for MGB construction. The current tools consist of a carry bar and a carry handle (Figure 16.6). The tools connect to the bridge components to allow the bridge crewmembers to manipulate components more easily than gripping them by hand. The current carry bar is used on the majority of the components and allows a two-handed grip. The current carry handle is used to move a few components and only allows a one-handed grip.

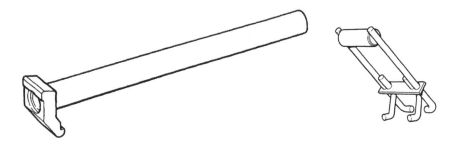

Figure 16.6   Current tools, carry bar (left) and carry handle (right)

Step 1—Identify the problem   The area of concern for this system is the repeated lifting, carrying, and positioning of large and heavy components. The heaviest, most lifted components of the MGB are the top panel, bottom panel, deck, long ramp, and junction panel. The task of lifting the bottom panel is split due to the significant weight distribution difference between the front and rear of the component. Table 16.2 summarizes the components under examination in this case study.

Table 16.2   MGB components

Component	Quantity per bridge	Weight (lb)	Weight per individual (lb)	Lift start height (in)
Bottom panel	24	425	154.00 (front) 66.00 (rear)	7
Deck (one-hand lift)	66	163	81.50	14
Junction panel	4	478	119.50	14
Long ramp (one-hand lift)	14	400	66.70	16
Top panel	32	385	96.25	9

Step 2—Analyze the severity   Although several individuals perform the lifts, the lifts were considered equivalent to a single-person lift of his or her proportion of the total weight.

Beyond this assumption, these tasks do not require any mechanical modeling before proceeding to compare against MMH standards. The users consist of a mixed gender population. The tasks include lifting from the ground to carry height, carrying, and lifting the components to a final placement height, which may be up to shoulder level. The most restrictive condition of the MIL-STD-1472F that applies for these components is the mixed gender lift to a height not exceeding 5 ft, which is 37 lb. These components all exceed that standard.

For simplification of analysis using the RNLE, a few additional assumptions were made. Since the lifting tasks involve little twisting of the trunk, occur infrequently (less than one lift per minute on average), and have a good hand-load grip, the lifts were assumed to have asymmetry, frequency, and coupling multipliers of 1. These assumptions result in a conservative analysis, that is, an underestimate of the physical demands. Applying the RNLE results in lifting indexes between 1.5 and 4.1 for the components previously identified (Figure 16.7).

Using postures obtained from video footage of the MGB being constructed, the University of Michigan 3DSSPP can also be applied to examine lower back (L5/S1 disc) compression at the start of each lift. For the 95th percentile stature male, who is at the greatest risk for low starting height lifts, the L5/S1 compression ranges from 830 lb to 1,667 lb for these components (Figure 16.8). These lifts all exceed the NIOSH design limit of 764 lb and one lift (bottom panel—front) exceeds the NIOSH upper limit of 1,438 lb.

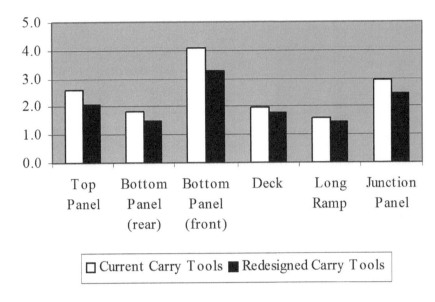

Figure 16.7   RNLE results — LI before and after redesign

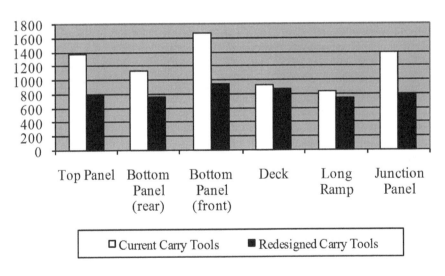

Figure 16.8   3DSSPP results—L5/S1 disc compression (lb) before and after redesign

Step 3—Determine suitable solutions With input from soldiers, the general design goals were developed: improve the posture of lifts using the carrying tools, develop a load-bearing system (for example, a cart) for carrying components, and eliminate the manual overhead positioning lifts by developing an integrated mechanical lifting system. To meet these goals, specific design solutions were identified. The hand tools should all use a two-handed grip and allow the maximum initial height for reducing lower back stress among tall individuals, but without restricting the lifting capabilities and carry height for shorter crewmembers. The cart system should be designed for pushing by one or two crewmembers and should have durable wheels to handle varied terrain. The hoist lifting system should be used to lift both top and bottom panels at a rate similar to manual lifting, should be able to secure and release components easily, as well as quickly advance to the next bridge section.

Step 4—Evaluate proposed solution The carry bar and carry handle were redesigned to raise the lift starting height by 14 in and 6 in respectively (Figure 16.9). The redesigned carry handle also allows for a two-handed grip. The hand tools were designed in a modular manner to reduce the weight and size of the required tools. Graphical results of the RNLE and 3DSSPP analyses for the redesigned tools are presented in Figures 16.7 and 16.8, respectively. Significant improvements were made for all tasks with the redesigned tools: the average RNLE lifting index was decreased by 16 percent and the average L5/S1 disc compression was reduced by 36 percent. For a 95th percentile stature male with the redesigned tools, there are no lifts that exceed the lower back compression upper limit and all the lifts are near the design limit. An all-terrain, six-wheeled cart and an electric hoist system were also designed to integrate into the build process and minimize the MMH tasks required of the soldiers (Figure 16.9). A large number of design trade-offs were evaluated when developing these mechanical assists, but that process is beyond the scope of this chapter.

Step 5—Define benefits and limitations for customer Analysis of the redesigned tools presents major ergonomic improvements, especially in the lower back stresses developed during lifting. An MCAM model of the change in musculoskeletal injury risk exposures between the old and new carrying tools identified that the cost of the new tools is offset by the reduction in long term medical care expenses for these soldiers. It is also theorized that efficient use of the cart and hoist systems will take most of the load off of the soldiers, as well as allowing the soldiers to use fewer personnel to complete a given task, increasing bridge building efficiency. To verify this benefit, an EEPP analysis was conducted to model the total savings in metabolic energy expenditure for the bridging unit by employing the new tools and assists. Across all elements of the build process that are affected by the redesigned tools, the total team energy expenditure for the bridge build was reduced by around 51 percent (Kumar 2010).

Figure 16.9   Redesigned hand tools, cart, and hoist systems

Conclusion

While more thorough procedures exist for the analysis of the physical ergonomics of musculoskeletal injury risk, the approach outlined in this chapter should enable the practitioner to rapidly identify, analyze, and propose solutions to a wide range of common MMH tasks given the information available at any phase of the system life cycle. Current military and civilian analysis procedures were reviewed for determining the safety of MMH tasks and case studies were provided to illustrate the application of the outlined rapid MMH analysis procedure.

References

Bratt, G.M., Kluchinsky, T.A., Coady, P., Jordan, N.N., Jones, B.H. and Spencer, C.O. 2010. The Army health hazard assessment program's medical cost-avoidance model. *American Journal of Preventive Medicine*, 38(1), Supplement, S34–S41.

Chaffin, D.B., Andersson, G.B.J. and Martin, B.J. (eds) 1999. *Occupational Biomechanics*. 3rd Edition. New York: John Wiley and Sons.

Garg, A., Chaffin, D.B. and Herrin, G.D. 1978. Prediction of metabolic rates for manual materials handling jobs. *AIHAJ*, 39(8), 661–74.

Kumar, K. 2010. *Energy Expenditure Analysis of Redesigned Mechanical Assists for Medium Girder Bridge*. Thesis. Missouri University of Science and Technology.

MIL-STD-1472F 2001. *Department of Defense Design Criteria Standard: Human Engineering*. Available at: http://www.public.navy.mil/navsafecen/ Documents/acquisition/ MILSTD1472F.pdf (accessed: July 26, 2012).

Sanders, M.S. and McCormick, E.J. 1993. *Human Factors in Engineering and Design*. New York: McGraw-Hill.

Waters, T.R., Putz-Anderson, V. and Garg, A. 1994. *Applications Manual for the Revised NIOSH Lifting Equation*. (DHHS Publication No. 94-110), January. Cincinnati, OH: US Department of Health and Human Services.

Chapter 17

Sizing up the Situation: A Three-dimensional Human Figure Modeling Framework for Military Systems Design

Richard W. Kozycki

Introduction

Digital 3D human figure models are now widely used in the design and evaluation of vehicles, furniture, assembly lines and other items and equipment with and within which people work The ability to model physical characteristics of the human body, simulate movement, and interact with these equipment items before physical prototypes are built are key features that have made these models almost as much a part of the design and analysis process as the use of computer-aided design (CAD) software itself. Anthropometrically and biomechanically accurate human figure models or body manikins can be used to visualize the geometric relationship between the human body and the physical environment. These visualizations include reach zones, field of view, visible ground intercepts, simulation of dynamic body motions for various operator and maintenance tasks, as well as the joint torques and the physical strength required to perform them.

Human figure modeling (HFM) can help identify shortcomings early in the design process when recommended changes to the design are easier to implement. Assessments such as the placement of displays and instrumentation, reach restrictions to operational controls, or the ability to perform a maintenance task within a confined space can be quickly assessed in relation to critical body dimensions of the target user population. These assessments would otherwise rely solely on physical mockups or prototypes with a large number of test subjects. Consequently, the use of human figure models can reduce the need for real test subjects as well as physical models and prototypes. This, in turn, generally saves time and cost, allowing the analysis to be performed earlier in the design cycle before reaching the system development stage, where the design baseline has been firmly established and changes are costly and time-consuming to implement.

However, like any other tool, HFM can produce misleading results if not applied properly. The use of this technique generally requires a skilled designer or analyst knowledgeable in the basic principles of human factors. For military systems design, this knowledge would also include a familiarity with the guidelines contained in sources such as DOD-HDBK-743A (1991),

MIL-HDBK-759C (1995), and MIL-STD-1472F (2001). Contributing variables that could lead to potential problems with performing an HFM analysis are anthropometry data, manikin construction, posture and positioning of the manikin, mission-essential clothing and equipment models, and CAD models of the system or equipment design.

This chapter covers the steps and techniques that can be used to improve the fidelity and accuracy of performing a design analysis using HFM. There is no standard way of performing a HFM analysis and the analysis itself can vary in its approach, based on the type and capabilities of the modeling software used, the requirements for the system design, and the type of analysis to be performed (operator task, maintenance task, ingress/egress, and so on). Lastly, context is provided for these steps and techniques by showing some examples of HFM analysis applied to actual systems.

HFM Software

Software tools used to simulate characteristics of the human body run the gamut of applications, with each one generally targeting a specific use or specialized area. One field that has seen widespread application of 3D HFM is that of the entertainment and video gaming industries. The pervasiveness of video games and computer animations has led designers to increasingly consider how to present real-world information flows and visualizations in the entertainment world to satisfy user demands for 3D environments or "virtual worlds." In fact, it is through these particular fields that the public has had the most exposure to simulated humans. The results, as observed in some of the more recent computer-generated animation movies and computer video game applications have been very realistic in both their appearance and body motions. The detail even extends to facial expressions that are very lifelike, as well as hair and clothing that flow and move, closely simulating their real-world counterparts. While these 3D human models look very realistic, they are generally not intended to be used for analysis or scientific purposes. There is, however, 3D HFM software that targets very specialized areas for use as an analysis tool. A common example of this is software intended for medical use such as for the development of orthopedic and prosthetic devices (Hauschild, Davoodi, and Loeb 2006). This type of software typically features detailed musculoskeletal and joint models. A recent development in the field of virtual medicine is 3D human modeling software used to assist with virtual surgery and training (Sourin, Sourina, and Sen 2000). Similarly, this software features highly detailed models of the internal organs along with nerve and tissue detail. Other modeling tools are used for vehicle crash simulation to examine resultant external forces on the human body (Dsouza and Bertocci 2010). Still other models specialize in biomechanics for areas such as sports performance simulation, lifting tasks for use in analyzing work-related injuries, or gait simulation to assist with physical rehabilitation (Cote and Hubbard 2001).

The discussion in this chapter focuses on HFM used for ergonomic and human factors engineering analysis as applied to military systems and equipment. These types of analyses can be considered workspace analyses and are used to examine how the warfighter would physically interact with an equipment or system design or within a specific workspace environment. These analyses generally examine fit, reach, and visual field of view of the intended user population and include not only static postures but also dynamic movements to investigate possible obstructions with components of the system design and body parts or with body-borne equipment.

Approach

Although it was mentioned previously that there is no standard way of performing a HFM analysis, it is still helpful to provide a list of basic steps. This list is not all-inclusive and is only meant to serve as a guideline for performing an analysis. Basic steps include the following:

1. Determine the questions to be answered with HFM in order to establish what aspects of the system design need to be analyzed and what data or supporting graphics detail will need to be provided.
2. Develop a procedure for sizing human figure model manikins from the anthropometric data to be used.
3. Investigate the clothing and equipment items worn and used by operators and maintainers and determine how they will be factored into the analysis.
4. Construct the human figure models.
5. Translate and optimize CAD files of the system design.
6. Posture and position the 3D human figure models within the system design model.
7. Perform the required analysis.
8. Provide recommendations for system design modifications.

Pre-modeling Analysis Questions

Before starting a HFM analysis of a system design, the analyst should begin by asking the following questions:

- What questions do I want to answer with HFM and is it the appropriate or best tool to use?
- What data or feedback from the modeling analysis is required to answer the questions?

These may seem like obvious questions to consider, but it is all too commonplace after obtaining the 3D CAD model of the system to insert the human figure models and begin investigating the design without giving any thought as to what information or data will need to be derived from the HFM analysis. For example, a typical question might be, Can the equipment design be safely and efficiently operated and maintained by the required percentage of the user population and what metrics do I use to make that assessment? These metrics could be a list of tasks and functions the operators and maintainers are required to perform. This might include, for example, a list of critical controls that must be accessed and activated or displays that must be viewed to determine optimal placement, or a method of checking for possible obscurations due to other components within the design. In order to make these assessments using HFM, the following additional considerations need to be made:

1. Is the available model the most recent 3D CAD model of the equipment design?
2. Is the model complete and accurate?
3. Does the model incorporate the range of motion for all articulated components?
4. Is the correct ground offset (that is, ground clearance for a vehicle design) known in order to assess a visual ground intercept?
5. What specific clothing and equipment items worn by the operator or maintainer must be considered for proper assessment of the design?
6. What tools and equipment will the maintainer need to use and are models for those available as well?
7. What posture should be used to assess the crewstation or workspace design?

By taking into account these considerations, an analyst is well primed to perform a HFM evaluation on the given system design. The caveat to this is that there are certain design activities that do not lend themselves to HFM. Consider the evaluation of emergency egress when there is a time standard involved. For instance, determining whether or not a squad could egress from the back of a vehicle in less than 30 seconds adds a layer of complexity that exposes the limitations of HFM. The complex body motions involved in an egress as well as predicting where possible snag or trip hazards would occur within the vehicle may be difficult for HFM to accurately simulate or detect, much less determine an approximate time to perform the egress. Another example where HFM may be of limited use might be in the assessment of small handheld devices that may need to be operated in a cold weather climate with a gloved hand. Many HFMs do not incorporate a hand model with the fidelity to accurately assess such devices, much less possess the capability to evaluate the impact of cold weather on manual dexterity. A third example might be the assessment of a display and whether a certain sized character font on that display could be read in a vehicle that is traveling over rough terrain. Again, many

HFMs do not incorporate a vision model with sufficient fidelity to provide an accurate assessment.

The point here is that the analyst should know the strengths and limitations of the HFM tool they are using to assess the system design and the design questions that need to be answered. Sometimes workarounds can be developed to overcome the limitations. However in some cases there may be no substitute for data that can be collected on actual human subjects.

Anthropometric Data

Anthropometric data is one of the basic building blocks of HFM. It must be given careful consideration when performing a workspace analysis of the system design for the target user population. This is the data that will be used to size a manikin or manikins for the analysis. The question that arises then is: what data should be used?

The overarching human factors engineering goal for most military systems designs is at a minimum to ensure the accommodation of the central 90 percent of the intended user population. This population normally includes all personnel who must safely operate, occupy, or maintain the system. A percentile approach stated in terms of 5th through 95th percentile body dimensions has typically been used as a means to accomplish a 90 percent accommodation goal. These percentiles refer to tables of anthropometric data where critical body dimensions are ranked relative to a sampled population. For example, a 95th percentile stature dimension is one that is taller than all but 5 percent of the sampled population. Likewise, a 5th percentile stature dimension is shorter than all but 5 percent of the sampled population.

Over the past 20 years, the most widely used source for body dimension data that has served as a guideline for sizing the designs of most US military systems has been the *1988 Anthropometric Survey of US Army Personnel* (Gordon et al. 1989). The publication contains anthropometric data from a sampled population of 1,774 male and 2,208 female soldiers that have been categorized into percentile values for each of the body dimensions listed. Designers and engineers have used these percentile values, typically 5th and 95th percentile values, to accommodate the range of user population specified in the system's requirements documentation.

When used as a means to define an accommodation range, it must be remembered that percentiles are only relevant to one specific body dimension. Problems arise when univariate percentiles are applied to systems and equipment that are multivariate in their function and design. The use of univariate percentile language inserted into a requirements document to define a desired accommodation target will likely result in a design that actually accommodates less of the population than the percentile range would imply. In fact, it has been shown that placing 5th and 95th percentile limits on all key body dimensions in a multivariate design could actually exclude 30 percent of the population instead of the 10 percent implied by the percentile range (Bittner 1974). This is aside from the fact that in real life, no human comprises all univariate percentile body dimensions such as 5th or 95th.

A more effective approach for defining an accommodation range is to use a multivariate statistical method such as principal component analysis (PCA) (for example, Bittner et al. 1987, Zehner, Meindl, and Huson 1992, Gordon, Corner, and Brantley 1997, Gordon 2002) that incorporates a set of critical body dimensions intrinsic to the system design from an existing or known anthropometric database. This method allows a desired range of a population to be accommodated in such a way that the size differences as well as body proportion variability are taken into account. Additionally, this method allows for a set number of manikins or forms to define the boundary or range of the desired population accommodation.

The design of a crewstation or workspace area using multivariate anthropometric data begins with the selection of body dimensions or body descriptors, known as "landmarks," which are essential to ensure accommodation of the target user population. The most crucial of these body dimensions are Sitting Height, Eye Height Sitting, Acromion Height, Knee Height Sitting, Popliteal Height, Chest Depth, Buttock–Knee Length, Bideltoid Breadth, Hip Breadth Sitting, Thumbtip Reach, and Functional Leg Length (Figure 17.1).

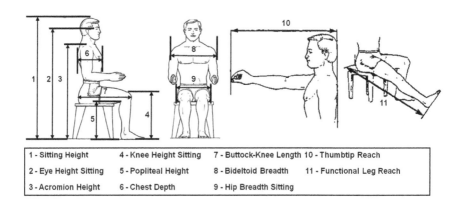

1 - Sitting Height	4 - Knee Height Sitting	7 - Buttock-Knee Length	10 - Thumbtip Reach
2 - Eye Height Sitting	5 - Popliteal Height	8 - Bideltoid Breadth	11 - Functional Leg Reach
3 - Acromion Height	6 - Chest Depth	9 - Hip Breadth Sitting	

Figure 17.1 Illustrations of crucial body dimensions for workspace design

PCA can then be used to derive an accommodation envelope by reducing this larger set of critical body dimensions to a more manageable number of dimensions (that is, two or three) that account for a large proportion of the variation by using linear combinations of the original measurements. After derivation of the principal components, a database of subjects can be scored and plotted in a new PCA space, and a 2D or 3D ellipse or ellipsoid can be fitted to the population distribution in order to capture the desired percentage of the user population. The surface of the generated ellipsoid then represents the accommodation envelope or "boundary" associated with the percentage of the selected subjects (Lockett et al. 2005).

A problem still remains, however, for the personnel who are faced with writing the requirements documents for military systems. If one of the goals for these systems is still to obtain 90 percent accommodation for the target user population, but the requirements writers are no longer supposed to insert univariate language to accomplish this, what are they to use instead? For the initial design phase, a table of anthropometric data matched to the subjects selected from the boundary of a specific accommodation ellipsoid could be used. However, as the design stage progresses and dynamic body movements in the workspace are factored in, HFM can play a key role. Boundary manikins, which are sized to the anthropometric data of the subjects selected from the accommodation ellipsoid, can be constructed and integrated into the CAD models of the system for design and evaluation purposes (Figure 17.2).

Figure 17.2   Illustration depicting the concept of a boundary manikin human figure model

Manikin Construction

After the anthropometric data for building the boundary manikin sets has been derived, it may seem that all that is required to produce the digital human models is to enter data into the HFM software and wait for it to automatically produce the individual models or set of models. While many HFM packages do have the ability to generate default or "library" figures, the creation of custom figures usually involves more effort. The procedure outlined below covers the method for generating these custom boundary figures using the Jack™ HFM software.

The Jack™ software includes an advanced body-scaling utility; the interface is shown in Figure 17.3. The interface features input for 26 specific body dimensions that can be used to build custom-sized figures. In addition to the advanced-scaling utility, a body part-scaling interface is also provided. This interface allows the user to scale individual body segments of the human figure.

The analyst should note that the scaling process involves more than just entering numerical values into the interface. Some of the dimensions are cumulative. For instance, Sitting Height includes Sitting Eye Height, Acromial Sitting Height, and Head Height. If one dimension is changed, the software changes the other dimensions by some proportion that may be unintended. A constant process of rechecking all changes is required to achieve the desired result. For example, if the manikin's Seated Height dimension is changed, the Seated Acromial Height and the Seated Eye Height dimensions will also change unless these dimensions have been locked. Likewise, if the Bideltoid Breadth dimension is changed, the Chest Breadth and Biacromial Breadth will also change unless these dimensions have been locked. Throughout this construction process, a series of reference markers (Figure 17.4) can be used to pinpoint the exterior body landmarks that represent anthropometric measurement points on the Jack manikin. These markers should be stored with the reference figure model as a visual check against unintentional changes in dimensions that should remain fixed. When each figure is completed, a final point-to-point measurement of all critical figure dimensions should be performed to ensure the highest degree of accuracy.

A subset of seven figures from a crewstation boundary manikin set is shown in Figure 17.5. The figure illustrates the variation in the body dimension proportions from within that set. These figures depict the population extremes with the small female and large male, as well as some of the widest variations in limb and torso dimensions.

Developing 3D Clothing and Equipment Models

The military services, through research funded by the US Department of Defense, were among the first to recognize the need for information on the decremental effect of clothing on mobility and performance (Paquette et al. 1999). An early investigation into the area of clothed anthropometry (Roberts 1945) looked at the

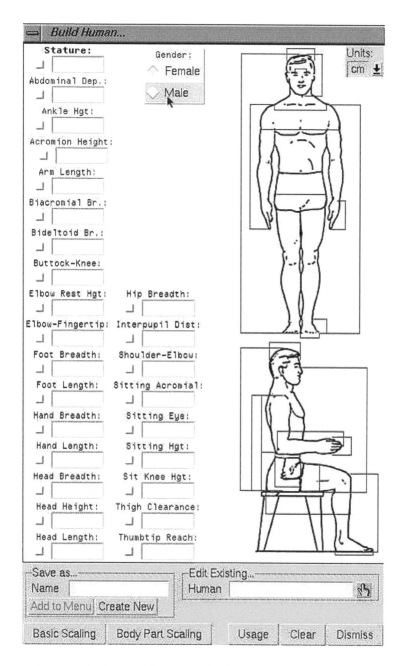

Figure 17.3   Jack advanced figure-scaling interface

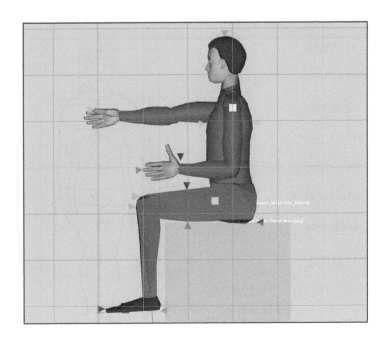

Figure 17.4   Large male figure with file reference markers

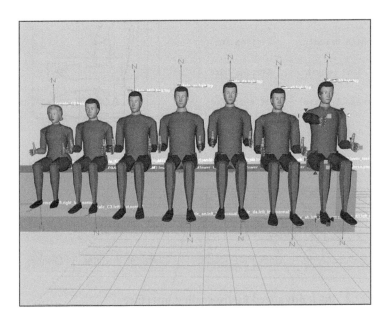

Figure 17.5   Selection of several boundary manikins from a crewstation manikin set

workspace requirements for nine different US Army clothing combinations. During the same period, a survey of seating dimensions for railway car design (Hooten 1945), in which the subjects were measured with various clothing combinations, was published. It was probably this survey, along with one performed on US Army Air Corps personnel (Randall et al. 1946), which ultimately led to the widespread use of applied anthropometry in the military. Randall and co-workers tested the effect of heavy flight clothing on the ability of fliers to fit into B-29 gun turrets.

During the 1950s and 1960s Kobrick (1956a, 1956b, 1957a, 1957b), Kobrick and Crist (1960), White, Kobrick, and Zimmerer (1964), and Johnson (1984) examined increases in body size due to wearing US Army clothing. Additional research on the effect of flight clothing on body size and function was also done by the US Air Force (Alexander et al. 1964, Alexander, Garrett, and Flannery 1969, Alexander, Garrett, and Robinette 1970, Garrett 1968, Middleton, Alexander, and Gillespie 1970).

While clothing and encumbrances are often disregarded in the design process for many types of office and commercial workspace designs, they are an important factor to consider for military systems where space is often at a premium, and the additional clothing and equipment can add significant weight and bulk to each individual. Some typical examples include multilayered ensembles that provide protection against nuclear, biological, and chemical threats; clothing to operate in extreme cold weather environments; and body armor for ballistic and fragmentation protection. Additionally, load-bearing vests and packs are worn to help transport sustainment supplies, along with advanced tactical equipment such as communication gear, components for night vision and thermal imaging capability, as well as lasers for range-finding and target designation.

Clearly then, when performing workspace analysis on military system designs, one must account not only for the physical body dimensions of the intended user population for the system design but also the clothing and equipment that will be worn when operating, occupying, or maintaining the system, as well as their effect on ingress/egress. When clothing and equipment models are added to the design considerations, a new layer of complexity is encountered. This is because specialized clothing and equipment have a significant impact on a person's range of motion, field of view, and ability to fit in a workspace, making it difficult for the wearer to complete required tasks successfully. For this reason, efforts using human figure models to analyze these workplaces and associated tasks should also include models of the same specialized clothing and equipment (Kozycki 1998, Kozycki and Gordon 2002). An example of clothing and equipment having an impact on operator performance is shown in Figure 17.6.

There are several methods in use today for constructing clothing models for 3D human figures. One of the popular methods employed is texture-mapping (Figure 17.7), which is analogous to wrapping a 2D image or picture around a 3D object. This method allows one to place intricate pattern designs and prints on the clothing model. For military analyses, camouflage patterns can be texture-mapped onto the human figure for an added realistic appearance. Texture-mapping is often used

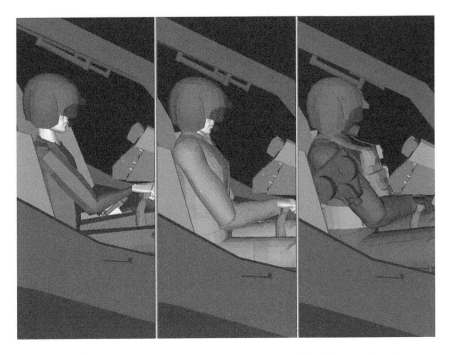

Figure 17.6 Human figure with different layers of clothing and equipment seated in an Apache helicopter cockpit model

for virtual reality and gaming applications, where real-time human figure motion is critical or an added degree of visual realism is desired. However, its use as a sole means of simulating clothing adds no bulk or thickness to the human figure. Also, it does not help to simulate any of the other physical properties of clothing material. For these reasons then, texture-mapping alone is usually not employed for clothing models when performing an ergonomic analysis except to enhance appearance.

A fairly simple method for representing clothing bulk is to expand the human figure body segments uniformly by some dimension. The expanded body segment method is shown in Figure 17.8.

A drawback to the expanded-segment method is that it usually leads to an exaggerated-looking figure with dimensions that are out of proportion in some areas of the body. In actuality, though, the clothing bulk thickness is not uniformly distributed over the entire body, such as the lower torso when the figure is in a seated posture. This incorrect addition of segment thickness to the lower portion of the upper leg and lower torso segments, for example, can lead to positioning of the human figure that would be slightly higher and more forward in a seat than would actually be the case. This would, of course, depend on the amount of uniform expansion applied to the body segments.

Figure 17.7   Human figure model with texture-mapped clothing segments

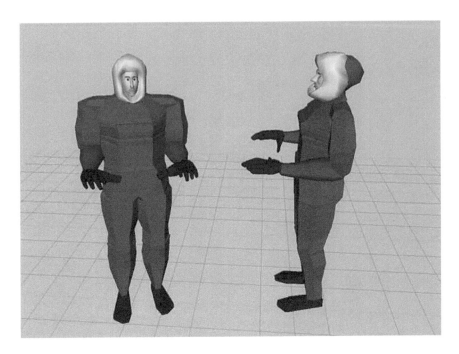

Figure 17.8   Human figure with cold-weather clothing, constructed using the expanded segment method

An advanced geometric modeling methodology for simulating clothing incorporates a deformable finite element mesh of polygons or use of B-spline surfaces to represent complex 3D shapes. This method, when coupled with collision detection to sense collision of the fabric with other body parts, provides one of the most powerful techniques for simulating 3D clothing. In fact, this method is commonly used for computer-generated animations and rendered motion sequences where the highest degree of realism is desired. The problem for the analyst is that these methods are often computationally expensive, making it difficult to obtain the simulated clothing effects in real time. Additionally, they usually require that motion sequences be scripted or captured first and then played back later. Depending on the complexity of the human motion involved and the power and speed of the graphics card being used, the sequence can take anywhere from a few minutes to a few hours to render and play back.

Probably the most widely used method for developing 3D models today involves using some type of 3D scanning equipment. An example of a digitized soldier clothing and equipment ensemble using a whole body scanner is shown in Figure 17.9.

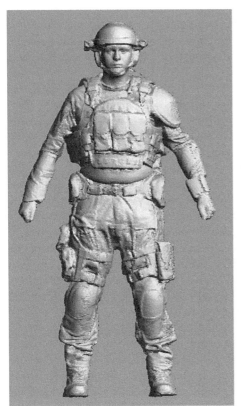

Figure 17.9 **Three-dimensional whole-body scans of soldier wearing a combat-equipped ensemble (scanned images courtesy of Brian Corner, Ph.D., US Army Natick Soldier Center)**

This equipment typically employs a laser scanning head or multiple scanning heads, depending on the size of the scanner. The advantages of using this method are that the clothing and equipment can be scanned quickly and captured at a very high level of the detail, which, in turn, produces very high-resolution models. The scans are capable of capturing clothing details such as the draping and folding of the fabric, as well as how weighted objects such as backpacks or body armor would fit and hang on the actual subject. This adds an element of realism and accuracy to 3D clothing and equipment models that was nearly impossible to capture before the advent of these scanning devices.

However, there are drawbacks as well to using scanning equipment to produce 3D models of clothing and equipment. For starters, when scanning a subject wearing the clothing and equipment, only the exterior surfaces are scanned and not the underlying interior surfaces, which makes it difficult to capture the

384 Designing Soldier Systems

bulk and thicknesses of the individual components that make up that ensemble. A possible solution to this may be to scan the subject in "layers." For example, perform multiple scans starting with the subject wearing undergarments, followed by a scan with combat coat and pants, then next with body armor, and then with a load-bearing vest, and so on until the complete clothing and equipment ensemble has been scanned. The problem with this solution is that for each scan, the subject would have to be in the exact same posture, otherwise the successive scans would be difficult to align. This still leaves the analyst with the task of building the interior surfaces of the clothing and equipment that were not captured by the scans. Also, since the clothing and equipment are scanned together, the analyst will have to develop a method for separating the clothing and equipment into separate components in order to produce individual models. Additionally, the detailed folding and draping of the clothing fabric captured by the scanner may only be representative of the clothing properties in the specific posture in which the subject was scanned. In this case, several body postures may have to be scanned. However, it would be difficult to anticipate what posture or postures may be needed to perform a design analysis. This high level of detail usually creates models that have a high polygon count and can be too complex for HFM use, particularly if a multilayered ensemble needs to be modeled.

One method that has been used by the US Army Research Laboratory to build 3D models of Army clothing and equipment ensembles involves using a combination approach. This method uses a portable coordinate measuring machine (CMM), such as a FaroArm™, to digitize the individual clothing and equipment components and then uses 3D whole body scans to help align the modeled ensemble on the human figure model. The use of a CMM to digitize the clothing and equipment is a more time-consuming process and the models usually do not capture the detailed folding and draping properties of the clothing. However, with the use of a CMM both the exterior and interior surfaces of the clothing can be captured to generate models that have thickness and represent the bulk of a cold weather coat, for example. Also, the use of this method allows layers of clothing and equipment to be built so that different combinations of equipment can be examined.

An important factor to consider when building clothing models is that the models must be digitized in segments so that they can be applied to the human figure in such a way that the model still retains the ability to move the limbs and torso and be placed into a variety of postures needed to perform an analysis. Figure 17.10 shows an example of digitized and segmented clothing.

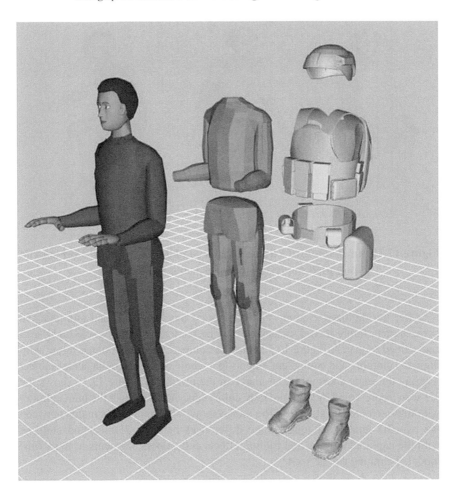

Figure 17.10 Digitized and segment clothing and equipment models to be attached to a human figure model

A human figure model in various postures with the attached clothing and equipment models is shown in Figure 17.11.

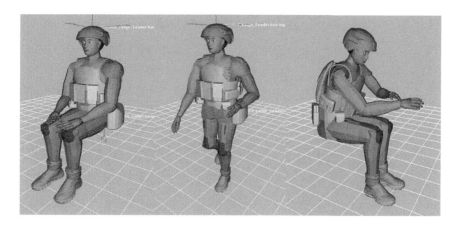

**Figure 17.11 An example of a human figure model with digitized clothing and
equipment models placed in various postures**

After the digitized clothing and equipment has been modeled and attached
to the human figure model, placement can be compared to the whole body scan
and refined to match. Figure 17.12 shows a comparison of the human figure with
digitized clothing and equipment models to a whole body scan of an actual subject
wearing the same ensemble. The image in the middle of Figure 17.12 shows the
3D scan (represented by the solid gray color) superimposed over the human figure
with digitized models attached. The image shows an overall close match between
the two 3D images, although some differences between 3D scanned clothing
with the captured draping and folding effects and smooth segments of the CMM
modeled clothing can be seen.

CAD File Preparation

An essential requirement for HFM is the availability of CAD files representing the
system design. This is often a significant challenge to analysts because system CAD
files, in many cases, may not exist, are unobtainable, or are only available in a form
that is not easily configurable for HFM. Most system designs originate as CAD
files created for the purpose of manufacturing the item. This results in having to use
CAD files that are solid geometry models with parametric features that incorporate
the material properties of the item to be manufactured. As compared to the solid
geometry models, CAD models typically used for HFM analysis are 3D surface
models. In other words, only the surface or boundary representation of the design
is required. This difference in CAD model types means that a translation process
is required to convert the solid model into a surface model or, more specifically,
faceted geometry when one is using a standalone HFM package (Figure 17.13).

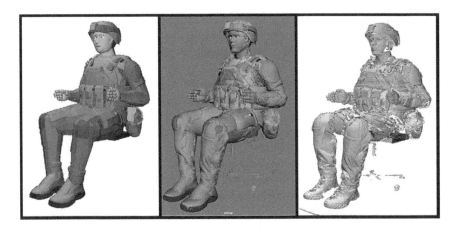

Figure 17.12 Comparison of a human figure with individually modeled clothing and equipment items to a 3D image generated from a whole body scan of an actual subject wearing the same clothing and equipment items

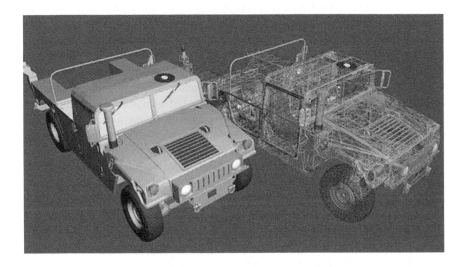

Figure 17.13 Surface model of a high-mobility multi-purpose wheeled vehicle (HMMWV) displayed in shaded and wire frame forms

Translated CAD geometry can also pose an additional challenge because of the number of polygons used to represent objects in the model. Often, models that have been translated from solids to surfaces are not optimized and the surface model has many more surface polygons than is required to efficiently define the object. For example, a single part in a vehicle design may have hundreds or even thousands of extra polygons after the translation of the file. If this problem is multiplied by the hundreds or even thousands of parts that comprise a vehicle design, computer graphics and computational resources can become overtaxed. Software known as "decimate" programs can be used to reduce the number of polygons on a surface model in order to optimize the file. However, care must be exercised because reducing the polygon count may sacrifice the fidelity of the model. Many of the decimate programs will reduce the polygon count but at a certain point, these programs will also begin to remove detail and features of the model that the user would prefer to keep. It is important for this step to be performed by someone who is familiar with the vehicle design; otherwise, highly desired design detail could be eliminated inadvertently.

Another problem with translated files manifests as nonessential detail that may be included with the desired model file. In many cases, when a CAD file for the design does exist, the human factors analyst receives a file of the entire system instead of just the components required for analysis. Often, these translated files contain nonessential detail included with the desired model file. A typical example of this is a vehicle file that contains all the major subassemblies including every bolt, washer, and screw thread as well as internal subcomponents. If the analyst has access to the CAD software that was used to create the original design, it may be possible to remove the unwanted detail by turning off the layers that these parts are assigned to and then exporting the file. In many cases, however, the analyst may not possess or have access to the software and the task of culling unwanted detail from the design can become quite challenging and time-consuming. One way of overcoming this challenge is direct communication between the analyst and the system CAD designer. In this way, the level of detail can be communicated directly to the CAD designer responsible for generating the files. The designer may be able to remove the unwanted detail in advance, thus saving the human factors analyst hours or even days of effort in performing the same task.

In some cases, the problem is missing detail rather than too much detail. For example, it is important that the CAD file include articulated joint definitions of movable parts that match the actual mechanical limits specified for that assembly. This information allows the human factors analyst to evaluate the effect of adjustable parts such as seats on accommodation. Joint definitions are frequently deleted from the files during the translation process or they may not have been defined. A recommended solution is to show movable parts at the extremes of displacement or rotation in the exported CAD file. Depicting a seat pan at its maximum fore and aft travel is a good example of this technique. If coordinates for centers of rotation (that is, pivot points along with rotation and displacement limits) are known, they can be manually added as joints in the translated model.

The last consideration for the translation and optimization process is that it must be performed in a rapid and timely manner in order not to lag behind the design effort. If human factors design analysis is to be effective, the process must keep pace with the ongoing design processes during the system development. Failure to do this severely limits the impact of human factors design considerations on total system design and integration as design baselines are established. Achievement of this objective is through the early and sustained interaction between the human factors engineer and members of the relevant integrated process team working with the same files simultaneously. This approach is more efficient and reduces the risk of redesign attributable to insufficient human factors engineering design considerations.

Human Figure Model Integration

Once the CAD files have been obtained and the appropriate human figure models have been developed, some thought now must be given as to how these figures are to be integrated into the design in order to perform the analysis. This requires a method for positioning and posturing the figures within the design.

Positioning

Positioning of the human figure model will depend on the role of the operator, occupant, or maintainer and how they will be interacting with the equipment design. One of the most common applications of human figure models is the evaluation of a workspace designed for a seated posture. However, realistically positioning seated human figure models can be difficult. A problem that analysts face is that the shape of the pelvis segment in the models remains fixed and does not change with the figure's posture. In many cases, some posterior point on the exterior of the pelvis segment is used to establish a relationship with the seat reference point (see Roebuck, Kroemer, and Thomson 1975 for a definition). This method, however, does not account for the stretching of muscles and compression of soft tissue on the bottoms of the thighs and buttocks for a seated posture. This, in turn, usually leads to a seated figure that is positioned too high and too far forward of the seat reference point. It is better, therefore, to use a skeletal landmark or joint center for positioning purposes instead of an exterior point on the surface of the figure's pelvis segment (Figure 17.14). An approach that has been employed is to use the location of the hip joint center to establish the seated position for the figures. This location method is adopted from Reynolds, Snow and Young (1981) and studies performed by the University of Michigan Transportation Research Institute (for example, Reed, Manary, and Schneider 1999). The method uses proportions of the pelvic bone and the ischial tuberosity and posterior–superior iliac spine landmarks that can be found on the Jack model pictured in Figure 17.14.

Posturing

After the figure has been positioned, the next step in the process is to determine a reasonable posture for the figure based on the tasks and location assigned to the occupant or operator of the design. These postures will vary, for example, between

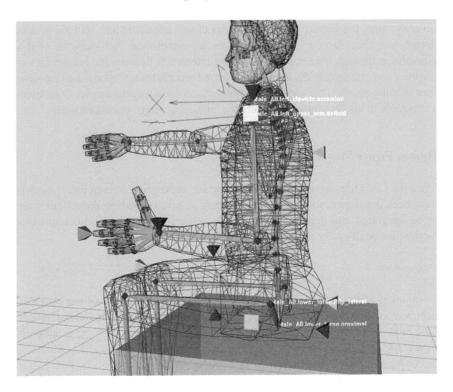

**Figure 17.14 Male figure displayed in wire frame form to show Jack internal
skeletal link model with joint locations**

operators assigned to crewstation positions and occupants who will be sitting in
the squad location of the vehicle. The crewstation operator may have to perform
several tasks simultaneously such as using foot pedals and hand-operated controls
as well as monitor displays and communication devices. In order to decide the best
locations for the placement of displays and operational controls in a crewstation,
a reasonable posture must be incorporated into the figures after they have been
positioned. A common mistake made when posturing a seated figure is to posture
the upper body in a straight upright-seated posture. This may be due in part to this
standardized posture being used to collect seated anthropometric dimensions such
as Sitting Height or Seated Eye Height. However, rarely does anyone sit in a seat
in a straight upright posture, but instead will sit with some degree of slump that
will affect placement of the hips, shoulders, and head. Specialized clothing and
equipment such as body armor, survival vests, and chemical protective ensembles
will also have an effect on the posture of the occupant or operator and must be
considered when posturing the human figures. In some cases it may be necessary
to collect posture data on actual subjects wearing the required specialized clothing
and equipment in order to determine an accurate seated posture. An example of

Figure 17.15 Collecting aviator posture data with portable 3D coordinate measuring devices

posture data being collected on pilots wearing mission-essential clothing and equipment, sitting in prototype seating, is shown in Figure 17.15.

The posture of an operator, occupant, or maintainer, however, is usually not static and therefore the modeling effort must also examine the system designs employing dynamic postures as well. The use of dynamic postures will help ensure that sufficient body clearance is available in the workspace area when reaches with the arms or legs are performed, as well as clearance for the head, shoulders, and upper torso as movement of those body parts occur. This is especially important when the human figures are encumbered with bulky clothing and equipment. A static posture, at first glance, may indicate that the design will accommodate the full range of the user population. However as dynamic postures are factored in, one may find that the design is actually lacking in space available for the occupant to perform all required tasks. This is why it is equally important that the designer or analyst have a solid understanding of the roles and tasks that the operator, occupant, or maintainer must perform or have access to a subject matter expert (SME) to provide guidance.

Evaluating the Design

Having completed the steps outlined in the HFM procedure discussed in the previous parts of the chapter, the analyst is ready to begin the investigation of the system design by evaluating the design against a set of known requirements, standards, or guidelines. For military vehicles and aviation platforms, the typical areas of analysis are crewstation or cockpit design, troop transport area design, emergency ingress and egress routes, fields of vision and maintenance tasks, and component accessibility. In the following parts of the chapter, examples of these design areas are presented, along with a discussion of some of the factors that must be considered for their design and analysis.

Operator Workspace Design

There are many factors to be considered in the process of designing the layout of a platform operator station. The most important of these considerations is a clear understanding of the tasks the operator must perform during a mission profile. As the layout begins, usually one element takes precedent over the others as a fixed starting point. It could be the foot pedals, the seat, a direction control, or a vision constraint. Other elements then must have built-in adjustability to account for human variation and task performance. For example, an operator station design may start with a fixed set of foot pedals. The seat, steering wheel, and vision must then be adjusted around that fixed element. A common starting point is the eye reference point or Design Eye Line (DEL). Once the eye reference point or DEL has been established, the distance down to the seat can be established if the Sitting Eye Height of the tallest operator is known. We can determine the vertical seat adjustment range by knowing the Sitting Eye Height of the shortest operator in the target audience description. The human factors analysts will need to convey to the design engineers the impact that human variations will have on hardware design and location. This is the point where several boundary manikins may be swapped into the crewstation to see where interferences might occur. If foot pedals or foot rests are part of the design, then the analyst needs to review the range of comfortable foot reach zones for a variety of operator sizes. If the comfort zones for small operators overlap with the comfort zones for large operators in different seat locations, then pedals placed within the overlapping area will likely be a good starting location for a common pedal. If the zones do not overlap, then adjustable pedals may be required. Again, the process includes reviewing several boundary manikins to check for clearance violations.

Fore and aft seat adjustment is usually a function of allowing all operators to comfortably reach or see controls or displays. Figure 17.16 shows an example of examining a helicopter crewstation design with a small female human figure integrated into the 3D CAD model. In this example, the model is placed on the DEL and reach to the foot pedals and critical flight controls are evaluated.

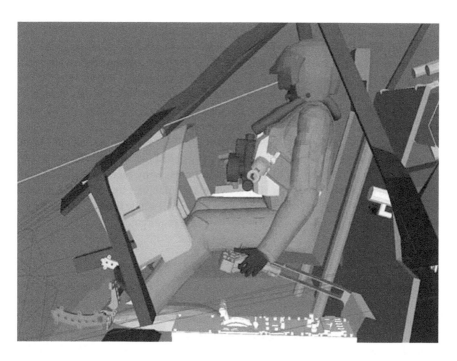

Figure 17.16 Small female human figure integrated into a 3D CAD helicopter crewstation model to evaluate design accommodation of the target pilot population

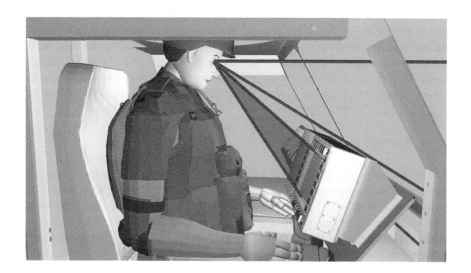

Figure 17.17 Evaluation of a display to see if placement falls within recommended guidelines for viewing angles

Informational displays are critical components within a crewstation design and placement of these displays must be evaluated not only for optimized viewing but also to ensure that they are not obscured by other components within the crewstation. The evaluation will also ensure that placement does not impinge on the operator, hamper body movement, or pose an obstacle for crewstation ingress and egress. An example of a display evaluation is depicted in Figure 17.17 with sight lines and viewing angles for the operator shown.

To ensure that the design accommodates the full range of the intended user population, each human figure from the boundary manikin set is evaluated in the model for fit, reach, and field of vision requirements. Any reach shortfalls, body interference or problems meeting display viewing guidelines for any of the figures in the set will likely mean that the target population will not be fully accommodated and may require a design modification to meet the requirement.

Field of View (FoV) Analysis

In addition to using HFM for the assessment of a vehicle interior design for optimum placement of displays and other critical components, it is a tool also used to perform exterior FoV analysis from various positions within the vehicle. Vehicle personnel must be vigilant for hazards such as concealed roadside explosives, enemy snipers, or badly deteriorated road conditions. HFM can be used to determine ground intercepts or the nearest ground point that can be seen from the driver's vantage point or any other crewstation within the vehicle. Also, polar plots can be derived using HFM to determine how wide a FoV the vehicle design provides and what blind spots might exist. FoV is often a key design parameter that is specified in a requirements document for military vehicles. HFM can be used early in the design process to check if this requirement will be met for the specified personnel population range that will be operating the vehicle. If the requirement cannot be met, then the analyst could recommend possible design modifications such as an optimal seat placement or adjustability that would help to meet the desired FoV goal or suggest where technology such as cameras or sensors could be placed to fill in any gaps.

Troop Transport Area Design

The troop transport section of a vehicle or aircraft is designed to transport multiple numbers of military personnel and is usually configured with the seats in an in-line arrangement, with seated troops facing the front (or occasionally the rear), or a side-by-side arrangement, with occupants facing toward the center of the transport area. In addition to transporting the required number of military personnel, the troop transport section must usually accommodate stowed equipment, gear, or weapons and may also contain fire suppression systems, radio racks, and communication equipment, as well as other specialized equipment depending on the intended mission of the vehicle or aircraft. This means that the space claim for seated passengers is limited and, in most cases, less than optimal. Most tactical transport vehicles, especially legacy platforms that carry troops in a side-by-side seating arrangement, tend to use a bench-type seat of a certain length that is designed to accommodate the desired number of seated passengers. However, to meet current evolving survivability requirements, transport vehicles are using individual seats that provide protection against exposure to land mines and can help to mitigate the effects of blast energy. The seat is designed to stroke downward, or it contains some sort of energy-absorbing device. This, in turn, means that equipment or gear that could be stowed under a bench seat in older designs can no longer be stowed under the newer individual-type seats, as it would interfere with designed capability to mitigate the effects of blast energy. Stowage for this equipment or gear will have to be found elsewhere inside an already crowded vehicle and someplace where it is not blocking an aisle or passageway that would impede ingress or egress.

These individual seats should have sufficient spacing to accommodate the larger end of the population, along with any of the specialized clothing and gear that they are expected to wear or carry. For this reason, then, it is equally important when using human figure models to analyze the design of this area that they be equipped with models of the same clothing and equipment. Likewise, models of the equipment or gear that must be stowed in this area should also be available in order to make sure that it will not impinge on the space claim for the seated troops.

Other important seat dimensions to consider are, for example, the Seatpan to Ceiling Height dimension. This dimension should be sufficient to accommodate the largest Seated Height dimension of the population, including space for a helmet and, nominally, an additional 2 in to allow adequate clearance for travel over rough terrain. The seat back height should provide proper back support; and the seat pan width and length should accommodate large seated hip dimensions, including clothing and gear worn by the seated occupants. The seat pan should also provide sufficient support to the upper thighs to avoid leg fatigue on extended rides.

Human figure models can be used to investigate the seating arrangements for the troops in this area and to analyze hip, back, and leg angles to see if the design can comfortably accommodate the specific number of passengers that the vehicle or aircraft is required to transport. Figure 17.18 shows an example of a possible seating arrangement for a troop transport vehicle. The body dimensions of the large end of the population are used to set the dimensions of the seats and seat spacing. However, a seatpan that is too long or sits too high off the floor could result in a design that may not accommodate some personnel at the small end of the population. It is a good practice, therefore, to examine the design with both large and small human figure models from the boundary manikin set.

Emergency Ingress and Egress Routes

Emergency ingress and egress for military vehicles and aircraft may be one of the most difficult categories to analyze using HFM. Emergency egress for most military platforms has an objective time requirement associated with it.

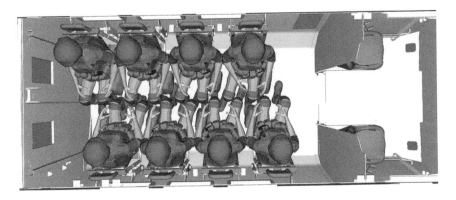

Figure 17.18 Seating configuration shown for the troop transport area of a tactical military vehicle

For example, a typical requirement might be: "All personnel must be able to safely egress from the vehicle in less than 30 seconds." While the analyst may be able to examine the postures, body movements, and route that the vehicle occupants may select, it is difficult to accurately predict the time it would actually take to complete the egress with HFM. There are many variables that could factor into performing the egress. A seated passenger may experience difficulty in unfastening a seat restraint or could become entangled in the harness. Trip or snag hazards can also degrade egress performance and may be difficult to spot or predict with modeling. Finally, personnel may have to enter or exit through a narrow interior passageway or through an escape hatch or emergency exit. Even if the analyst can conclude that there is sufficient space for occupants to pass through the passageway, hatch, or emergency exit, it is difficult to accurately predict the time that it would take. For these reasons, evaluation of emergency ingress and egress time requirements should be performed with human subjects with the actual equipment or a prototype of the design.

These limitations do not exclude HFM from consideration when attempting to examine ingress and egress issues. In fact, HFM is commonly used to examine the space or clearance issues associated with ingress and egress. The analyst can also use human figure models to examine the posture that one might have to assume in order to traverse an egress route or passageway. However, any use of human figure models to examine ingress or egress clearance issues must also include models of the mission-essential clothing and equipment items. As has been mentioned, items such as body armor, load-bearing vests, hydration packs, and protective clothing can add significant bulk to personnel that have to perform emergency ingress or egress procedures. Without models of these clothing and equipment items, it would be difficult to properly assess ingress and egress issues using HFM. Figure 17.19 shows an example of HFM use to examine an emergency egress scenario.

Figure 17.19 Human figure model used to examine an emergency egress scenario through a restricted area inside a helicopter

A common application of HFM in emergency egress situations involves the examination of access to hatch openings. The analyst is interested in evaluating, for example, whether all personnel can reach a hatch opening or whether they would be required to stand on a seat or other interior structure to be able to egress through the hatch. Also, if a vehicle has a moveable or rotating component mounted on the exterior, one must consider the possibility that its orientation could block or restrict space for ingress and egress through a hatch opening. HFM could be used to answer such questions and help identify any design modifications that may be required.

HFM can also be used to examine egress for a vehicle or aircraft in a rollover condition. Consideration must be given for emergency egress in rollover situations because the vehicle or aircraft is not in its normal upright state of operation. If a vehicle has rolled over onto the roof, all emergency escape hatches may be blocked. The path to the other emergency exits would certainly be different than when the vehicle was upright. Unfortunately, it may be difficult and hazardous to simulate rollover with an actual vehicle or prototype design with live subjects. However, the analyst can easily rotate the CAD model of the vehicle design and use HFM to look at the egress paths that may or may not be available for this condition and examine what obstacles for egress might be present. Additionally, HFM could be used to examine optimal locations for possible alternative escape openings in the floor or side of the vehicle.

Lastly, consideration must also be given to the extraction of injured or incapacitated personnel. HFM can also be used in this instance to examine space requirements if the patient must be transported into or out of the vehicle or aircraft on a stretcher or litter. In this case, sufficient space must exist not only for the patient and litter but also for the personnel that will need to lift and carry the patient. Also if, for example, a vehicle has rolled over on to its left side and the driver is incapacitated, it may need to be determined if emergency responders would be able to gain access to the injured personnel and if room exists to perform the extraction. HFM can be used to examine this scenario and flag possible design problems that could pose potential problems.

Equipment Maintenance Issues

Maintenance requirements and procedures must always be given careful and thorough consideration for any system design. Access to components that will need to be serviced along with their installation and removal paths are areas where HFM is frequently used when examining maintenance requirements. When examining a component that will need to be accessed to perform a maintenance procedure, the design should provide sufficient space to accommodate the hand or arm to reach it. Also, maintenance personnel may be required to wear cold weather or other protective clothing that would add extra bulk and increase the space required to perform the maintenance procedure. Tools must also be considered—not only the space to accommodate the tool itself but also any space required to operate a tool such as a wrench or pry bar. Visual access is almost equally important to gaining physical access to a part or component. HFM can also be used to examine the visibility for a maintenance procedure. This may require checking to see if space is available for the maintainer to move his head to a strategic position. Figure 17.20 shows an example of a small female maintainer trying to gain access to a component for removal.

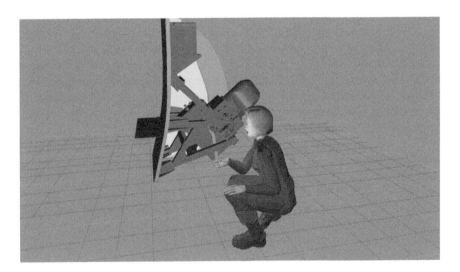

Figure 17.20 Access to component removal shown for small female figure

Modeling Analysis Feedback

As human factors analysts identify design problems, they also have to develop recommendations for addressing the problems. It is not sufficient to merely point out the problem (that is, inadequate seat adjustment or an inaccessible hatch).

Design problems must be quantified and feedback provided to the designer. Analyses must be performed in a timely and efficient manner. An effective technique is to model suggested changes via the vehicle or aircraft CAD files and export them back to the design engineers for consideration. The process of analyzing the designs must keep pace with the design process for the recommendations to have an impact. Feedback on design aspects that do not meet requirements gives the decision-makers valuable information that they can use in looking at possible design options. Often, certain design modifications may be very costly to implement and decision-makers may need to weigh the cost versus the percent of the population that will not be accommodated due to flaws or problems with the current design. HFM allows for several design options to be examined and early in the design process when flexibility in the design permits these modifications to be implemented.

Conclusions

The steps and methodology presented in this chapter focused on a very narrow application for HFM, specifically, workspace evaluation with regard to military vehicles and aircraft. Although many of the human factors design principles can be applied to other workspace environments, military systems designs face unique challenges due to the specialized clothing and equipment that must be accommodated along with the personnel who will operate, occupy, or maintain the system. Also, the hazardous conditions and environments in which military systems must operate put added requirements on the design to ensure that the personnel can safely interact with the system and exposure to injury is minimized.

As has been pointed out earlier in this chapter, models of the specialized clothing and equipment, along with models of weapons, ammunition, tools, and other stowage items, are critical to an accurate design analysis of military systems. However, these models are scarce. Most of the models of these items that can be located have been developed for 3D animation or video gaming use and are not suitable for military workspace design analysis. The intended purpose for these models are cosmetic, that is to say, they must look realistic but do not have to represent the actual size and proportions of the original clothing and equipment. Care must also be exercised when a dimensionally accurate model of a clothing or equipment item is built, so that it is not simply scaled to fit a family of boundary manikin figures. Doing this can distort the proportions of the model, making the model no longer reliable for workspace analysis. Models must be "custom-fitted" to each human figure model to maintain accurate proportions.

The importance of posture data has also been emphasized in this chapter. While considerable resources have been spent on evaluating postures for the automotive industry and office workspaces designs, this data may not be applicable to military systems because of the specialized clothing and equipment, as well as other unique requirements such as indirect vision driving, emergency ingress and egress, and specialized mission requirements. These clothing and equipment items not only require the designs to accommodate the additional bulk but also place restrictions on range of motion and alter the posture of the person. The restricted range of motion should also be incorporated into the human figure models along with posture data that would accurately reflect personnel in the encumbered condition. Note also that the restricted range of motion and changes to the posture for an encumbered person is likely not generalizable across the entire population. The weight and bulk of an armored vest combined with a hydration pack, for example, will affect the seated posture of a large male differently than a small female occupant in a military vehicle design. If models of the specialized clothing and equipment are scarce, range of motion restrictions and posture data related to wearing these items are still more scarce. This does not mean that HFM should be abandoned, but rather that these physical influences on the human must be considered when performing the design analysis.

Lastly, while significant strides have been made toward improving human figure models and the software used for workspace design, their pace has been slow in comparison to the pace of improvements and investment made in modeling software targeted for the entertainment industry. One of the most obvious improvements in most of the ergonomic human figure models is the appearance. Gone are the days of primitive looking "bubble-men" or low-resolution segmented figures. These human models are more contoured and more human-like in their appearance and are beginning to rival those characters used in video games. Caution must be applied here as well, as some of these newer figures do not yet incorporate full anthropometric scaling and only offer a limited library of pre-constructed figures or limited scaling of body dimensions. Without full anthropometric scaling it will be difficult to create an accurate set of boundary manikin figures. In addition to the more realistic appearances of the figures, software vendors are moving beyond the limited use they may have originally served and are adding new capabilities. For example, a software package that may have been previously used only for workspace design analysis now offers the ability to perform biomechanical analysis or crash simulation as well. Still, it may be wise to be wary of vendors who claim that one model can serve all purposes until a detailed evaluation of these claims can be made.

As the figures and software continue to improve and progress, so too must models of specialized clothing and equipment be developed in order to perform accurate analysis of military system designs. The current process is slow and expensive, and requires almost a customized hand-built process. Even these models are limited, in that they do not incorporate deformable properties to simulate compressibility or flexibility. Future scanning and software technology will hopefully speed up the process, lower the cost, and offer models that can simulate actual material properties in near-real time. Additionally, future studies and data collection efforts can help provide some of the missing physical impacts that these items have on the human in order to improve modeling assessment capabilities as well.

References

Alexander, M., Garrett, J.W. and Flannery, M.P. 1969. *Anthropometric Dimensions of Air Force Pressure-Suited Personnel for Workspace and Design Criteria.* (Technical Report AMRL-TR-69-6, AD697 022). Wright-Patterson Air Force Base, OH: Aerospace Medical Research Laboratory.

Alexander, M., Garrett, J.W. and Robinette, J.C. 1970. *Anthropological Applications in High Altitude Flight Systems.* (Technical Report AMRL-TR-70-3, AD706 888). Wright-Patterson Air Force Base, OH: Aerospace Medical Research Laboratory.

Alexander, M., McConville, J.T. Kramer, J.H. and Fritz, E. 1964. *Height–Weight Sizing of Protective Garments, Based on Japanese Air Self-Defense Force Pilot Data, with Fit-Test Results.* (Technical Report AMRL-TDR-64-66, AD606039). Wright Patterson Air Force Base, OH: Aerospace Medical Research Laboratory.

Bittner, A.C. Jr. 1974. *Reduction in Potential User Population as the Result of Imposed Anthropometric Limits: Monte Carlo Estimation.* (Technical Publication TP-74-6). Point Mugu, CA: Naval Missile Center.

Bittner, A.C., Glenn, F.A., Harris, R.M., Iavecchia, H.P. and Wherry, R.J. 1987. CADRE: A family of manikins for workstation design, in *Trends in Ergonomics/Human Factors IV*, edited by S.S. Asfour. North Holland: Elsevier, 733–40.

Cote, D. and Hubbard, M. 2001. *Musculoskeletal Dynamic Computer Simulation of the Shoulder During Baseball Pitching. Proceedings of the 8th International Symposium on Simulation in Biomechanics*, Milan, Italy, July 2001, 117–22.

DOD-HDBK-743A 1991. *Department of Defense Military Handbook: Anthropometry of US Military Personnel (Metric).* Available at: http://www.everyspec.com/DoD/DoD-HDBK/DOD-HDBK-743A_16856/ (accessed: July 25, 2012).

Dsouza, R. and Bertocci, G. 2010. Development and validation of a computer crash simulation model of an occupied adult manual wheelchair subjected to a frontal impact, *Medical Engineering and Physics,* 32(3), 272–79.

Garrett, J.W. 1968. *Clearance and Performance Values of the Bare-Handed and Pressure-Gloved Operator.* (Technical Report AMRL-TR-68-24, AD681457). Wright-Patterson Air Force Base, OH: Aerospace Medical Research Laboratory.

Gordon, C.C. 2002. Multivariate anthropometric models for seated workstation design, in *Contemporary Ergonomics 2002*, edited by P.T. McCabe. London: Taylor & Francis, 582–89.

Gordon, C.C., Churchill, T., Clauser, C.E., Bradtmiller, B., McConville, J.T., Tebbetts, I. and Walker, R.A. 1989. *1988 Anthropometric Survey of US Army Personnel: Methods and Summary Statistics.* (NATICK/TR-89/044, AD A225 094), Natick, MA: US Army Natick Research, Development, and Engineering Center.

Gordon, C.C., Corner, B.D. and Brantley, J.D. 1997. *Defining Extreme Forms for Designing Body Armor and Load Bearing Systems: Multivariate Analysis of Army*

*Torso Data. (*NATICK/TR-97/012, ADA324730). Natick, MA: US Army Soldier Systems Command, Natick Research, Development, and Engineering Center.

Hauschild, M., Davoodi, R. and Loeb, G. 2006. Designing and fitting FES and prosthetic systems in a virtual reality environment. *IEEE 2006 International Workshop on Virtual Rehabilitation*, New York, August 2006, 167–73.

Hooton, E.A. 1945. *A Survey in Seating.* Westport, CT: Greenwood Press.

Johnson, R.F. 1984. *Anthropometry of the Clothed US Army Ground Troop and Combat Vehicle Crewman.* (Technical Report NATICK TR-84/034). Natick, MA: US Army Natick Research and Development Center.

Kobrick, J.L. 1956a. *Quartermaster Human Engineering Handbook Series: I. Spatial Dimensions of the 95th Percentile Arctic Soldier.* (Environmental Protection Research Division Technical Report EP-39). Natick, MA: US Army Quartermaster Research and Development Center.

——. 1956b. *Quartermaster Human Engineering Handbook Series: II. Dimensions of the Upper Limit of Gloved Hand Size.* (Environmental Protection Research Division Technical Report EP-41). Natick, MA: US Army Quartermaster Research and Development Center.

——. 1957a. *Quartermaster Human Engineering Handbook Series: III. Dimensions of the Lower Limit of Gloved Hand Size.* (Environmental Protection Research Division Technical Report EP-43). Natick, MA: US Army Quartermaster Research and Development Center.

——. 1957b. *Quartermaster Human Engineering Handbook Series: IV. Dimensions of the Lower Limit of the Body Size of the Arctic Soldier.* (Environmental Protection Research Division Technical Report EP-5 1). Natick, MA: US Army Quartermaster Research and Development Center.

Kobrick, J.L. and B. Crist 1960. *Quartermaster Human Engineering Handbook Series: VII. The Size and Shape of the Available Visual Field during the Wearing of Army Headgear.* (Environmental Protection Research Division Technical Report EP-133). Natick, MA: US Army Quartermaster Research and Engineering Center.

Kozycki, R.W. 1998. *Developing a Modeling and Simulation Paradigm for Assessing the Encumbrance of Helicopter Aircrew Clothing and Equipment. Proceedings of the 1998 SAFE Symposium*, Phoenix, AZ, September 1998.

Kozycki, R.W. and Gordon, C.C. 2002. *Applying Human Figure Modeling Tools to the RAH-66 Comanche Crewstation Design. Proceedings of the 2002 SAE Digital Human Modeling Conference*, Munich, Germany, June 2002, 191–200.

Lockett, J.F., Kozycki, R.W., Gordon, C.C. and Bellandi, E. 2005. *An Integrated Human Figure Modeling Analysis Approach for the Army's Future Combat Systems.* (SAE Technical Paper 05MV-5). Warrendale, PA: SAE International.

Middleton, R.H., Alexander, M. and Gillespie, K.W. 1970. *Cockpit Compatibility Studies Conducted with Aircrew Members Wearing High Altitude Flying Outfits in B-57D, B-57F, F-104B and F-106B Aircraft.* (Technical Report ASD-TR-70-25, AD880672). Wright-Patterson Air Force Base, OH: Aeronautical Systems Division.

MIL-HDBK-759C 1995. *Department of Defense Handbook for Human Engineering Design Guidelines*. Available at: http://www.hf.faa.gov/docs/508/docs/milhdbk759C.pdf (accessed: July 25, 2012).

MIL-STD-1472F 2001. *Department of Defense Design Criteria Standard: Human Engineering*. Available at: http://www.public.navy.mil/navsafecen/Documents/acquisition/ MILSTD1472F.pdf (accessed: July 25, 2012).

Paquette, S.P., Case, H.W., Annis, J.F., Mayfield, T.L., Kristensen, S. and Mountjoy, D.N. 1999. *The Effects of Multilayered Military Clothing Ensembles on Body Size: A Pilot Study*. (Technical Report Natick TR-99/012, ADA359792). Natick, MA: US Army Soldier and Biological Chemical Command Soldier Systems Center.

Randall, F.E., Damon, A., Benton, R.S. and Patt, D.I. 1946. *Human Body Size in Military Aircraft and Personal Equipment*. (Technical Report AAF-TR-5501 (ATI 25 419)). Dayton, OH: Air Materiel Command, Wright Field.

Reed, M.P., Manary, M.A. and Schneider, L.W. 1999. *Methods for Measuring and Representing Automobile Occupant Posture*. (SAE Technical Paper 1999-01-0959). Warrendale, PA: Society of Automotive Engineers.

Reynolds, H.M., Snow, C.C. and Young, J.W. 1981. *Spatial Geometry of the Human Pelvis*. (Memorandum Report AAC-119-81-5). Oklahoma City, OK: Federal Aviation Administration, Civil Aeromedical Institute.

Roberts, L.B. 1945. *Size Increase of Men Wearing Various Clothing Combinations*. (Project No. 9, SPMEA 741-3). Fort Knox, KY: Armored Medical Research Laboratory.

Roebuck, J.A., Kroemer, K.H.E. and Thomson, W.G. 1975. *Engineering Anthropometry Methods*. New York: John Wiley and Sons.

Sourin, A., Sourina, O. and Sen, H.T. 2000. Virtual orthopedic surgery training. *IEEE Computer Graphics and Applications Magazine*, May/June, 6–9.

White, R.M., Kobrick, J.L. and Zimmerer, T.R. 1964. *Reference Anthropometry of the Arctic-Equipped Soldier*. (Technical Report EPT-2). Natick, MA: US Army Natick Laboratories, Pioneering Research Division.

Zehner, G.F., Meindl, R.S. and Hudson, J.A. 1992. *A Multivariate Anthropometric Method for Crew Station Design: Abridged*. (Technical Report AL-TR-1992-0164, ADA274588). Wright-Patterson Air Force Base, OH: Armstrong Laboratory.

Chapter 18

Interface Design: Dynamic Interfaces and Cognitive Resource Management

Thomas W. Davis, Michael Sage Jessee, and Anthony W. Morris

Introduction

Advancements in technology have greatly improved the capability of military systems, often changing the dynamics of the human–machine relationship when accomplishing mission tasks. These high-tech, intelligent systems give soldiers a technological edge on the battlefield; however, in some instances they impose greater demands on the soldiers' mental resources available for managing attention. Technological solutions are conceived, designed, and developed often with the purpose of providing support for a single task. The way in which a soldier works toward mission success, however, is through the management of multiple priorities and the distribution of attention to many tasks. The single-mindedness and focus required to design and engineer technology to support the soldier proliferate many independent solutions, which can quickly become at odds with the task-rich environment in which the soldiers operate. For example, soldiers may need to maintain visual surveillance to identify targets in a geographical area while making radio contact to communicate current position coordinates to their commanding officer. These three macro-activities, visual surveillance, radio operation, and communicating, illustrate the degree to which multiple, concurrent task-management requirements occur in field operations. Different, independent technologies exist for supporting each of these common field functions. Visual scanning may be enhanced with low-light optical amplification and magnification equipment. The radio itself includes multifunctional interface screens for mode selection and tuning. The speech acts associated with conveying location may realistically involve reading a paper or a digital map.

This chapter focuses on design principles gathered from scientific research and operational tests that can help the human factors researcher and designer identify and alleviate some of the soldier strain that arises from managing the complex technology, as well as the many concurrent goals of mission effectiveness and combat. Through disciplined application of human factors engineering (HFE) principles, technology can transparently facilitate the operational needs of soldiers. In particular, this chapter first identifies design analysis principles for understanding the effectiveness of human attention processes. Next, research techniques for eye-movement studies and system control and operation are discussed in the context of system interface design and analysis.

Theoretical Motivations

Soldier-System Domains and Boundaries

The boundary that soldiers manage between their capabilities and those of their mission support system is a dynamic one. Human-centric technology amplifies a soldier's capability. When technology is well designed, it affords the soldier an adaptive and dynamic interface with which to manage mission-critical information and responses. Engineering this soldier–system boundary requires that human-centered design analyses and processes take into account transactions that take place in domains that share a diverse array of relationships and interface agreements. The persistent pursuit that a soldier makes toward the successful completion of a mission's goals may be the clearest window through which to view the assistive roles that technology has for soldier performance. Soldiers perform goal-directed behaviors. They hunt. They fight. They learn. Each of these behaviors reflects purposeful behaviors directed toward other actors and environmental constraints. Mission goals distribute across domains composed of environmental circumstances, soldier capability, and other actor capability. For example, a hunt is an ecosystem between a soldier and the local environment or between a soldier and an enemy, where each participant encounters unique, egocentrically defined environmental constraints. A fight might be a defensive act against a hostile opponent. Learning is the integration of past-pending and future-tending challenges. Encounters such as these specify the myriad dispositional states active in the operational theater of a mission's successful deployment and execution. It is against the backdrop of the transactions from these different domains that we seek to extract principles and lessons learned for understanding how to technologically support the soldier's initiation of intentions and their successful finalization. In what remains of this introduction, we discuss human factors design principles that have emerged for us through the observation of relationships that soldiers encounter across different mission domains. Relationships that exist between the soldier and the environment have implications for how the soldier maintains contact with the mission-relevant information and energy. Soldier attention and how tasks get executed influence whether different types of information, for example, visual versus tactile, constructively support or destructively interfere with a soldier's execution of a mission's goals. A soldier's visual perception pattern interacts with task process structure, system layout, and environmental events. A soldier's control performance relates to the style of change that a monitored system displays.

Relationships of Soldier to Environment

Most agree that a soldier's mission can be conservatively characterized as concurrent task management. Although any given mission may have a single objective, for example, transporting surveillance equipment to a hilltop, the soldier must manage a multitude of subgoals. The soldier's abilities of perception and action when coupled to environmental circumstances make possible an indefinite number of events and encounters. Dual task-processing, where someone attempts to simultaneously do two tasks like driving a car while carrying on a conversation, is, to say the least, a lower bound on the typical set of tasks facing the soldier, especially in combat. The relationships between the soldier and the environment serve the human factors design analyst by bounding and quantifying the effects of behaviors that soldiers believe will accomplish their goals. The soldier–environment coupling defines the dynamic set of events that a soldier's intent sets in motion. These events entail the distribution and uptake of energetic and informational resources relevant to the goal an actor intends to successfully complete (Shaw and Kinsella-Shaw 1988). Pursuing a mission objective or its subgoal is not a free lunch. Soldiers must carefully plan when and how to distribute goal-relevant resources in order for a sequence of behaviors to succeed. By studying the dynamics of an intention (the time variations in performance that an intention initiates), a design analyst can objectify the behavioral steps that a soldier believes will accomplish a goal. These goal-directed behaviors reveal what the soldier believes will be successful. By means of intended, incremental changes in behavior, a soldier learns what works and what does not work.

How a soldier is coupled to the local constraints provided by the environment shapes how the soldier pursues goals. This coupling in its most basic form describes a mutual and reciprocal energy transaction that soldiers manage. To illustrate, consider a person's actions when encountering ice on the sidewalk while walking to lunch. Black ice poses a hazard when a pedestrian encounters it on a sidewalk; it interrupts a walker's grip. Walking is a goal-directed activity that typically accompanies a superseding goal like getting to the diner for lunch. When successfully taking a step, the soles of a walker's shoes and the surface of the sidewalk grip one another. They fit together to form a relation called "gripping." Grip is a dispositional property in that it emerges out of the relationship formed between two systems, in this instance, shoe soles and a sidewalk surface. Grip is an essential and taken-for-granted ingredient for successful walking. Simultaneously, the sole of the shoe is pushed in one direction by the force of the leg while the sidewalk's surface resists, in the opposite direction, the leg's push. The friction of a grip between the sole of the shoe and the sidewalk surface converts these two pushes into walking. The moral of this story is that grip is the local, emergent constraint between an actor and the environment that black ice can dangerously disrupt. Why? Because black ice is slippery and it does not reliably reflect light so as to inform a walker about the change in the sidewalk's surface texture. Quickly, a person's safe passage can be disrupted by unknowingly stepping onto black ice. The intended trip to the diner may be thwarted and an uneventful walk reconsidered.

Relationships of Soldier Attention to Task Process Structures

Simultaneous task requirements and performance choices influence a soldier's distribution of mental resources (Wickens 1980). How effective that distribution is for accomplishing a goal can be influenced by analyzing the ways in which a task may have similar perceptual or movement process structures involved in its completion (Wickens 2008). Consider the example of driving a car while using a dash, mounted global positioning system (GPS) map system. To drive safely, a driver must look out the windshield and monitor the rearview mirrors using vision. These activities of seeing the optical flow out the window are the task process structure of detecting visual information for safely driving. Similarly, looking at the GPS display involves detecting with the eyes the position of the car relative to an upcoming turn, for instance. In this example, the task process structures for the visual detection of navigationally relevant information out the window and on the GPS map are similar. Each detection activity draws heavily on the same metabolic potential underwriting the visual detection of goal relevant information. Designing a task or procedure that requires performing two similar, but different visual detection tasks close in time typically degrades one of the tasks. In contrast, if the GPS had an auditory mode that played aloud the relative position information, then different metabolic potentials, one for looking out the window and one for listening would be involved, and hence would reduce the total draw-down on a single resources that supports the visual detection task process.

Relationships of Soldier Visual Perception to Performance Control

A central concern of information display design is the control of mission-relevant information and the minimization of mission-irrelevant information. Mission goals specify target information requirements necessary for soldiers to execute mission-relevant performance goals. Mission goals define the desired distribution of a soldier's resources with respect to the information and performance conditions specified by the task environment. The goal establishes the criteria for successful performance, acceptable progress (suboptimal success), and non-success for a given task. In order to successfully respond to the old adage, "When I say jump," the jumper does need to know when and how high. Target parameters (what) and manner parameters (how) of a task's goal are specific to how task-relevant resources should be distributed with respect to detecting goal-relevant information and executing goal-relevant behaviors. These two sets of parameters are generalized constraints that apply to most intentional or goal-directed tasks. Target parameters involve information specific to time-to-contact with a target, direction-of-contact toward a target, and distance-to-contact with a target. Manner parameters involve performance constraints specific to impulse force, which initiates movement toward a target; torque, which controls bearing toward a target; and work, which sustains moving to a target (Shaw and Kinsella-Shaw 1988).

The primary interest of performance control is specifying the manner in which a behavior is to be executed. Practically speaking, a motion that "hits its mark" is one that effectively and efficiently coordinates the effort generating the motion with the information that guides it. Whether a motion is quick and short or slow and long describes the manner or style of the motion. The manner in which one moves entails controlling several energy-relevant variables. Impulse forces initiate a movement. Work sustains the movement until the target is reached. Torque maintains the motion's heading toward the target. The style or manner of movement also entails detecting information about the target or the final conditions of the movement. This information includes when to move, how far to move, and which way to move. These information variables are specific to the manner properties. A person's time of arrival at a target depends on when impulse forces are applied (time-to-contact). A target's location specifies the direction in which forces are needed (torque-to-contact), as well as how far the forces need to be applied (distance-to-contact). These manner and target parameters combine under the influence of a person's goal or intent to form a unique trajectory.

Practical Motivations

Also common to the human factors practitioner's (HFP) approach is to solicit user feedback whenever possible. However, there are several limitations inherent in subjective feedback, namely, an operator's willingness and ability to provide responses. Thus, the HFP must adapt techniques to mitigate these issues. Furthermore, technological advances in military systems increasingly focus on visual dynamic displays. This increase in visual demand often produces an overload of finite visual resources available to the soldier. In order to measure and design systems around these finite resources, a model of the resources must first be introduced.

The Multiple Resource Model (Wickens 1980, 2008) consists of three dimensions each of which is a dichotomy. Each dimension outlines the different perceptual, processing, and response modes required of the task and the soldier. According to this view, decrements in performance may occur because the pairing of certain resources can interfere and thereby create a resource shortage. Conversely, through proper task analysis and system design, the HFP can pair resource dimensions and modes so as to avoid a resource shortfall and enhance the collaborative interaction of the dimensions. For example, driving a vehicle or flying an aircraft draws heavily on the visual perceptual system. A design decision that pairs another visual process with the driving task will compete for visual processing resources. A display that requires a driver or aviator to read, for instance, sets up a competition among resources. In contrast, a design decision that augments information exchange through an audio channel may avoid the resource competition and may, in fact, create a collaborative gain in the operator's ability to manage the multiple and simultaneous demands on visual attention. A car radio is a case in point.

Car drivers can see out the window and listen to the radio more effectively than they can see out the window and read text-based messages. Seeing to navigate and seeing to read draw on the same resource, whereas seeing to navigate and listening to navigate or be entertained distributes the resource draw-down across Wickens' independent resource pools. Thus, Army systems must be designed such that the perceiving–acting cycle that the system enables does not overly tax one of the resource pools. Since vision is of primary importance to many military systems, measuring the amount of visual resources a system demands from the soldier can provide insights into how to reduce the cost of perceiving information (visual workload), increase the efficiency of which information is processed (mental workload), and correlate the perception and action exchange that is produced by the overall system. The following parts of this chapter outline the proper selection of eye-tracking equipment, previous research that has correlated mental workload with ocular activity metrics, and specific case studies in aviation applications.

Selecting the Proper Eye-tracking Equipment

Several eye-tracking solutions are available in slightly different configurations. The most prominently used method is video-based eye-tracking. However, another broad category of eye-tracking methods that should noted by the well-versed practitioner is called "electrooculography." (EOG). EOG is a technique that implements a pair of electrodes, either above and below the eye or on either side of the eye, to measure the resting potential of the retina. Another EOG technique is to attach search coils directly to the eye, via contacts, a rubber ring, or surgical implantation, that measure eye movements with the same electromagnetic principles. These techniques are generally more intrusive than camera-based solutions and are often used to measure rapid eye movements during sleep and eye movements of animals, or to diagnose nystagmus. For this reason, HFPs almost always select a camera-based solution.

Camera-based eye trackers are often infrared (IR)-based, which allows for a more accurate and reliable capture of the two primary objects the eye-tracker system uses to calculate point of gaze, the center of the pupil, and cornea reflection. However, non-IR cameras are often favorable to implement in outdoor scenarios where ambient light interferes with the IR light that is used to illuminate the cornea reflection. In this case, the edge of the iris and the sclera can be used to track eye movement via light/dark field boundary detection algorithms. This would be a preferred method under circumstances in which unmanned systems were tested in a realistic field environment. For example, an evaluator would use this data to quantify the frequency and duration of operator gazes at a hand controller in order to infer quick shifts in mental workload, visual workload, and operator efficiency level. For most HFE applications, however, the preferred eye-tracking method is IR-based. The next consideration is the location of the cameras.

For applications in which maximum head movement is required, a head- or helmet mounted eye-tracking system is recommended. This allows the researcher to connect a motion-sensing device (magnetic, optical, or laser-based devices can be integrated with eye-tracking systems) to the helmet or headband in order to implement eye–head integration, which allows for free head movement of the soldier while collecting point-of-gaze data. The advantage of this method is that it allows for a large range of motion, which presents the most natural environment for the soldier to operate. For an operations test, a non-intrusive measurement is imperative. The trade-off, however, is that the helmet-mounted optics can slightly alter how the weight shifts the helmet. Typically, a small weight can be attached to the back of the helmet in order to counterbalance this effect. Although the counterweight slightly alters the feel of the helmet, it allows the soldier to have maximum range of head movement and does not constrain how the soldier operates the equipment, alleviating a potential confound to the overall operational test data sources. For tests in which point-of-gaze data only need be collected in regard to a specific area or display, a remote optics solution may be implemented.

Desktop remote optics tend to allow for a much smaller range of head motion, but do not require that the soldier be fitted with optics. This is the least obtrusive method of collecting eye-movement data, but it does not allow for a highly dispersed scene in which eye-tracker data will be consistently recorded. In addition, remote eye-tracking optics can be either relatively unnoticeable, in cases where only one camera is integrated into the monitor, or relatively obtrusive to the environment, in cases where multiple cameras are directed at the operator in order to increase the detectable range of head movement. In the latter case, multiple cameras can slightly alter the presentation of the environment that is being evaluated, as well as alter operators' general response when multiple noticeable cameras are directed at them. Although remote desktop eye-tracker solutions tend to be cheaper than complete eye-head integration systems, head-mounted eye trackers with eye–head integration are much more scalable to a variety of research and evaluation contexts. Thus, given the flexibility, it is most often advised to acquire an eye–head integration system if the researcher/evaluator must implement it in a variety of scenarios. Given a head-mounted eye-tracker system, a compatible motion-sensing solution must be employed.

As suggested above, three broad categories of compatible head trackers are available: magnetic, optical, and laser based. Magnetic trackers tend to be most cost effective, but if data are going to be collected in environments where ferrous metals are abundant, such as a cockpit, then a laser or optical solution should be used because of issues regarding interference that will yield inaccurate data. Here, the primary trade-off is that laser and optical solutions require that the sensor be in direct line of sight of the camera or scanner and tend to have a more limited range of accurate sensing. In addition, the cost tends to be higher, but laser or optical head-tracking with helmet-mounted optics tends to provide the most flexible head-tracking solution. Although this system is flexible, it does not allow the operator to walk around freely, which would be required when evaluating a system that is used during a dismounted field test, as with the Raven unmanned aerial system.

For these situations, a mobile eye-tracking device can be used when operators need to move freely across large spaces. The data from these solutions tend to be much more limited in scope since data on the motion of the head is not captured. Essentially, pupil and blink activity is captured, but point of gaze data is bound to a head-mounted scene camera as opposed to preselected scene planes. This can require a somewhat cumbersome post hoc video-coding process, but often the data yield invaluable information regarding gaze patterns in a realistic setting. The efficiency of eye-tracker data analysis tools should also be considered in order to provide an initial look at the results during overall after action reviews (AARs). This is very important in order to gain operator feedback regarding gaze-pattern explanations to facilitate discussion.

Analysis Techniques

Head- and eye-tracker systems often provide the HFP with a plethora of ocular activity variables. Given this large data set, it is import for the human factors engineer to understand which variables are correlated with different operator states. A considerable amount of research has been conducted correlating various ocular activity variables with mental workload (MWL). Table 18.1 provides a list of variables that are positively correlated with MWL along with their respective research authorship.

Although these variables have been correlated with other measures of MWL, we propose that some of them actually measure a correlated component of MWL, specifically, visual workload (VWL). Compared to MWL, the VWL construct has received considerably less attention, but has been mentioned in some studies (Backs and Walrath 1992, Hancock and Desmond 2001, van der Horst 2004, Verwey and Veltman 1996, Wickens, Helleberg, and Xu 2002). In addition, VWL has received even less consideration regarding its definition and theoretical underpinnings. Here, VWL is defined as the effort that is required to detect visual information, while MWL represents the *servicing* of information. Both of these detection and servicing events require effort from the operator and have various implications regarding the overall design of a system. As has previously been mentioned, the Multiple Resource Model can be used to conceptualize soldiers in terms of their attentional resource structures. The importance of this theoretical position is that measuring each resource pool separately leads to different design implications. For example, if the visual system has to exert too much effort in order to detect information that is broadly dispersed or buried within layers of pages on a dynamic display, then the operator runs the risk of not detecting other prominent visual information. In this situation, VWL is too high, and the system must be redesigned such that visual information is easily accessible. This may be achieved in a number of ways including, but not limited to, optimizing font, font size, color, and the dispersion of visual information.

Table 18.1 Ocular activity variables correlated with MWL

Authors	Positively Correlated with MWL					Negatively Correlated with MWL			
	Fixation Duration	Pupil Diameter	Blink Interval	Fixation Frequency	Saccadic Extent	Blink Rates	Blink Frequency	Blink Duration	Fixation Duration
Ahlstrom & Friedman-Berg (2006)		*						*	
Brookings, Wilson, and Swain (1996)						*			
Ryu and Myung (2005)			*						
Tole et al. (1983)	*								
Van Orden, Limbert, Makeig, and Jung (2001)				*	*		*	*	
Van Orden, Jung, and Makeig (2000)							*		
Veltman & Gaillard (1998)						*		*	
Willems, Allen, and Stein (1999)									*
Wilson (2002)						*			

Aviation Case Studies

To illustrate a specific example, the following demonstrates an overall system analysis and specific task analysis of a rotary wing aircraft. The overall system analysis was conducted in a simulator during a Limited User Test for an upgrade of the Black Hawk helicopter. Here, the design was evaluated in light of several upgrades in order to provide a human factors data package that informed decision-makers about the overall system performance. For this analysis, overall dwell times were recorded as a percentage of total flight time. Once all dwell times were calculated, they were compared to similar data that were collected for the earlier baseline model. Figure 18.1 outlines the percentage of the dwell times for pilots flying the UH-60M Upgrade helicopter.

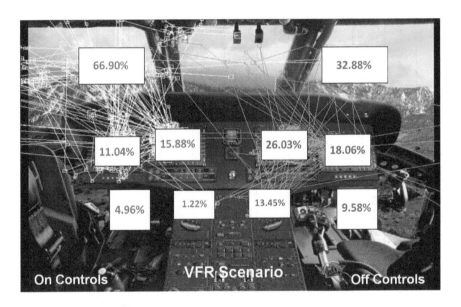

Figure 18.1   Dwell time of pilots while flying the simulated helicopter mission

These data will be compared to similar flight scenarios of the previous aircraft in order to quantify the differences in heads-down data versus heads-up data. If enough instances were recorded to ensure adequate statistical power, then parametric statistics should be conducted. However, low operator populations do not always allow for that level of analysis. In these cases, if the data are within a close range, then the modifications did not bring the pilot's visual attention into the cockpit more than the previous model. To date we have not identified research conducted that demonstrates a threshold requirement of heads-up time, but typically greater than 75 percent is preferred.

The next example illustrates how to handle eye-tracker data in a similar fashion but with a more specific task for a developmental test. During an instrument panel software upgrade, in which a new navigation capability was introduced, only eye-tracker data segments specific to tasks related to that capability were analyzed. The primary task of the pilot on the controls is to aviate. The pilot off the controls is primarily tasked with navigation and communication in most circumstances. In order to evaluate the demands of the task related to the new capability being introduced, pilot eye-tracker data were observed to look for gaze patterns, which can be viewed in Figure 18.2.

Figure 18.2   **Pilot (gray scan pattern) and copilot (white scan pattern) fixations during an emergency tactical approach under instrument flight rules**

These patterns indicate the pilot on the controls had to support the copilot task, which can be viewed by the frequency of fixations of the pilot on controls (gray circles) that transitioned to both the pilot and copilot Flight Management System (FMS). The pilot off the controls (illustrated in white) scanning path resembles a typical pattern of an operator conducting an FMS-related task. The pilot on the controls, however, demonstrated that, momentarily, visual information from the FMS captured his attention. This could indicate an unacceptable crew workload if the pilot on the controls is required to support the pilot off the controls. Typically, before any conclusions can be drawn, these data will be compared to a review of the video captured during the event, subject matter expert (SME) observations captured by the tactical steering committee, and subjective measures capturing workload during the task. This allows the practitioner to evaluate and identify issues with several different data sources offering convergent data points to a particular problem or success.

A complementary example of this type of analysis is also demonstrated in Figure 18.3. These data were collected during a holding pattern, also during instrument meteorological conditions, but illustrate the pilot on controls focusing solely on the inboard and outboard multi-functional displays, indicating that no navigation support was required by the pilot off the controls. Figure 18.3 shows that the pilot on controls was free to primarily focus on the task of aviating, indicating that the workload required for navigation was handled by one pilot during the holding procedure. Typically, this would be considered a success of the program if additional capability is provided to the pilot with no added operator costs.

Figure 18.3   Pilot (gray scan pattern) and copilot (white scan pattern) eye-tracker data during a holding procedure

In conclusion, it should be noted that there are some frequently used procedures for analyzing eye-tracker data, but there are no "cookie cutter" strategies that apply to all systems and capabilities. Thus, eye-tracker data from the specific aspects of the system under scrutiny should be focused on using several different analysis strategies. For example, Figure 18.2 demonstrates the high, non-primary task-relevant, visual workload of the pilot, but other ocular activity variables such as pupil diameter (MWL), blink inhibition (VWL), and fixation duration variability should be analyzed during those periods as well to offer converging or discriminate evidence of the practitioners' claims. Given the breadth of literature surrounding ocular activity and its correlates, the human factors practitioner should study and apply various analyses and strategies in a way that answers specific questions about the system being evaluated.

References

Ahlstrom, U. and Friedman-Berg, F.J. 2006. Using eye movement as a correlate of cognitive workload. *International Journal of Industrial Ergonomics*, 36, 623–36.

Backs, R. and Walrath, L. 1992. Eye movement and papillary response indices of mental workload during visual search of symbolic displays. *Applied Ergonomics*, 243–54.

Brookings, J.B., Wilson, G.F., and Swain, C.R. 1996. Psycho-physiological responses to changes in workload during simulated air traffic control. *Biological Psychology*, 42, 361–77.

Hancock, P.A. and Desmond, P.A. (eds) 2001. *Stress, Workload, and Fatigue*. Mahwah, NJ: Lawrence Erlbaum.

Ryu, K. and Myung, R. 2005. Evaluation of mental workload with a combined measure based on physiological indices during a dual task of tracking and mental arithmetic. *International Journal of Industrial Ergonomics*, 35, 991–1009.

Shaw, R.E. and Kinsella-Shaw, J. 1988. Ecological mechanics: A physical geometry for intentional constraints. *Human Movement Science*, 7, 155–200.

Tole, J.R., Stephens, A.T., Vivaudou, M., Ephrath, A.R. and Young, L.R. 1983. *Visual Scanning Behavior and Pilot Workload*. (NASA Contractor Rep. No. 3717). Hampton, VA: Langley Research Center.

van der Horst, A.R.A. 2004. Occlusion as a measure for visual workload: An overview of TNO occlusion research in car driving. *Applied Ergonomics*, 35(3), 189–96.

Van Orden, K.F., Jung, T-P. and Makeig, S. 2000. Combined eye activity measures accurately estimate changes in sustained visual task performance. *Biological Psychology*, 52, 221–40.

Van Orden, K.F., Limbert, W., Makeig, S. and Jung, T. 2001. Eye activity correlates of workload during a visuospatial memory task. *Human Factors*, 43(1), 111–21.

Veltman, J.A. and Gaillard, A.W.K. 1998. Physiological workload reactions to increasing levels of task difficulty. *Ergonomics*, 41(5), 656–69.

Verwey, W.B. and Veltman, J.A. 1996. Detecting short periods of elevated workload. A comparison of nine common workload assessment techniques. *Journal of Experimental Psychology: Applied*, 2, 270–85.

Wickens, C.D. 1980. The structure of attentional resources, in *Attention and Performance VIII*, edited by R. Nickerson. Hillsdale, NJ: Lawrence Erlbaum, 239–57.

——. 2008. Multiple resources and mental workload. *Human Factors*, 50, 449–55.

Wickens, C.D., Helleberg, J. and Xu, X. 2002. Pilot maneuver choice and workload in free flight. *Human Factors*, 44, 171–88.

Willems, B., Allen, R.C. and Stein, E.S. 1999. *Air Traffic Control Specialist Visual Scanning II: Task Load, Visual Noise, and Intrusions into Controlled Airspace.* (DOT/FAA/CT-TN99/23). Atlantic City International Airport, NJ: Federal Aviation Administration, William J. Hughes Technical Center.

Wilson, G.F. 2002. An analysis of mental workload in pilots during flight using multiple psychophysiological measures. *International Journal of Aviation Psychology*, 12(1), 3–18.

Chapter 19

Using a Holistic Analytical Approach to Meet the Challenges of Influencing Conceptual System Design

Diane Kuhl Mitchell and Charneta L. Samms

When a new system is being developed, human system analysts are one of several groups, representing different disciplines, who want to influence the system design to meet their required objectives. System engineers, for example, want to ensure the design meets system requirements. Cost analysts, on the other hand, want to ensure the design meets budget constraints. Human system analysts (HSAs) want to ensure the system is designed so human operators can use it effectively. Because the objectives of the various groups may conflict, the HSAs' design objectives are more likely to be implemented if a holistic approach is used to incorporate the objectives of the other disciplines as well.

In addition to integrating the objectives of multiple disciplines, HSAs must apply an integrated approach within their own analyses. This holistic approach is critical because it is possible for the design of one system to influence performance with another system that is being used concurrently. If, for example, an in-vehicle navigation aid is being designed, the HSA professional would need to evaluate how this system interacts with any other driving displays. This holistic approach makes it more likely that human operators will be able to use the system effectively.

Analysts working at the US Army Research Laboratory (ARL) have successfully implemented both aspects of this holistic HSA approach. Using the Improved Performance Research Integration Tool (IMPRINT)[1] and the integrated approach, they have successfully modified the conceptual designs of several military systems. In addition, they have enhanced the human performance modeling-tool, IMPRINT, to provide other HSAs with the capability to effectively implement this approach.

To understand the development and benefits of the ARL holistic approach, analysts must first understand the benefits and limitations of more simplistic analytical approaches and their implementation in IMPRINT.

---

1   Available at: http://www.arl.army.mil/IMPRINT (accessed: July 27, 2012).

Simplistic Human System Analysis Approach

HSAs view system design from the perspective of the human operators of the system. For this reason, their primary concern is the effectiveness of the design in meeting the goals of the human operators. To evaluate this aspect of design, they can use the simple technique of observing a human operator interacting with a mock-up of the system design. They might, for example, evaluate how the position of an in-vehicle navigation aid influences truck driver performance by observing truck drivers using a mock-up of the navigation aid within the driver's compartment of the truck (Tijerina 1996). Within this simple analytical approach, a basic technique used to capture the interaction of the human operators with the system is task analysis (Kirwan and Ainsworth 1992). A task analysis is a record of the sequence of activities of the user with the system. It can represent a diverse range of human systems interactions, from human–computer interactions (Diaper and Stanton 2004) to the mental workload associated with driving (Wojciechowski 2004).

The level of detail in a task analysis depends upon the hypothesis or question that the HSAs are investigating relative to the system design (Mitchell and Samms 2007). By analyzing the task analysis data as they would experimental data, HSAs can identify potential design problems early in the system development cycle. Identifying potential problems early reduces the number of costly design changes. In addition, they can identify and eliminate unsafe interactions inherent in the design, and identify and reduce human errors within the system (Stanton and Baber 2005). Ultimately, this ensures that the system design is safer and increases the productivity of the human operators of the system (Kirwan and Ainsworth 1992).

One of the limitations to this simple approach, however, is that mock-ups of the system may not be available early in the design cycle when changes are less costly to make. Human performance modeling is a technique that HSAs can use in this situation. Specifically, human performance modeling tools provide them with the capability to conduct a task analysis of the system and make recommendations for design changes early in the design cycle. The Improved Performance Research Integration Tool (IMPRINT) is one of these tools.

Improved Performance Research Integration Tool

Developed by the US Army Research Laboratory, IMPRINT is a dynamic, stochastic, discrete event simulation tool that can be used to quantify the effect of human operator performance on system performance. HSAs use IMPRINT throughout the acquisition lifecycle, from concept design through field-testing and system upgrades to estimate system performance as a function of warfighter performance. This is achieved by building a model of the system mission decomposed into functions and tasks of the warfighter–system interaction. The parameters of each task, such as predicted task time and mental demand, are then set. Once the models are built and run, the IMPRINT reports provide output data that allow HSAs to understand how long the mission will take to perform and what task combinations will likely cause mental overload, thereby increasing the likelihood that the system operators will make mistakes. By building models and conducting analyses across different variations of function allocation and/or technology incorporation, HSAs can provide quantitative data that can be used to support trade-off decisions early in the design process to meet their objectives. Since IMPRINT was developed by the US Army, earlier versions focused on the analysis of Army systems and soldier performance. The latest version of IMPRINT, known as IMPRINT Pro, expands the analysis capability of the tool to support all of the services.

Implementation of the Simplistic Approach with IMPRINT

Wojciechowski (2004) illustrates that HSAs using human performance modeling, specifically IMPRINT, can represent the interactions of the human operator with a system and accurately predict operator performance. The goal of her 2004 project was to validate that HSAs could use IMPRINT to represent three driving control modes that were still in the early design phase and predict the operator mental workload associated with each control mode. The three driving modes modeled were direct, teleoperation, and semi-autonomous. For each of these three modes, the researcher developed four functions in an IMPRINT model representing driving: move, see, maintain situational awareness (SA), and communicate.

In the IMPRINT model, moving included the tasks associated with steering and controlling the speed of the vehicle. Seeing included the tasks associated with looking for obstacles or threats, as well as keeping the vehicle on the route. Maintaining SA included the tasks associated with being aware of the status of the vehicle such as traction and direction of movement. Finally, communications represented the vehicle driver speaking with other individuals.

Based on the results of the IMPRINT model, the HSA concluded that all three driving modes lead to high workload. When communications tasks are combined with driving tasks, mental workload increases to an overload level where there is great potential for human error. Numerous open literature studies of driving combined with communications tasks support this IMPRINT analytical result (Strayer, Drews, and Johnston 2003, Redelmeier and Tibshirani 1997, Tijerina 1996). Because the predicted results match experimental results, this analysis demonstrates IMPRINT's capability to provide HSAs with the ability to accurately represent human–system interactions.

This example is a demonstration of the simplistic analysis approach and how results of such an approach can potentially identify human performance issues associated with a system. If, however, the ultimate goal of HSAs is to identify issues early in the design cycle, they are more likely to achieve this goal if they broaden their analytic approach to include the perspectives of other disciplines. By identifying issues early in the design cycle, HSAs have expanded their focus to support the concerns of cost analysts who want to ensure that budget constraints are met. However, the focus of the simplistic human analysis approach is the human ability to use the system. To have a greater impact on design process they need to expand their analytical approach even more to include the concerns of other disciplines. They can, for example, include system engineering and the system engineers (SEs) who are responsible for designing the system.

Multidisciplinary Approach

Whereas HSAs want to ensure a proposed system is designed most effectively for the human operators, SEs want to ensure that the system meets engineering design requirements. For example, consider SEs developing an unmanned ground vehicle (UGV) for the US Army. Soldiers will control this UGV via a display inside the vehicle in which they are riding. Although the system requirements specify that the UGV should be capable of moving at a certain speed and weigh a specific amount, they do not specify which soldier inside the vehicle will control the UGV. Therefore, the SEs place the UGV display in a location within the vehicle that is accessible only by the gunner of the vehicle. They select this location because it is the most cost-effective location for the display. They do not consider whether a soldier could perform gunnery tasks while also operating a UGV. HSAs, however, have analyzed the workload and performance of soldiers who perform gunning concurrent with UGV control and determined they would be mentally overloaded and perform ineffectively. The SEs, however, have no requirement to consider this information when selecting the display position and, therefore, may be unaware of the HSA analyses. If the HSAs include the SEs in their analysis approach, then this situation may be resolved. Furthermore, the design is more effective from both a system and an operator perspective.

One approach for integrating the perspective of the system engineering discipline with the human factors engineering discipline is to integrate traditional SE techniques with traditional HSA techniques. This integration, while including the concerns of both groups, provides analysts with results in a format they can understand and consider valid. Because the techniques have validity for both groups, the results will be more likely to be accepted and implemented. One way of achieving this integration is by combining unified modeling language activity diagrams with IMPRINT task networks.

IMPRINT and Unified Modeling Language (UML)

As SEs design a system, they use unified modeling language (UML) products to document design requirements and exchange design concepts (Strandrud 2004). The UML use cases are products that document how the system will be used by the human operator. Associated with the use cases are activity diagrams showing the interactions of the user and the system. These activity diagrams are analogous to the task analyses HSAs complete to record human–system interaction. The difference between the two techniques is their focus. Whereas UML emphasizes system requirements and system actions, task analyses emphasize human operator goals and actions. A UML activity, for example, might be "*display* friendly unit locations." The comparable task analysis activity might be "*read* friendly unit locations."

ARL analysts recognized that a multidimensional analysis approach could combine the focus of each discipline and present the information in a format that was familiar to both SEs and HSAs. Furthermore, presenting information in familiar formats should make it easier for design recommendations to be implemented. Therefore, in 2005, they completed a proof of concept pilot project demonstrating that UML activity diagrams could be incorporated into IMPRINT task network diagrams (Mitchell 2005a).

Implementation of Multidisciplinary Approach with IMPRINT

For the 2005 pilot project, the ARL analyst started with a model of a future concept tank. This model was part of an analysis conducted specifically to predict crew performance rather than system performance (Mitchell et al. 2003). Because the focus was crew performance, the analyst included all the activities of the tank crew in the task network but none of the activities of the tank. When the project was briefed to the SEs designing the tank, they pointed out that they had certain design constraints to meet when building the crew compartment that did not appear in the task network model. The ARL analysis needed to demonstrate the interrelationship between the crew performance requirements and the SE design requirements for the SEs to address the potential performance issues identified by the IMPRINT analyst.

To demonstrate the interrelationship between the human and system requirements, the ARL analyst included some of the UML activities into the IMPRINT task network model. For example, the tank crew was required to fire 3 rounds then move quickly to another position in 3 minutes. The crew could push the button to fire the rounds and then hit the gas pedal, but ultimately, whether they can move in 3 minutes depends on how fast the mechanical systems in the tank are initialized and ready to move. The tank use case and activity diagram contained the startup time requirement for the system. The ARL analyst incorporated this information into the IMPRINT model to show the SEs the impact of the tank system startup initialization time and the human 3-minute mission requirement. As a result of the pilot study, the SE community within this program recognized IMPRINT as an effective tool for identifying potential issues with system requirements.

Another result of the pilot study was the addition of a new IMPRINT feature, which would not have happened without the multidisciplinary approach. Because the SE community recognized the effectiveness of IMPRINT analyses for identifying potential design issues, SEs asked the ARL analysts to recommend design modifications that could resolve the issues. IMPRINT, however, did not provide analysts with this kind of information. It would be up to IMPRINT analysts to translate IMPRINT results into design recommendations based on their experiences or previous research. Recognizing the variability in results that would be created by requiring analysts to develop design recommendations and that this capability would further strengthen the analytical relationship between the HSA and SE disciplines, the ARL analysts added this capability to IMPRINT via a plug-in developed from another tool.

Multimodal Information Design Support Tool (MIDS) Plug-in to IMPRINT

Developed by Design Interactive, Inc., the Multimodal Information Design Support Tool (MIDS) is a human performance modeling tool that allows users to calculate overall estimated workload by inputting task and user interface specifications such as task type, display type, and mental workload demand to identify points of mental overload. While this functionality of MIDS is similar to IMPRINT, the unique capability of MIDS is that the tasks that are identified with the instances of mental overload are then connected to specific multimodal design guidelines and heuristics that when implemented into the system design could help to mitigate the overload. MIDS as a standalone tool is powerful in its capability to connect mental overload to potential mitigation strategies; however, it was developed to examine only command, control, communications, computers, intelligence, surveillance, and reconnaissance tasks and did not include stochastic task execution. Recognizing the similarities, strengths, and challenges in each tool, ARL engaged Design Interactive, Inc., to develop a MIDS plug-in to bring the design guideline identification capability into IMPRINT (Samms et al. 2009). The MIDS plug-in, once installed with IMPRINT, allows the IMPRINT analyst to run IMPRINT models and have MIDS use IMPRINT workload data to identify the instances of mental overload and then suggest design guidelines and heuristics for potential workload mitigation.

By connecting the capability of IMPRINT and MIDS, ARL provided analysts with the capability of implementing a multidisciplinary approach. However, analysts can be even more effective if they incorporate a multi-systems approach into the multidisciplinary analysis.

Multi-systems Analytical Approach

With the multi-systems approach, analysts take into consideration the system under development and all the systems the humans will use with it. This approach is beneficial because when multiple systems are used simultaneously, one system can influence the performance of another. For example, two radio systems operating via the same frequency may work fine independently but jam each other when used together. Another benefit from this approach is that analysts can integrate the displays and controls the human operators use concurrently to ensure more effective human performance. A tank driver, for example, may use a navigation aid while driving. If the navigational aid requires the driver to look at it for directions, then it will take the driver's attention away from the road. An auditory navigation aid, on the other hand, would allow the driver to continue looking at the road while listening to directions.

Analysts can use human performance modeling to implement this multi-system approach. They can, for example, model the car driver performing all driving-related tasks with all in-vehicle controls and displays. An analysis that incorporates all of the interfaces that the driver will need to interact with would provide a more accurate picture of the human–system interaction and therefore lead to a better prediction of human and system performance. To assist analysts in representing all the controls and interfaces the human operator may use and thereby better interpret the impacts on system performance, ARL modified IMPRINT to support modeling a variety of interfaces and the associated mental resources required to use the interfaces.

Mental Resources and Interfaces in IMPRINT

Earlier versions of IMPRINT offered two methods of estimating mental workload: the visual, auditory, cognitive, and psychomotor (VACP workload) or the advanced workload (Mitchell 2000). The VACP workload option required users to use four mental resources scales (visual, auditory, cognitive, and psychomotor) to estimate the mental demand of each task. The Advanced workload option also used mental resource scales but was based on Multiple Resource Theory (MRT) (Wickens 2002), which proposes that there are penalties associated with using more than one mental resource at a time. Therefore, when using this option in IMPRINT, analysts had to identify the interface associated with the mental resource pairs that each task required and estimate the mental demand of each pair using the provided visual, auditory, cognitive, motor, and speech mental resource scales. While both of these approaches are effective, ARL analysts found them limiting for the purposes of supporting the examination of new theories or approaches in mental workload prediction using IMPRINT. Therefore when IMPRINT Pro was developed, the concept of two separate workload modeling approaches was eliminated and the interface and subsequent reports were modified to support a more flexible mental workload modeling structure. This new workload modeling structure gives analysts the power to develop and test their own mental workload theories while still allowing novice users to use the research-supported MRT option (previously called the "Advanced workload" option). IMPRINT Pro allows users to use either the validated mental resource scales available by default or enter their own values for each resource. Also, an analyst can create a new mental resource that can be added to the workload calculation.

In addition, new workload scales have been developed to expand the collection of default mental resource scales available and allow for more specific calculation of mental workload in IMPRINT. The previous mental workload scale of motor was expanded into two separate scales: fine motor and gross motor. The fine motor scale resembles what was previously the motor scale in IMPRINT. The gross motor scale was developed to support the need to estimate the mental workload of dismounted operations such as walking, climbing, and running. A tactile mental resource scale was also added to allow analysts to examine the effect of haptic or tactile technology on operator and system performance. To pair up with the mental resources, IMPRINT analysts can add any number of interfaces that represent the controls and displays the operators will use within the system. These additions to IMPRINT provide analysts with the flexibility to model multiple systems within a given analysis.

Implementation of Multi-system Approach

The ARL incorporation of their driving model (Wojciechowski 2004) into a model representing a tank crew (Mitchell et al. 2003) is an example of the application of the multi-system approach. This analysis included the controls the driver needed for driving, such as the yoke, brake pedal, accelerator, and forward terrain view, but it also included the weapons and communications equipment the driver uses such as the intercom, radios, and gun sights. Furthermore, the analysis included all the tank crewmembers and their equipment and tasks. The system design for this tank included a requirement for a crew of two soldiers to engage targets while the platform was moving. By incorporating all the systems, tasks, and crewmembers into the analysis, the analysts demonstrated that this requirement would not be met due to the requirement to engage multiple targets. The IMPRINT analysis demonstrated that in order for this requirement to be met, the driver would need to line up a target for the tank commander while also driving. Subsequently, the design requirement was changed to a three-soldier crew requirement.

After successfully achieving a multidisciplinary, multi-system analytical approach, the final step analysis need for a holistic analytical approach is consideration of the human operators and the proposed system as a part of a larger system of systems.

System of Systems (SoS) Approach

To implement a SoS approach, analysts need to consider how operator interactions with one system may affect other systems that are interrelated, a tank crew allocated a proposed system, for example, may be part of a tank platoon consisting of three tanks, each with its own crew. In addition to the tank vehicles being part of the tank platoon, the tank platoon is a part of a larger company mission. The company mission may include, for example, the tanks protecting an infantry platoon. The survivability, then of the infantry platoon is partially dependent upon how well the tank platoon performs in protecting them. The analysis of the proposed system, therefore, will have greater impact if the analysts can demonstrate how design recommendations will impact the tank platoon and its ability to accomplish the company mission. Analysts can achieve this goal by generalizing the results of the analysis of one system to the other systems in the platoon. How effectively analysts can make this generalization depends upon the quality of the analysis results available to them. In order to provide the analysts with the information they need to implement the SoS approach, ARL enhanced the format of the IMPRINT output reports.

Development of New IMPRINT Output Reports

While the identification of mental overload is important, the true value of this identification is to understand what tasks are contributing to the overload in order to develop mitigation strategies. Previous versions of IMPRINT reports identified high mental workload, but no corresponding information was available to determine what task combinations caused the high workload. In order to address this issue, new IMPRINT mental workload reports were developed. The Operator Workload Summary and Operator Workload Detail reports were designed and implemented into IMPRINT to provide the analyst with more detailed information about the tasks the operators are performing over the mission associated with the mental demand of the task combinations. The Operator Workload Summary report provides the overall workload of each operator across the entire mission compared to the workload threshold set by the analyst. With this report, the analyst can quickly view the time periods when the operators are experiencing workload over the desired threshold. By referencing those time periods, the analyst can then go to the Operator Workload Detail report to identify what tasks are being performed by that operator during that time period to better understand what task combinations are contributing to the mental overload. This report also lists the task demand for each mental resource, the overall workload and the total conflict value (how much of the overall workload is caused by mental resource channel conflict). In addition, the Operator Overload report was recently added to support a new feature that allows analysts to set a workload threshold for each individual resource, not just overall workload. The addition of these reports makes it much easier for the analyst to understand the tasks being performed by the operators and the mental demand of these tasks over the entire system mission.

Implementation of the SoS Approach

The ARL analysts used IMPRINT to extend the impacts of human and system performance into impacts on a SoS mission (Mitchell and Samms 2009). For this SoS approach, the ARL analysts started with the output reports from an analysis of a tank platoon that has been allocated unmanned assets (Mitchell 2005b). These output reports indicated when a particular tank crewmember would be mentally overloaded when performing concurrent tasks. In addition, the reports displayed which task combinations created mental overload for each crewmember. For example, the reports indicated that the tank gunners would be overloaded when concurrently monitoring an unmanned vehicle while scanning for threats.

Based on this information from the IMPRINT reports, the ARL analysts recommended against gunners being allocated unmanned asset control because they might stop scanning for threats in order to maintain unmanned asset control. An ARL experiment (Chen and Joyner 2009) validated that gunners do indeed reduce scanning in this situation. Furthermore, the experiment indicated that gunners might reduce scanning by 56 percent. ARL analysts transitioned this result into a SoS analytical result by incorporating it into a statement about the impacts on the company mission. Specifically, they stated that the reduction in gunner scanning increased the probability that the gunner would miss a potential threat and the tank would be hit. If the tank is hit it may be destroyed and. as a consequence, be unavailable to protect the infantry platoon and the company mission may fail. Presented in this way the ARL results had a greater impact on system design and, ultimately, the crew workload became a major evaluation issue with the advanced concept tank program.

Presenting analytical results in the SoS context, especially when the impacts on overall operator goals are demonstrated, has the additional advantage of presenting analysis in a way that test and evaluators can use in test and evaluation plans. Therefore, the SoS approach supports the multidisciplinary analytical approach to include another discipline, test, and evaluation. Furthermore, it supports the cost analysts by helping to focus testing on the critical human–system interactions.

References

Chen, J.Y. and Joyner, C.T. 2009. Concurrent Performance of Gunner's and Robotics Operator's Tasks in a Multitasking Environment. *Military Psychology*, 21(1), 98–113.

Diaper, D. and Stanton, N. (eds) 2004. *The Handbook of Task Analysis*. Hillsdale, NJ: Lawrence Erlbaum.

Kirwan, B. and Ainsworth, L.K. (eds) 1992. *A Guide to Task Analysis*. Washington, DC: Taylor and Francis.

Mitchell, D.K. 2000. *Mental Workload and ARL Workload Modeling Tools*. (Technical Report ARL-TN-161). Aberdeen Proving Ground, MD: US Army Research Laboratory.

Mitchell, D.K. 2005a. *Enhancing System Design by Modeling IMPRINT Task Workload Analysis Results in the Unified Modeling Language (UML)*. Paper to the Naval Engineers Human System Integration Conference: Arlington, VA.

Mitchell, D.K. 2005b. *Soldier Workload Analysis of the Mounted Combat System (MCS) Platoon's Use of Unmanned Assets*. (Technical Report ARL-TR-3476). Aberdeen Proving Ground, MD: US Army Research Laboratory.

Mitchell, D.K. and Samms, C.L. 2007. Please don't abuse the models. *Proceedings of the 51st HFES Annual Meeting*, Baltimore, MD, October 2007. 1454–1457.

Mitchell, D.K. and Samms, C.L. 2009. *Using IMPRINT to Translate Human Performance into System and Mission Effectiveness*. Presentation to the NDIA

25th Annual National Test and Evaluation Conference, Atlantic City, NJ, March 2009.

Mitchell, D.K., Samms, C.L., Henthorn, T. and Wojciechowski, J.Q. 2003. *Trade Study: A Two- Versus Three-Soldier Crew for the Mounted Combat System (MCS) and Other Future Combat System Platforms*. (Technical Report ARL-TR-3026). Aberdeen Proving Ground, MD: US Army Research Laboratory.

Redelmeier, D.A. and Tibshirani, R.J. 1997. Association between cellular-telephone calls and motor vehicle collisions. *The New England Journal of Medicine*, 336(7), 453–58.

Samms, C., Jones, D., Hale, K. and Mitchell, D. 2009. Harnessing the power of multiple tools to predict and mitigate mental overload, in *Engineering Psychology and Cognitive Ergonomics*, edited by D. Harris. Berlin and Heidelberg: Springer, 279–88.

Stanton, N.A. and Baber, C. 2005. Task analysis for error identification, in *Handbook of Human Factors and Ergonomics Methods,* edited by N. Stanton, A. Hedge, K. Brookhuis, E. Salas and H. Hendrick. Washington, DC: CRC Press, 38, 1–9.

Strandrud, T.S. 2004. *UML Modeling Guidelines (D950-10677-1)*. (Phantom Works Engineering and Information Technology-Modeling and Simulation (8GK0) and the Boeing Company). St. Louis, MO: The Boeing Company.

Strayer, D.L., Drews, F.A. and Johnston, W.A. 2003. Cell phone induced failure of visual attention during simulated driving. *Journal of Experimental Psychology: Applied*, 9(1), 23–32.

Tijerina, L. 1996. *Heavy Vehicle Driver Workload Assessment*. (DOT HS 808 466). Washington, DC: National Highway Traffic Safety Administration.

Wickens, C.D. 2002. Multiple resources and performance prediction. *Theoretical Issues in Ergonomics Science*, 3(2), 159–77.

Wojciechowski, J.Q. 2004. *Validation of Improved Research Integration Tool (IMPRINT) Driving Model for Workload Analysis*. (Technical Report ARL-TR-3145). Aberdeen Proving Ground, MD: US Army Research Laboratory.

Chapter 20

C3TRACE: Modeling Information Flow and Operator Performance

Jennifer C. Swoboda and Beth Plott

The future of military command and control (C2) processes will continue to be affected by proposed organizational designs, reconfiguration of personnel, as well as advances and changes to information technologies. The advances in technology promise a significant increase in the rate of information flow. Therefore, the C2 processes are continually in a state of change. These processes must be amenable to dynamic changes in organizational structure as well as to the addition or elimination of information technologies. To predict how these changes will impact soldier performance, the Human Research and Engineering Directorate (HRED) of the US Army Research Laboratory (ARL) sponsored the development of a modeling environment in which one can develop multiple concept models for any number and size of organizations, staffed by any number of people, performing any number of functions and tasks, under various communication and information loads. The modeling environment that was developed is called Command, Control, and Communications—Techniques for the Reliable Assessment of Concept Execution (C3TRACE).[1] C3TRACE can be used to model representations of C2 tasks and functions performed by a group of people engaged in various military missions, within conceptual and actual military organizational designs, and using various types of communication and information technologies. C3TRACE has been used to analyze organizations ranging in size from platoon to brigade level, with the inclusion of message flow from platoons to a division staff. C3TRACE enables the rapid construction of models to analyze multiple organizational, personnel, and systems architectures as they emerge through concept exploration for the Army Future Force. This chapter discusses the background, components, and application of the C3TRACE tool.

Background

ARL-HRED has a history in the development of C2 human performance models (HPMs) and in the sponsorship of tools to support easier HPM development.

---

[1] C3TRACE is controlled by the ARL's Human Research and Engineering Directorate. For more information about the tool, contact ARL-HRED (410-278-5800)

Some of the initial HPMs developed by ARL were based on a Tactical Operations Center (TOC) operating under current doctrine using only analog communications and information processing methods (Knapp et al. 1997a). It was these HPMs that established a baseline for the efficiency and effectiveness of the proposed battalion TOC organizational design. The model was further enhanced in order to examine the effects of introducing digital equipment that the TOC personnel would use to conduct their missions and communicate among staff cells (Knapp et al. 1997b). Finally, these models were modified again in order to represent one possible "Army After Next configuration," where all C2 positions were equipped with fully integrated C2 systems (Knapp et al. 1998). These modeling efforts led to the development of the Computer Modeling of Human Operator System Tasks (CoHOST) project (Middlebrooks et al. 1999). The CoHOST modeling methodology was built and exercised to perform an analysis for comparing different personnel and equipment configurations for operations in the Army of today and tomorrow.

An HPM was also developed by ARL for evaluating alternatives of a conceptual organization depicting the manner in which the US Army would conduct its field artillery missions (Wojciechowski et al. 2000). The goal was to move toward effects-based fire decisions, centralized fire resources, and opportunities to better accomplish the commander's intent. To this end, the US Army Depth and Simultaneous Attack Battlelab at Fort Sill, OK, developed a conceptual organization consisting of a Fires and Effects Coordination Cell (FECC) and a field artillery (FA) TOC. In conjunction with Micro Analysis and Design (MA&D), ARL designed an HPM of an FECC and an FA TOC to analyze operator performance within those organizational concepts. The model represented individuals performing different tasks and functions on a prototype C2 system in co-located and distributed environments. This modeling effort provided an economical means to examine efficiency and effectiveness for the alternate organizational concepts. The development of the FECC and FA TOC models provided the basis for the development of "sensor-to-shooter" task flow logic. From this logic, the FECC and FA TOC hybrid sensor-to-shooter model, also called "Performance of the Virtual Soldier," was developed. This tool created a more extensive capability for conceptual brigade-level fires and effects FA organizational designs (Wojciechowski 2001). It also provided the capability to assess the quality of decisions made by the personnel in the brigade-level organization (Wojciechowski et al. 2000) by embedding an information decision algorithm into the model. The algorithm calculated the probability that the decision-maker would have the information needed to make a good decision. Model output included operator utilization, task drops and interrupts, task completions, and sensor-to-shooter timelines. Researchers were able to use these measures to examine the process of battle command.

Due to a large group of novice modelers, the user community expressed the need for a modeling environment with greater flexibility and an easier-to-use graphical user interface (GUI). This redesign resulted in the modeling environment

C3TRACE. C3TRACE (Kilduff, Swoboda, and Barnette 2005) provides an environment for targeted evaluation of the effects of different configurations of soldiers and information technology on performance. This tool has an embedded discrete event simulation engine, Micro Saint Sharp, and a graphical, point-and-click, Windows standard interface that facilitates the model development process.

The background provided here is a developmental timeline of the creation of C3TRACE. Further details about any of the aforementioned modeling efforts can be obtained from the references provided.

Model Development

A C3TRACE model represents the tasks that different, resource-bound operators perform as they process various messages or communication events. This allows the analyst to study how resource requirements, task flows, and operator capabilities interact and impact system performance. The C3TRACE computer simulation works according to the "input–throughput–output" process. Specifically, the inputs to the model are the communication events, which in turn form an information events stream in a frequency-driven time sequence. As the communication events enter the model, tasks are triggered and performed in a pattern that reflects the logic from the task branching and interrupt priorities. Communication events can be triggered at certain times in the simulation or by tasks in the model.

The development of a C3TRACE model requires three basic levels of information and input: the organization, the functions and tasks, and the communication events. The tool includes a user-configurable GUI that allows easy manipulation of information that drives the model. A detailed description of how to use C3TRACE can be found in the Software User's Manual for C3TRACE (Plott 2009a). The following parts of this chapter describe the process used to build a C3TRACE model. Data required to build the models can be obtained from multiple sources depending on the project. Subject matter experts provide an invaluable source of data and validation during model development.

Tool Input

Define the organization   The first step in C3TRACE model development is to consider what level of organizational entities and personnel are to be included in the analysis. C3TRACE handles this by breaking an organization down into sections and operators in a hierarchical tree diagram. The organization can be broken down into sections, then each section can be further broken down into personnel and sections within that section. C3TRACE does not limit the number of levels in the organizational hierarchy or the number of personnel represented.

At this point, the user can also define the personnel in terms of their personnel characteristics or attributes. Some of these attributes are military rank, military training level, length of service, aptitude level, age, battle command experience, and military occupational specialty. It is here where the analyst is able to specify the various model constraints such as the number of simultaneous tasks an operator can perform and the amount of time that a task can be suspended before being dropped permanently.

Define the functions and tasks   The second step in C3TRACE model development is to decompose the functions and tasks to include sequencing of tasks, decision tasks, collaboration tasks, and queues. Each function is represented by a sequence of tasks corresponding to the order in which an operator performs them. Figure 20.1 shows an example of a task-level network diagram for processing a message.

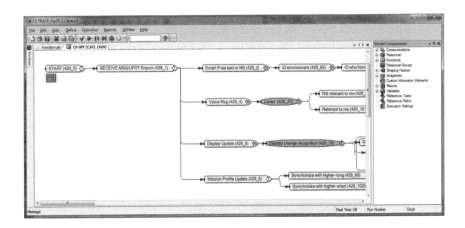

Figure 20.1   Sample C3TRACE task network diagram and model components

Each task is defined by the analyst and it is the tasks that are the basic building blocks of any C3TRACE model. The analyst-defined attributes for each task include priority of task; situational awareness (SA) level (perception, comprehension, prediction) (Endsley 1995); mode (manual, automatic, or both); specification of decision task, collaborative task, or both; task time information; operator task assignment (primary operator, alternate primary operator, supporting operators); workload level (Visual, Auditory, Cognitive and Psychomotor); and decision task information element weighting. The information element weighting is discussed under the communication events heading.

Decision branches and queues are included in the task flow as is appropriate for the tasks. Decision branches include single, multiple, probabilistic, communication-based tactical, and user-defined tactical. A single decision is when there is only one task following the decision task and it never varies. In a multiple-decision branch, all the tasked following the decision execute simultaneously after execution of the previous task. A probabilistic decision has a probability assigned to each task that immediately follows the decision. As the model is executed, C3TRACE will determine which one of the tasks will be executed, based on its assigned probability and a random number generator. A communication-based tactical decision type means that a specific communication event type determines the task execution order. Lastly, a user-defined tactical decision type is represented by a mathematical Boolean expression that determines which task will be executed next.

Communication events   The third major category of input for C3TRACE model development is message traffic. The types of communication events that can be defined are face to face, digital messages (common relevant operating picture updates, email, "whiteboard" messages, sensor updates), voice (radio or intercom), haptic, and written. The communication events provide the information necessary to initiate or drive the analysis. The following list is the data needed to describe each communication event: type of communication; time communication entered the input stream; incoming communication frequency; initial information quality; tasks that are triggered as a result of a communication; and communication priority by communication type. The initial information quality is the actual quality of the incoming message when the operator receives it. This score can be changed to reflect the degradation of transmission quality of a message or degraded sensor capability. Unless there is reason to believe that there should be degradation in the initial information quality, the score defaults to 100 percent, that is, perfect information quality.

Tool Processes

Information-driven decision algorithm   A successful model of a process provides predictive measures that can be examined to determine whether the design of the organization or information flow is correct. In order to establish an accurate assessment of the decisions being made, "information-driven decision-making" architecture was embedded into C3TRACE (Wojciechowski et al. 2001). The basic premise behind this architecture is that information is collected in processing tasks, shared in collaborative tasks, and used in decision-making tasks. This architecture takes into account which operator knows what elements of information, how recent that information is for each operator, and whether the information is sufficient to support the decisions. The five components of the decision architecture used in C3TRACE, which are briefly described in the following subsections, include: specification and weighting of what information elements are used to make a decision; identification of the information elements; initial information quality; decay rates; and calculation of a decision-quality score.

> *Information elements used to make a decision*   The user defines the nature of the information required for each decision task. Any or all of the information elements can be weighted to reflect their relative importance to

a particular decision. Based on importance, the weighting scale is 0 to 10, where 0 has no importance and 10 is the most important.

The match between available information and the information required to make a decision occurs whenever a decision task is executed: does the decision-maker have the right information, either processed directly from a message or received from a collaborator, to make a decision when the time comes? In the end, the "quality" of a decision, that is, the probability of making a good decision, is based on the match between the information received by the decision-maker and the information required to make a decision, and also by how much the "value" of the information has decayed over time. This technique can help to identify system and organizational inefficiencies, bottlenecks, or obstacles relevant to the high quality and recent information required for effective decision-making.

Information elements The information taxonomy within C3TRACE has expanded over the years. In the first version of C3TRACE, the information was defined according to the Army's accelerated decision-making process for developing a successful plan (referred to collectively as the "course of action" (COA) development, or information elements). Five aspects of a successful plan were identified, and it was determined that these aspects contained 24 information elements. These 24 elements included information about enemy force and actions, information about friendly force and actions, feasibility of the current plan, suitability of the current plan, and information on the enemy course of action and potential course of actions.

- *Completeness* includes the parameters *Who, What, When, Where* and *How* for both *friendly* units and *enemy* units in addition to the parameter *Why*, which describes why the friendly and enemy units are there.
- *Feasibility* includes *Time* and *Means*.
- *Suitability* explicitly separates *Accomplish the Mission* from *Comply with the Commander's Guidance* and *Comply with the Commander's Intent*.
- *Risk Assessment* is assessed in terms of *Time, Soldiers, Equipment*, and *Position Losses*.
- *Flexibility* allows modifications to the success of the COA based on three different enemy COAs: *Enemy COA1, Enemy COA2*, and *Enemy COA3*.

For version 2.0, aspects of Politics, Military, Economic, Social issues, Information, and Intelligence were added to C3TRACE. The 29 new information elements that were added are spread across the following six domains:

- *Political:* Own Government, Coalition Government, Enemy Government, Host Nation Government, Insurgent Leadership

- *Military:* Air Force, Ground Forces, Naval Forces, Special Forces, WMD Forces, Guerilla Forces, Terrorists
- *Economic:* Manufacturing, Agrarian, Labor, Banking, Black Market
- *Social:* Religious, Cultural, Ideological
- *Information:* Media, Psychological Operations
- *Infrastructure:* Roads, Airport, Railroads, Ports, Waterways, Terrain, Water Supply, Waste, Communications.

In Version 3.0 the ability for users to define their own Custom Information Elements was added. Using this feature, users can define whatever information elements relate to their analysis. Once a new element is added, it is available to attach to communication events, use in decision tasks, and to assign decay rates.

Initial information quality   The initial information quality represents the probability that the information contained in each message is valid at the time it is collected. A value of 100 percent would indicate that if the soldier were to act immediately, the information would be completely reliable and correct. In other words, the decision would not fail based on this information element alone. The initial information quality defaults to 100 percent but can be changed by the user to reflect things like a lower transmission quality of a message or diminished sensor capability.

Decay Rate

The initial information quality level is adjusted as a function of decay rate and time since last update or when the message is initially executed in the model. As the model executes and messages are generated and forwarded to personnel to be processed, the "age" of each information element is computed, based on when the assigned person first read the message and how much time elapsed before the person used the information in a decision task.

It is a basic assumption in C3TRACE that the older the information, the less useful it is. It is also assumed that the rate of decay is not the same for all types of information. For example, information about the location of the enemy is likely to change frequently and rapidly and, thus, that information element will decay quickly. By contrast, information about a location of a building is likely to remain stable over a given period of time. C3TRACE provides a decay algorithm to capture this differential decay. For each information element, the information quality decay rate can be determined by a quantification of the information's volatility and frequency of change. The user determines if the information has a high volatility and frequency of change, and assigns the appropriate category.

There are four categories of decay rates built into C3TRACE: high frequency–high volatility, high frequency–low volatility, low frequency–high volatility, and low frequency–low volatility (the categories are named A, B, C, and D, respectively as shown on Figure 20.2). By default, information that changes slowly (rate of change) with a small change in the size of the data (volatility) has a time-dependent decay rate in its probability of 1 percent per minute. Information that changes rapidly, and the size of the change is large, has a decay rate of 5 percent per minute. Information that either changes rapidly, or the size of the change is large, but not both, has a decay rate of 2.5 percent. However, users are allowed to change the percentage rate for each category of decay as well as the default decay rate category that is assigned to each information element.

Effect of Time on Info Quality

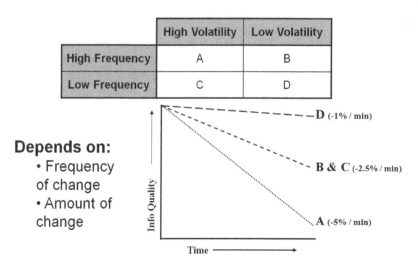

	High Volatility	Low Volatility
High Frequency	A	B
Low Frequency	C	D

Depends on:
• Frequency of change
• Amount of change

D (-1% / min)

B & C (-2.5% / min)

A (-5% / min)

Info Quality

Time ⟶

Figure 20.2   Decay rates

Obviously, for each decision, the order in which the various aspects and elements are collected will vary. The time for each element for the purposes of information decay should be the difference between the time when the information is collected and when it is used to make a decision. For example, information may be collected on Enemy Who with an information quality score of 90 percent 10 minutes before a decision is made (Figure 20.3). Since this information is rather stable, it has a low decay rate of 1 percent. This is compared to the information collected on Enemy Where, which is much more recent (2 minutes before the decision) but it is more volatile and decays at 2.5 percent.

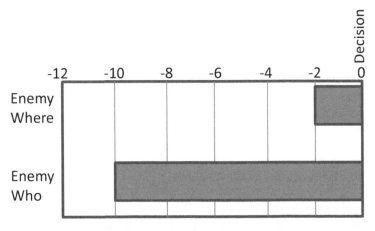

	Enemy Where Information Decay	Enemy Who Information Decay
Initial Information Quality	90	90
Decay Rate	2.5%	1%
Time Since Update	2	10
Information Quality at Time of Decision	85.5	81

Figure 20.3 Information decay example

Calculation of Decision Score A unique abstraction of the C3TRACE approach is that the decision quality can be predicted without having to consider the actual content of the communications. The time since update will determine the information accuracy based on the volatility and frequency of change of each element of information. This is then used to determine the present state of the information elements for a particular operator based on a defined rate of change (the decay rate). The time since the operator last updated his knowledge (in this case, the information elements) is multiplied by the rate of change of each information element and subtracted from the initial quality to produce an information quality score:

$$\text{Information Quality} = [(\text{Initial Quality}) - (\text{Initial Quality} \times \text{Decay Rate} \times \text{Time Since Update})] \times \text{Weight}$$

Probability of Making a Good Decision = 1−[SUM (Information Quality)/
(total weight of all information elements used)]

This information quality score is used to compute the probability of a good decision for the operator performing a decision task. The decision probability score allows an analyst to determine the likelihood that the operator made an appropriate decision. In order to obtain an accurate estimate of whether good decisions are being made, C3TRACE compares the decision score to a user-settable, decision-value criterion (0–100). If the decision score is greater than the criteria it is marked as a "good" decision; otherwise it is labeled as a "bad" decision.

Output files   The outputs provided by C3TRACE include a number of useful human performance metrics, including operator workload, task drops and suspensions, operator task completions, basic SA, and the decision scores. These measures together provide analysts with a tool that can be used to examine the process of battle command. Specifically, the available C3TRACE reports are as follows:

- *Operator Utilization:* This report shows what the total average utilization was for each operator. In C3TRACE, "utilization" is defined as the percentage of time an operator spends performing tasks. An important output of C3TRACE is the capability to see which operators are over- or under-utilized. The results are shown both in graphical and tabular format.
- *Utilization over Time:* This report shows what the job utilization was over the length of the scenario. Specifically, utilization at a given time is defined as the percentage of time on task divided by the total time available (simulation clock time). It is useful to see at what points certain operators are over- and/or under-utilized. Utilization is computed each time a task begins or ends.
- *Operator Performance:* This report lists the run number, operator name, percentage of time busy, total time busy on tasks, total number of tasks completed (by mode), total number of tasks dropped, total number of tasks interrupted, and total messages received.
- *Sensor to Shooter:* This report displays the time it took from when a target was detected to the time it was fired upon.
- *Decision Data:* This report displays a large array of data including function name, decision task name, whether the decision was good, and the calculated decision quality.
- *SA:* Each task in C3TRACE can possibly contribute to the operator's SA. The three levels of SA included in C3TRACE are:
  - Level I Perception of elements in the current situation;
  - Level II Comprehension of the current situation; and
  - Level III Projection of future status.

- This report displays data on what SA levels (I, II, or III) were reached by which operator and the number of tasks that contributed to the operator's SA.
- *Operator Workload:* This report allows users to view each operator's workload data. This report can be used to quickly identify points of high workload. The workload is displayed as a line graph on an x–y axis. The x (horizontal) axis represents time and the y (vertical) axis represents the workload value. Each resource as well as the conflict value and overall workload score is displayed in a different color.
- *Interrupted Tasks:* This report provides information regarding what tasks were interrupted during the simulation. When two tasks of different priorities are competing for the same resources (that is, manpower), the lower priority task is suspended so that the higher priority task can be executed. If these suspended tasks can be resumed or restarted, and can eventually be completed, they are considered interrupted tasks, rather than dropped tasks.
- *Dropped Tasks:* This report provides information regarding what tasks were dropped during the simulation. Tasks are interrupted when a higher priority task is scheduled that competes for limited resources (that is, manpower). Tasks that have been suspended for a period of time longer than the user set limit for the lead operator are dropped.
- *Collaboration Performance:* This report provides information about all of the collaborative tasks that were performed during the simulation. Information presented in this report includes lead operator or personnel group name (depending on which is applicable for the given task), the type of communication event being processed (for example, digital, radio, and so on), the name and ID of the task, the time at which the collaborative task was completed, and a list of any other operators involved in the collaboration.
- *Collaboration Overall:* This report lists all of the operators, the amount of time they spent collaborating during the simulation, and the percentage of total time they spent collaborating.
- *Message Processing Times:* This report shows the amount of time spent on each communication event, including the run number, communication ID, start and end time of the communication event, total processing time, and total waiting time.
- *Operator Activity:* This report shows what tasks each operator performed and when the tasks were performed.
- *Task Summary:* This report records how many times each task was executed, the mean duration, the standard deviation, the minimum execution time, the maximum execution time, and the run number.
- *Task Timeline:* This report is a graphical depiction of task execution in the form of a Gantt chart. It shows the order of task execution and the duration for each task that executed in the simulation.

• *Communication Details Report:* This report shows each message that
  started, who sent the message, who received the message, and how long the
  message was in the system. The analyst can define their own data collection
  "snapshots" and view the collected data after execution. C3TRACE
  provides three methods to view the data:
  – View the raw data in C3TRACE;
  – Export the result data to a HTML formatted file;
  – Open and view the data in a spreadsheet application.

Model Applications

Since its inception, C3TRACE has been used in a wide variety of applications at
many different organizational levels. Military models have been built from the
smaller platoon organizational level to the larger battalion organizational level.
Although ARL models deal specifically with military applications, this tool is
appropriate for evaluation of any type of organizational configuration in which the
analyst wants to investigate the effect of information flow on operator performance.

Initial use of the tool was to represent and evaluate performance differences
between two different organizational concepts in support of the Future Combat
System for the US military. The models represented the unit of action Mounted
Combat System Company Headquarters. Two conceptual models were
built representing the same information technology but different personnel
configurations and vehicles. Results of data from the comparison of these two
configurations can be obtained from Plott et al. 2004.

Whereas the aforementioned model varied personnel configuration and
vehicles, other models focused on the manipulation of information technology. A
C3TRACE platoon-level model was built to examine the mounted and dismounted
phases of a mission and the types of information technology used by the soldiers.
The mounted soldiers were assumed to be using laptop-type displays and the
dismounted soldiers were assumed to be using wrist-mounted personal digital
assistants. Results for the comparison of these two models comparing the use of
the different information technologies can be obtained from Kilduff, Swoboda,
and Barnette (2006).

Recently, another study using the C3TRACE modeling tool investigated a
reconnaissance platoon and what happens when different assets assigned to that
particular platoon are eliminated. Two models were built to represent equipment
attrition and personnel attrition without replacement, respectively. The first model
indicated the effects of one of the primary operators losing digital communications
and how that impacted the other operator's performance. Similarly, in the second
model, one of the scout personnel was permanently eliminated, and the results
showed the effects on the remaining operators (Swoboda 2009).

Governmental agencies outside the C3TRACE-sponsoring organization
ARL have also taken advantage of this tool. The National Aeronautics and

Space Administration (NASA) used C3TRACE on the Kennedy Space Center's Simulation and Analysis of Launch Teams effort. Human performance models of shuttle launch functions and tasks were developed. The models simulated the last 20 minutes of a shuttle launch as well as off-nominal events. These off-nominal events included system problems and the processes and procedures that the launch team applied to solve the issue. The C3TRACE models were used to quickly and inexpensively measure system performance, improve operations, analyze alternate variations of system design, and conduct trade-off analyses.

Modeling efforts have also provided support to the work of the Joint Improved Explosive Device Defeat Organization (Plott 2009b). The Sentinel Hawk is a system of unmanned aerial vehicles (UAVs) used to monitor a specific road or path. The goal of the Sentinel Hawk system is to detect the emplacement of improvised explosive devices. A HPM effort was performed using C3TRACE in order to provide the Sentinel Hawk Team an understanding of the state of the system with respect to task allocation and the impact of task assignment on overall system performance (for example, throughput and errors). The primary objective was to document the tasks and use simulation modeling to analyze how fatigue, vigilance, crew size, and crew rotation schedule might affect operator workload and performance during use of the Sentinel Hawk system. A C3TRACE model was developed to simulate the tasks that the operators perform when monitoring up to four UAVs. The model included a fatigue and vigilance function to predict performance effects for different crew rotations. The model also allowed different detection rates to be simulated.

Steps have been taken to integrate task network and cognitive modeling within a unified development and simulation environment (Lebiere, Biefield et al. 2002, Lebiere, Archer et al. 2005, Lebiere, Best et al. 2005). This has resulted in the introduction of the Human Behavior Architecture (HBA) by the Alion/MA&D Operation (Lebiere, Archer et al. 2005, Lebiere, Best et al. 2005, Warwick et al. 2008). In particular, Alion/MA&D has taken a C3TRACE model that was developed by ARL to study the flow of communication in a Future Combat System and attempted a "cognitive retrofit." This exercise provided a new opportunity to verify the HBA function under the load of a very complicated, independently developed task network model. It also demonstrated how additional cognitive fidelity could make a marked but plausible impact over the predictions made by the unmodified model. This exercise laid the groundwork for further integration work we are currently performing under the Army's Communications-Electronics Research, Development, and Engineering Center Tactical Human Integration of Networked Knowledge Army Technology Objective (THINK ATO). This effort will take the integration one level higher, where the HBA itself serves as a component to be integrated with social network analysis tools and techniques for assessing team performance.

As part of the THINK ATO, work has already begun to identify specific points of contact between the analysis of a social network or team performance and the predictions that can be made using a C3TRACE task network model.

These points of contact will allow us to provide additional, empirically derived details about the communications that drive the scenario in a C3TRACE model (for example, actual vs. predicted communication links, frequency of communication), thereby producing better predictions about operator performance. Conversely, improved predictions about operator performance can provide detail to social network analysis by providing timing and task information to establish context for analyzing communication traffic. In short, we see the potential for a "virtuous cycle" of analysis.

Conclusions

The C3TRACE tool provides an easy-to-use environment to investigate information flow within any organizational structure. For the purposes of the current modeling efforts, most fall into the military arena. Regardless, C3TRACE can easily represent any organizational level, the people assigned to that organization, the tasks and functions they perform, and a communications pattern within and outside the organization. This tool allows for rapid "what-if" evaluations of conceptual organizations without the need for live exercises or experiments. This capability will save time and money, and will provide for many more evaluations than could be conducted by "human-in-the-loop" experiments alone.

References

Endsley, M.R. 1995. Measurement of situation awareness in dynamic systems. *Human Factors*, 37(1), 65–84.

Kilduff, P., Swoboda, J. and Barnette, D. 2005. *Command, Control, and Communications: Techniques for the Reliable Assessment of Concept Execution (C3TRACE) Modeling Environment: The Tool*. (Technical Report ARL-MR-0617). Aberdeen Proving Ground, MD: US Army Research Laboratory.

Knapp, B.G., Johnson, J., Barnette, D.B., Wojciechowski, J., Kilduff, P., Swoboda, J. and Bird, C. 1997a. *Modeling Maneuver Battalion C2 Operations of a Force XXI Equipped Army Command Post for a Force on Force Scenario* (Traditional Model Delivery Paper to US Army Armor Center (USAARMC)). Aberdeen Proving Ground, MD: US Army Research Laboratory.

Knapp, B.G., Johnson, J., Barnette, D.B., Wojciechowski, J., Kilduff, P., Swoboda, J., Bird, C. and Plott, B. 1997b. *Modeling Maneuver Battalion C2 Operations of a Current Army Command Post for a Force on Force* Scenario (Integrated C2V Battalion TOC Model Delivery Paper to USAARMC). Aberdeen Proving Ground, MD: US Army Research Laboratory.

Knapp, B.G., Johnson, J., Barnette, D.B., Wojciechowski, J., Kilduff, P., Swoboda, J., Bird, C., Middlebrooks, S.E. and Plott, B. 1998. *Modeling Maneuver Battalion C2 Operations of a Force XXI Equipped Army Command Post for a*

Force on Force Scenario (Revolutionary C2V Battalion TOC Model Delivery Paper to USAARMC). Aberdeen Proving Ground, MD: US Army Research Laboratory.

Lebiere, C., Archer, R., Warwick, W. and Schunk, D. 2005. Integrating modeling and simulation into a general-purpose tool. *Proceedings of the 11th International Conference on Human–Computer Interaction*, Las Vegas, NA, July 2005.

Lebiere, C., Best, B. J., Archer, R. and Warwick, W. 2005. *Integrating Task Network Models and Cognitive Architectures in dynamic, information-rich Environments*. Human Systems Integration Symposium, Arlington, VA, June 2005.

Lebiere, C., Biefeld, E., Archer, R., Archer, S., Allender, L. and Kelley, T.D. 2002. Imprint/ACT-R: Integration of a task network modeling architecture with a cognitive architecture and its application to human error modeling. *Military, Government and Aerospace Simulation, Society for Modeling and Simulation International*, 34(3), 13–19.

Middlebrooks, S.E., Knapp, B.G., Barnette, B.D., Bird, C.A., Johnson, J.M., Kilduff, P.W., Schipani, S.P., Swoboda, J.C., Wojciechowski, J.Q., Tillman, B.W., Ensing, A.R., Archer, S.G., Archer, R.D. and Plott, B.M. 1999. *CoHOST (Computer Modeling of Human Operator System Tasks) Computer Simulation Models to Investigate Human Performance Task and Workload Conditions in a US Army Heavy Maneuver Battalion Tactical Operations Center*. (Technical Report ARL-TR-1999). Aberdeen Proving Ground, MD: US Army Research Laboratory.

Plott, B. 2009a. *Software User's Manual for Command, Control and Communication—Techniques for Reliable Assessment of Concept Execution (C3TRACE), Version 3.5*. Boulder, CO: Alion Science and Technology.

Plott, B. 2009b. *Sentinel Hawk Human Performance Model—Final Report*. (Contract DAAD19-01-C-0065- TO 112). Aberdeen Proving Ground, MD: US Army Research Laboratory.

Plott, B., Quesada, S., Kilduff, P., Swoboda, J. and Allender, L. 2004. Using an information-driven decision-making human performance tool to assess US Army command, control, and communication issues. *Proceedings from the HFES 48th Annual Meeting*, New Orleands, LA, September 2004, 48(20), 2396–2400.

Swoboda, J. 2009. Alternative FCS-Approved Reconnaissance Platoon Models. (Unpublished Customer Report). Aberdeen Proving Ground, MD: US Army Research Laboratory.

Warwick, W., Archer, R., Hamilton, A., Matessa, M., Santamaria, A., Chong, R., Allender, L. and Kelley, T. 2008. Integrating architectures: Dovetailing task network and cognitive models. Proceedings for the Seventeenth Conference on Behavior Representation and Simulation.

Wojciechowski, J.Q. 2001. *PERVISO: A Tool for Representing Decision-Making in Command and Control*. Paper presented at the DOD HFE TAG Meeting 46, Colorado Springs, CO, May 2001.

Wojciechowski, J.Q., Knapp, B.G., Archer, S., Wojcik, T. and Dittman, S. 2000. Modeling human command and control performance sensor to shooter. *Proceedings of the Human Performance, Situation Awareness, and Automation Conference*, Savannah, GA, October 2000.

Wojciechowski, J.Q., Wojcik, T., Archer, S. and Dittman, S. 2001. Information-driven decision-making human performance modeling. *Proceedings of the 2001 Military, Government and Aerospace Simulation Symposium: Advanced Simulation Technologies (ASTC) Conference*, Seattle, WA, April 2001.

Chapter 21

The Dynamic Network Evolution of C2 Communications over Time

Jeffrey T. Hansberger

Military command and control (C2) relies heavily on efficient communication through the chain of command (for example, Kahan, Worley, and Stasz 1989, McCann and Pigeau 2000). In order to systematically assess patterns of communication within a C2 experimental structure or exercise, analyses that focus on social interactions must be used. One method that systematically analyzes communication patterns between individuals within a C2 structure is social network analysis (SNA).

SNA is based on network theory, which uses graphs as a representation of symmetric or asymmetric relations between discrete objects (Scott 2000). The graph is a mathematical structure to represent pairwise relations between objects. Placed within a social context of humans and their interactions, a social network is a set of individuals (nodes) connected through social interactions like face-to-face or email communication (links). The analysis of these social networks consists of a family of relational methods to systematically uncover patterns of people's interconnectedness.

The examination of interaction patterns within a C2 environment can provide information on a wide range of organizational and individual factors (Wasserman and Faust 1994). The nature and speed of information flow within a C2 structure can be examined through various measures, like the network path lengths and density. Important structural characteristics can be identified through cluster analyses, along with the ability to quantitatively compare networks over time or before/after an experimental manipulation. SNA also allows the identification of key individuals in the network and the ability to systematically focus the analysis to greater levels of depth and detail within the organization. The analyses presented in this chapter leverage the quantitative and empirically based analyses of social networks to provide insight to the changes of the C2 structure within a highly complex exercise at the Joint Forces Command (JFCOM).

The National Academy of Sciences recently assembled a committee of experts to investigate the nature of networks and network research both in general and specific to future US Army applications (NRC 2005). Based on analysis of existing literature and extensive interviews and questionnaires from active network researchers, the Committee on Network Science supports the finding of "the pervasive influence of networks in all aspects of life—biological, physical, and social—and concluded that they are indispensable to the workings of the global economy and to the defense of the United States against both conventional military threats and the threat of terrorism" (NRC 2005: 2). The analyses presented in this chapter attempt to leverage the quantitative and empirically based analyses of social networks to provide insight to the changes of the C2 structure within a highly complex exercise at JFCOM.

JFCOM Exercise Overview

JFCOM conducted two human-in-the-loop (HITL) exercises to explore the effect of new technology and organizational changes on communication and planning effectiveness within an urban environment. The HITL 1 purpose was to establish a baseline using today's technology and organizational configurations in a modified Joint Task Force (JTF) structure that focused its functional components on urban operations. HITL 2 introduced the technological solutions to solve problems similar to those faced by the experimental subjects in HITL 1. These technical solutions were used to supplement or replace the technology used by some of the role players within the JTF staff and components. A minority of positions received one or more technical solutions. However, the majority of the positions maintained their HITL 1 baseline and the HITL 1 organizational structure.

The objective of this SNA approach was to aid in the evaluation of C2 information flow and social network structure changes across the exercise. The SNAs presented here examine network changes *within* HITL 1 and 2 along with any changes *across* these two HITLs. The general hypothesis was that due to the technological benefits provided to the players in HITL 2, improvements would be evident in the information flow and C2 network structure characteristics from HITL 1 to HITL 2.

Method

Participants

Fifteen experimental participants were selected to complete the surveys. Participants included both active duty and retired military. The participants were selected as a key core process group, representing the sensor-to-commander pathway. This group was responsible for sensor data collection, analysis, information integration, and the sharing of raw and processed data throughout the JTF to support the JTF's many functions and tasks. This core process group was selected as a reasonable representative information path through the JTF chain of command. The group consisted of a Joint Force Air Component Commander Intelligence, Surveillance, and Reconnaissance (ISR) Officer, Joint Forces Land Component Command ISR Officer, JTF Headquarters Chief of Staff, JTF personnel (Deputy Commander,

ISR Collection Manager, Info Superiority Chief, Joint Urban Resolve 2 (JURS2) Chemical, Biological, Radiological, and Nuclear (CBRN) Chief, JURS2 Insurgent Analyst, JURS2 CBRN Analyst, JURS2 Insurgent Chief, Deputy Info Superiority Chief, Operations Chief, Current Intel Integrator, and the ISR Operations). Due to the similarity in duties, data for the two JURS2 Insurgent Analysts were combined.

The focus of this chapter is the subpopulation that has the fidelity of the larger organization and may be a reasonable representative of the JTF. Key personnel were selected along the sensor-to-commander pathway as discussed above. Using Figure 21.1 as a reference, the core process group's relationships and tasks are described. The JTF Deputy Commander was the senior decision-maker for the JTF. He was supported directly by the Chief of Staff. The Deputy Commander met with the highest-level military and civilian authorities and received input from and provided direction to his department chiefs and component commanders. The Chief of Staff provided assurance that the next tier of the organization is operating on the intent provided by the Deputy Commander and assumed tasks as assigned.

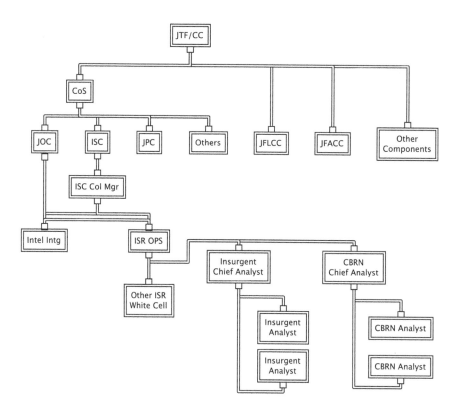

Figure 21.1   Basic organizational structures for HITL 1 and 2

In the next management tier, the ISC Chief was in charge of all Intel operations, and the Joint Operations Chief directed ongoing operational activities. The other sections did not complete a survey because they were outside the sensor-to-commander chain.

The Information Systems Command (ISC) Chief and Deputy ISC own the staff assets for Collection Management, Intel Integration, ISR Ops, the Intel Analysts (including CBRN and Insurgent Analysts) and other intelligence personnel. The Collection Manager, Intel Integrator, and ISR Operations (Ops) Manager had peer relationships with each other and each had their own staff. There were two specific ISR Ops staffing elements: the Chemical, Biological, Radiological, Nuclear and Enhanced Conventional Weapons Chief, his two CBRN analysts, the Insurgent Chief and his two Insurgent Analysts. The Ops Chief has operational, but not administrative, control over the Intelligence Integrator and ISR Ops Manager.

With respect to the Components, the Joint Forces Land Component Command's and Joint Force Air Component Commander's ISR Ops Officers maintained an informal relationship with the ISC Chief, Operations Chief, the Collection Manager, Intel Integrator and ISR Ops Manager.

Measures

The network survey was based upon a tested social network elicitation methodology (Cross and Parker 2004). The survey was composed of two parts, an egocentric set of questions and a bounded question set. This chapter focuses on the results from the first of four bounded questions: "Please indicate the frequency with which you typically turn to each person below for information on work-related topics." This question was restricted to relationships between the 15 surveyed subjects and measured the frequency of information exchange between the targeted 15 players in the group.

During each two-week HITL, the network survey was administered every Tuesday and Thursday. These particular days were chosen to allow for sufficient time for the scenario to cause interactions between players. Most of the 15 participants returned completed surveys by the end of the same day. Some surveys were not returned or incomplete and therefore were treated as missing data.

Results

The results are organized in two primary sections that represent the analyses of the SNA (1) within and (2) across HITL analyses. The analyses began at a high level addressing overall network structure characteristics. The analysis then progressed to the lower detailed levels of the network.

Social Network Analysis

Base network analysis   The base network is derived from a question that elicited the frequency in which each person contacted others for work-related issues. The base network is defined as responses of three to five on a five-point scale, which represents communication rates with others ranging from "sometimes" (3) to "very often" (5). The base networks represent most of the communication traffic among the 15 targeted players. The base network results represent the macro analysis level covered in this chapter, where within and across HITL changes are examined.

Within HITL 1 base changes   Social network structure changes were examined within HITL 1's two-week period and the four administrations of the network surveys (day 2, 4, 7, and 9 of the 10-day exercise). The density of a network is the number of links in the network, expressed as a percentage of the maximum number of links in the graph (Scott 2000). Higher density in a network typically indicates shorter path lengths and faster information flow within an organization. Figure 21.2 displays the network densities within HITL 1 over time. The important interpretation is the relative differences between the measured networks or the overall trend. Statistical analysis of the significance of a linear trend is only done for trends including data from both HITLs due to such a small sample with any one HITL (that is, four). The trend for HITL 1 base networks is generally positive, with increases from Trials 1 through 3 and little change from Trials 3 and 4. This result suggests a learning effect occurring within the network as the individuals involved explore and develop the needed information connections at the beginning of the HITL exercise.

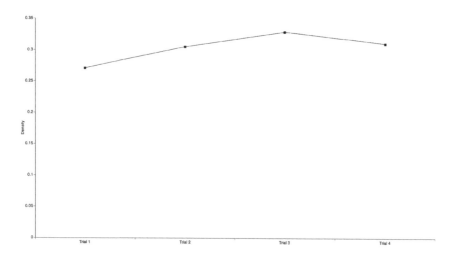

Figure 21.2   Density results for the Base network within HITL 1

A more direct measure of information flow within a network is the average path distance of the network. This measures the average distance between all pairs of nodes. Networks with high average path distances take more time to transmit information to all individuals in the network and are an indication of reachability. The average distance results for HITL 1 are shown in Figure 21.3. Again, the key factor to note is the relative differences over time and the general decrease in path distances within the network over time. Combined with the density results above, players in HITL 1 established more connections with other players (density), which helped to reduce the distance information must travel in the overall network (average distance).

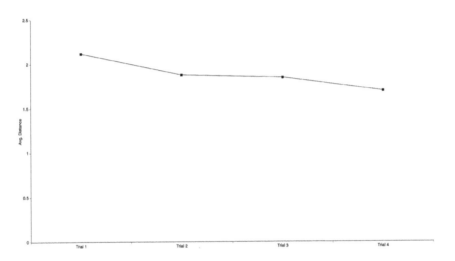

Figure 21.3   Average distance results for the Base Network within HITL 1

Another aspect of reachability within a network is the absence or presence of organized clusters of individuals within the network. The presence of distinct clusters in an organization may create vulnerable bottlenecks tying those clusters together. A hierarchical cluster analysis identifies subsets of individuals in a network that are relatively similar to one another (Wasserman and Faust 1994). Intuitively, a cluster is a relatively dense area in the network. A new and efficient hierarchical cluster algorithm developed by Newman (2003) was used to identify existing clusters in the networks. Among the four base networks for HITL 1 (Figure 21.4), three clusters were consistently found for the first three networks. The final network in HITL 1 (that is, Trial 4 in Figure 21.4) had the number of clusters reduced from three to two.

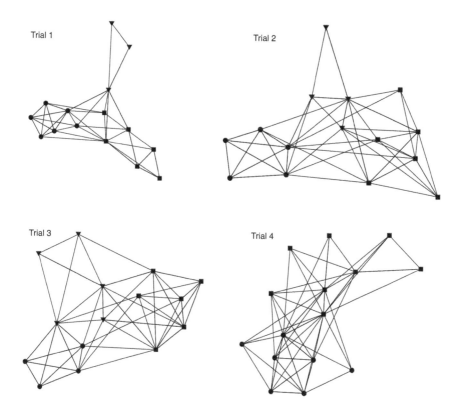

Figure 21.4   Social base networks for HITL 1 with clusters shape-coded

There are a number of significant observations to be made from the social network graphs and clusters found. The Newman cluster analysis defines and computes a Q, or quality function, that measures how meaningful the divisions it computed are (Newman 2003). The Q values for the components found across the networks for HITL 1 showed a steady decrease ($Q=0.3, 0.23, 0.26, 0.16$). This trend mirrors the basic increase in network density observed above and suggests that, as the players made more connections with the people they deemed as necessary information sources, they were able to reduce communication boundaries and form a more cohesive group overall. The most distinctive clustering occurred in Trial 1, whereas the largest drop in Q occurred with the reduction of 3 to 2 clusters in Trial 4. Another interesting observation is the consistently strong grouping among the JURS group throughout HITL 1.

Within HITL 2 base changes   HITL exercise 2 consisted of the same exercise structure over the same amount of time, but several technology changes were made that were not present in HITL 1. The density of the HITL 2 base networks at the four administrations of the network survey is shown in Figure 21.5. Unlike the HITL 1 density results, there is no evident trend over time for the density of the networks.

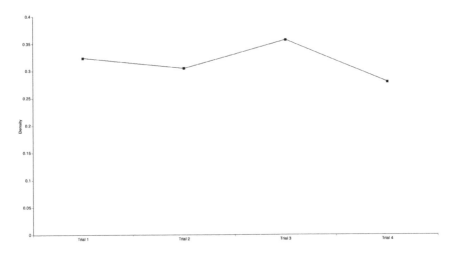

Figure 21.5   Density results for the base network within HITL 2

The average path distances in HITL 2 (Figure 21.6) mirrors the changes seen in the network density. As the density increases for Trial 3, the average distance decreases but increases at Trial 4 as density drops for the same trial. These changes in the network structure suggest significant change during this period.

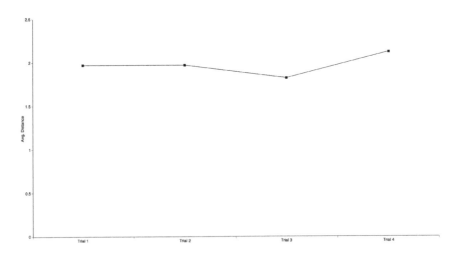

Figure 21.6   Average distance results for the base network within HITL 1

An examination of the clusters found in HITL 2 shows a slightly different picture than the findings from HITL 1. Clusters of three were found for both Trials 1 and 2 of HITL 2, with a reduction to two clusters occurring at Trial 3 and remaining for Trial 4 (Figure 21.7). The clusters between Trials 1 and 2 consist of the exact same members with only slight differences in the network structure itself, as the high correlation ($r = 0.83$, $p < 0.01$) between structures indicate using the quadratic assignment procedure (Krackhardt 1987). The shift from three clusters to two coincides with the jumps in network density and path distances observed above. The Q, or quality, function that measures how meaningful the cluster divisions are reinforces a significant shift in the information network occurring at trial 3 ($Q = 0.33$, 0.31, 0.21, and 0.32, respectively, across trials).

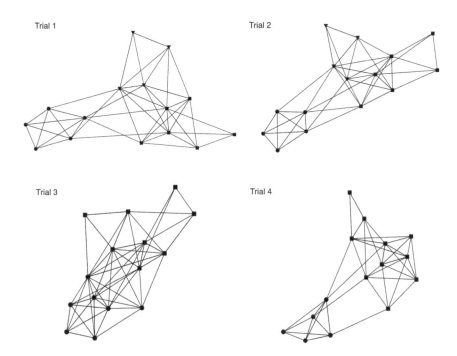

Figure 21.7　Social base networks for HITL 1 with clusters shape-coded

Across HITL base changes　The previous two subsections examined the differences across time within each of the HITLs. This section examines the networks for any differences or trends across the HITLs. Following the same approach as the within analyses, the density across HITLs can be examined. By combining the density line graphs for HITL 1 and 2, relative differences and trends are evident (Figure 21.8). The trials for HITL 1 and 2 have been assigned continuous number assignments where Trials 1–4 consist of HITL 1 measures and Trials 5–8 consist of HITL 2 measures. There is no overall significant linear trend overtime and similar values between the last trial of HITL 1 and the first of HITL 2. There is a similar increase followed by a dip in density at Trial 3 for both HITLs, but with much greater movement in HITL 2, where we saw the two-cluster formation develop earlier and with greater definition.

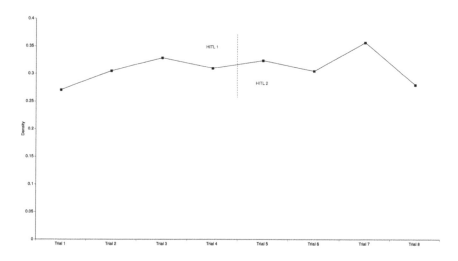

Figure 21.8   Density results for the base networks for HITL 1 and 2 with linear trend fit

A similar analysis of the average path distances across HITLs suggests a different phenomenon occurring between the HITLs (Figure 21.9). There is no significant linear trend across the HITLs, but there is a considerable jump between HITLs (Trials 4 and 5). Because there was no consistent manipulation throughout HITL 1, it is expected that the players go through a learning process to discover who and how to obtain information through the most efficient and shortest path. Given this expected learning effect and some amount of carryover to HITL 2, it would be expected that the average path distance at the beginning of HITL 2 be similar to the end of HITL 1 and continue improving or asymptote. This suggests that a factor or change between HITLs 1 and 2 caused the HITL 2 base network to start roughly where it began during HITL 1, losing its progress gained from HITL 1. There are number of possible changes between HITLs that could have caused this effect such as the minor player personnel changes that occurred between HITLs, the time between HITLs to reestablish or rediscover information links, the new technology changes introduced to HITL 2, or other reasons not listed here.

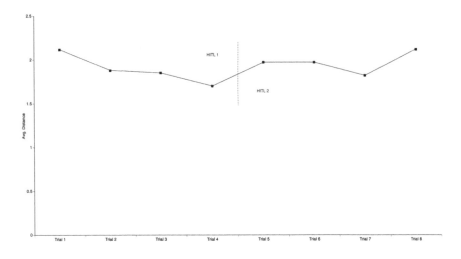

Figure 21.9   Average path distance results for the base networks for HITL 1 and 2

Comparisons between HITLs can also be made from the hierarchical cluster analyses done in the previous subsections. Both HITLs began with three identified clusters. During HITL 1, it was not until Trial 4 that those clusters were reduced to two. In HITL 2, that process of cluster reduction from three to two occurred at a faster rate exhibiting a two-cluster formation at Trial 3. It is worth noting that the individuals comprising the two clusters in Trial 4 of HITL 1 and Trial 3 of HITL 2 were exactly the same (Figures 21.4 and 21.7). However, the network structure (for example, the links between nodes) of these two networks is different as evidenced visually and through the moderate correlation level ($r = 0.51$, $p < 0.01$) between the two (Krackhardt 1987). These findings suggest that the development process the players are using within this local network were applied during both HITLs, but they were able to do it an accelerated rate in HITL 2. The cause for this acceleration cannot be clearly identified due to too many uncontrolled variables between HITL 1 and 2, but it is an excellent area for future investigation.

A phenomenon potentially as interesting as this shared development path and structure for Trial 4 of HITL 1 and Trial 3 of HITL 2 is the structure following this shared state in HITL 2. The structure the network evolved to at Trial 4 of HITL 2 is unlike any other network seen through HITL 1 or 2 and is the clearest representation of the two-cluster structure seen earlier. The accompanying factors of the network in HITL 2 prior to the formation of the Trial 4 network also suggested changes within the network. The increased network density observed at Trial 3 of HITL 2 is evidence of increased network activity. A similar pattern of increased cognitive activity or variability in behavior before a major strategy discovery or an "ah-ha" event has been observed at the individual level in past research (Siegler 1996).

It has been hypothesized in past research that a network of individuals working together with supporting technology can create a distributed cognitive system (for example, Hutchins 1995 and Hutchins 2000). The network results here may be suggesting that such a system is at work and is displaying strategy development behavior seen at the individual cognitive level (Hansberger, Schunn, and Holt 2006).

Core network analysis

The past analyses examined the base networks that consisted of most communication links among the targeted players. To analyze the network changes in greater detail, the lens that the networks are viewed through can be magnified to examine greater detail to the core of each network. The core network is defined as a response of "5," which represents only communication links with others that occur "very often." The core networks represent the foundation of the base networks examined in the above analyses and elucidate the heavy traffic areas (Figure 21.10). Analysis of these core networks focuses on both within and across HITL changes in order to complement the analyses done on the base networks.

Across HITL core changes   Some differences or changes across the HITL 1 and 2 core networks are evident. By combining the density line graphs for HITL 1 and 2, trends can be examined (Figure 21.10). The trials for HITL 1 and 2 have been assigned continuous number assignments where Trials 1–4 consist of HITL 1 measures and Trials 5–8 consist of HITL 2 measures. There is no overall significant linear trend over time but intriguing dynamics exist between Trials 3 and 4 for both HITLs as discussed in the within HITL sections.

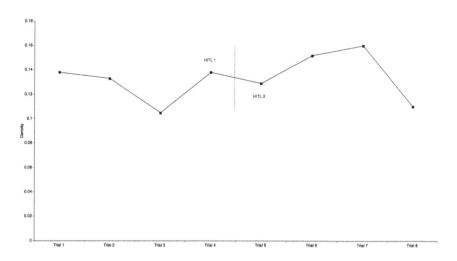

Figure 21.10 Density results for HITL 1 and 2 core networks with linear trend fit

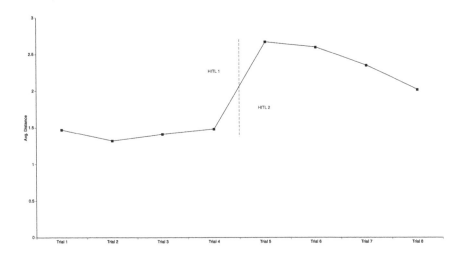

Figure 21.11 Average path distance results for HITL 1 and 2 core networks

Analysis of the average path distances across the two HITLs (Figure 21.11) shows a large discrepancy between the two HITLs. HITL 2 is showing considerably longer average path distances compared to HITL 1. When considered with the connectedness results (Figure 21.12), it is evident that the significantly increasing connectedness eliminated most of the smaller non-connected components found in HITL 1 (for example, the JTF Commander and Chief of Staff components). The linking of these components to the larger core network improved the connectedness but also created a larger core network and therefore greater distances for information to travel to reach all individuals in this core network. The connectedness scores show a strong positive linear trend overall ($R^2 = 0.65$, $p < 0.01$) across time (Figure 21.12).

When these results are tied back to the base network results, possible explanations begin to surface for the patterns and network behaviors seen there. There was a discrepancy between the base networks of HITL 1 and 2 for the average path distances (Figure 21.9), where a possible change in the environment was hypothesized to have caused the network to reset its average path distance progress between the HITLs. If that hypothesis were true, we would expect a similar pattern of networks being developed at the core level for HITL 1 and 2, as they would essentially be reset to begin their development path again due to an environmental factor rather than the technology manipulation. However, the network patterns are distinctly different between the HITLs. The HITL 1 core networks consisted of several separate components and the HITL 2 core networks consisted primarily of a single component, which increases its connectedness over time. This suggests continued internal development of the networks across HITLs or the influence of one or some of the manipulations made between HITL 1 and 2.

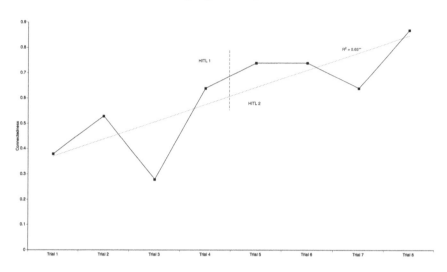

Figure 21.12 Connectedness results for HITL 1 and 2 core networks with linear trend fit

If the cause was due to continued development of the network, the greater degree of connectedness and the elimination of the separate components at the core level causing a spike in the core average path distance, possibly also created a ripple effect that influenced the base level network in a similar manner. In other words, what looked to be an outside environmental influence at the base network level now has some evidence to be influenced by internal factors at the core network level.

Discussion

The SNA applied to a limited but potentially representative subsample of the overall information flow from sensor to decision-maker. It also uncovered several dynamic network behaviors over and across HITL 1 and 2. Of particular interest was the evidence for the development of the networks within each HITL at the base network level and the accelerated network evolution in HITL 2. Because the exercise was not designed to test the cause and effect relationship in regard to network development, the cause for these changes cannot be explicitly identified. However, as has been mentioned previously, there is evidence of similar patterns of cognitive strategy and idea formation at the individual level (for example, Siegler 1996). This is an excellent area for further network analysis and research. At the core level, it was demonstrated that additional insights are possible by peeling back the layers of the network in order to uncover behaviors seen at more macro levels of the network. The significant trend increase in network connectedness at the core level across HITLs definitely shows improvement over time and across HITLs, but the cause again cannot be directly attributed to either the natural development process of the social network or the changes implemented in HITL 2.

References

Cross, R.L. and Parker, A. 2004. *The Hidden Power of Social Networks: Understanding How Work Really Gets Done in Organizations.* Boston, MA: Harvard Business School.

Hansberger, J.T., Schunn, C.D. and Holt, R.W. 2006. Strategy variability: How too much of a good thing can hurt performance. *Memory and Cognition,* 34(8), 1652–1666.

Hutchins, E. 1995. How a cockpit remembers its speeds. *Cognitive Science,* 19, 265–88.

——. 2000. *Cognition in the Wild.* Cambridge, MA: MIT Press.

Kahan, J.P., Worley, D.R. and Stasz, C. 1989. *Understanding Commanders' Information Needs.* (Report R-3761 of the RAND Corporation). Santa Monica, CA: Rand Corporation.

Krackhardt, D. 1987. Cognitive social structures. *Social Networks,* 9, 109–34.

McCann, C. and Pigeau, R. (eds) 2000. *The Human in Command: Exploring the Modern Military Experience.* NY: Kluwer Academic.

Newman, M.E.J. 2003. Fast algorithm for detecting community structure in networks. Physical Review E, 69(2), 066133/1–5.

NRC (National Research Council) 2005. *Network Science.* Washington, DC: National Academies Press.

Scott, J.P. 2000. *Social Network Analysis: A Handbook.* London: Sage Publications.

Siegler, R.S. 1996. *Emerging Minds.* New York: Oxford University Press.

Wasserman, S. and Faust, K. 1994. *Social Network Analysis: Methods and Applications.* Cambridge, UK: Cambridge University Press.

Index